Kenya: A Natural Outlook

DEVELOPMENTS IN EARTH SURFACE PROCESSES, 16

SERIES EDITOR -J.F. SHRODER JR

1. PALEOKARST: A SYSTEMATIC STUDY AND REGIONAL REVIEW
   P. BOSÁK, D. FORD, J. GLAZEK and I. HORÁCEK (Editors)
2. WEATHERING, SOILS & PALEOSOLS
   I.P. MARTINI and W. CHESWORTH (Editors)
3. GEOMORPHOLOGICAL RECORD OF THE QUATERNARYOROGENY IN THE HIMALAYA AND THE KARAKORAM
   JAN KALVODA (Editor)
4. ENVIRONMENTAL GEOMORPHOLOGY
   M. PANIZZA
5. GEOMORPHOLOGICAL HAZARDS OF EUROPE
   C. EMBLETON and C. EMBLETON-HAMANN (Editors)
6. ROCK COATINGS
   R.I. DORN
7. CATCHMENT DYNAMICS AND RIVER PROCESSES
   C. GARCIA and R.J. BATALLA (Editors)
8. CLIMATIC GEOMORPHOLOGY
   M. GUTIÉRREZ
9. PEATLANDS: EVOLUTION AND RECORDS OF ENVIRONMENTAL AND CLIMATE CHANGES
   MARTINI, A. MARTINEZ CORTIZAS and CHESWORTH (Editors)
10. MOUNTAINS WITNESSES OF GLOBAL CHANGES RESEARCH IN THE HIMALAYA AND KARAKORAM: SHARE-ASIA PROJECT
    RENATO BAUDO, GIANNI TARTARI and ELISA VUILLERMOZ (Editors)
11. GRAVEL-BED RIVERS VI: FROM PROCESS UNDERSTANDING TO RIVER RESTORATION
    HELMUT HABERSACK, HERVÉ PIÉGAY and MASSIMO RINALDI (Editors)
12. THE CHANGING ALPINE TREELINE: THE EXAMPLE OF GLACIER NATIONAL PARK, MT, USA
    DAVID R. BUTLER, GEORGE P. MALANSON, STEPHEN J. WALSH and DANIEL B. FAGRE (Editors)
13. NATURAL HAZARDS AND HUMAN-EXACERBATED DISASTERS IN LATIN AMERICA: SPECIAL VOLUMES OF GEOMORPHOLOGY
    EDGARDO M. LATRUBESSE (Editor)
14. THE WESTERN ALPS, FROM RIFT TO PASSIVE MARGIN TO OROGENIC BELT: AN INTEGRATED GEOSCIENCE OVERVIEW
    PIERRE-CHARLES DE GRACIANSKY, DAVID G. ROBERTS and PIERRE TRICART
15. GEOMORPHOLOGICAL MAPPING: METHODS AND APPLICATIONS
    MIKE J. SMITH, PAOLO PARON and JAMES S. GRIFFITHS (Editors)

Developments in Earth Surface Processes

Volume 16

Kenya: A Natural Outlook

Geo-Environmental Resources and Hazards

Edited by

Paolo Paron
UNESCO-IHE, Institute for Water Education,
Department of Water Science & Engineering,
P.O. Box 3015 - 2601DA,
Delft, The Netherlands

Daniel O. Olago
Department of Geology,
Institute for Climate Change and Adaptation,
College of Biological and Physical Sciences (CBPS),
University of Nairobi, P.O. Box 30197,
Nairobi, Kenya

Christian Thine Omuto
Department of Environmental and Biosystems Engineering,
University of Nairobi,
P.O. Box 30197-00100,
Nairobi, Kenya

ELSEVIER

AMSTERDAM • BOSTON • HEIDELBERG • LONDON • NEW YORK • OXFORD
PARIS • SAN DIEGO • SAN FRANCISCO • SINGAPORE • SYDNEY • TOKYO

Elsevier
The Boulevard, Langford Lane, Kidlington, Oxford, OX5 1 GB, UK
Radarweg 29, PO Box 211, 1000 AE Amsterdam, The Netherlands

First edition **2013**

British Library Cataloguing in Publication Data
A catalogue record for this book is available from the British Library

Library of Congress Cataloging-in-Publication Data
A catalog record for this book is available from the Library of Congress

ISBN: 978-0-444-59559-1
ISSN: 0928-2025

Contents

Foreword

Both Editor Dr. Paolo Paron and I began our postdoctoral careers in East Africa, he in the past decade and I half a century ago in the 1960s. The region is certainly the sort of primal place that geoscientists of all varieties can really discover for themselves the enthusiasm and means to utilize many dimensions of their training for its most beneficial effects to the local people and their environment. Kenya, of course, is well known as one of the most iconic of African locales, with its towering volcanoes above the huge African Rift and great grasslands covered with migrating herds that support the remarkable predators so beloved of the tourists, but not much by the local people. Africa is also well known, however, as being a place where the diverse human populations have been sorely beset and exploited for centuries by the largely external forces of slavery and colonialism visited upon them, as well as being riven by internal divisiveness, brutish tribal politics, and endemic corruption. The result was that the continent has been unfortunately left out of many of the world's more recent advances in industry, trade, economic development, higher education, and newer understandings of the natural environment. This essential realm of Kenya, however, is critical to protect for the future progress of the country, and must be better understood to be managed properly.

When Dr. Paron first noticed in Kenya the unhappy lack of so much essential environmental information about what most of the more developed world considers fundamental data, he resolved to do his best to find a way to alleviate some of the information shortages. Thus, this new book of edited chapters by different authors, *Kenya: A Natural Outlook*, is the direct result of his efforts to produce a general account of the physical geography and general geology of the country that might be useful in future in some pragmatic and beneficial ways.

As an example of the most common use of a book such as this, the volume is replete with exactly the sort of material one seeks to know when initiating any sort of new development project in a poorly developed place like East Africa. These essential projects commonly help out people in isolated rural communities or even in the fairly hopeless city slums that are so known for major deficiencies of such services as reliable transport, communication, potable water, power supply, and so forth. The booming populations, coupled with unknown changes that will likely arrive as world climates shift, cause demand for background works such as this to teach people about what exists at present and to provide certain baseline sources of information for planning. Any United Nations or World Bank project or even those development projects coming from the new donor countries in East and South Asia must rely upon works like this. All development officers need basic information sources on the local natural resources in the particular country, as well as to find out much of what they need to know about the most common natural hazards that threaten the people and any particular development project.

At the same time, however, where the reader may detect some less-developed sources of basic information, such areas of thin research can serve as an indicative guide to content areas where new work needs to be done. This would be most true, especially if such work could solve some urgent development problems or provide a thoughtful focus for new assessments by government ministries, or even just serve as an information source to provide local university staff personnel with good new, thought-provoking ideas for student or faculty projects.

In another useful fashion, this book also has been done relatively inexpensively by a major publishing house, with a full-colour version available online of some of the important graphics, so that the book might actually become more easily or cheaply available in Kenya itself. No guarantee of this possible low publication cost actually exists, of course, in these uncertain times for publishers, but at least the thought has been there in the preparation of this book.

The two sections of this book—the first part on natural resources and the second on natural hazards—are notable in having a plethora of Kenyan authors. With only three chapters having all foreigners for authors, the other 21 chapters have some 39 Kenyans as their writers. This amounts to well over half of the book as made up of contributions by Kenyans, which is an exemplary way to involve indigenous experts. The 18 chapters of the natural resources section include all of the expected topics of geology and physical geography that one would expect, with such topics as basic geology, mineral resources, relief, drainage, climate, soils, vegetation, and surface and ground waters. Also included are more specialized discussions of such unusual topics as landscape evolution in the Quaternary, renewable energy, remote-sensing applications, fire ecology, and environmental impact. Five chapters on natural hazards in Kenya include details related to earthquakes, landslides, floods and droughts, as well as coastal erosion.

This book, *Kenya: A Natural Outlook*, is thus a useful new view of the land of one of the most important countries of East Africa. The information given here about a leading African country provides a guide for other nations on a continent that most definitely needs other such examples of background information for use in those countries as well. Dr. Paron has succeeded in locating the best people in Kenya who could contribute to such a volume and has brought the book to fruition for use by the development community and the local secondary schools and university students. His professional post-doctoral career is thus off to a fine start from Africa, with a book that offers important sources of information for those who might need it. In his introductory comments, he notes also the need for a companion volume that details the human geography of the country, a suggestion to which we concur. In this it would be important to emphasize details, less on the quaint ancient tribal aspects that safari tourists may desire, but instead to focus upon the complex modern social developments which will control the future of Kenya.

John F. Shroder Jr.
Editor-in-Chief
Developments in Earth Surface Processes

Introduction

It is exactly 40 years since the first comprehensive synthesis of physical and human geography of Kenya appeared in one single book by Ojany and Ogendo (1973), titled 'Kenya, a study in physical and human geography'.

At that time, Kenya was a young independent nation with an already bright path of growth and development. That book, beautifully illustrated, described the environmental and human links and underlined implicitly the needs for an integrated social and environmental sustainable development of the country. Since then, this remained the only book that synthesized in one volume the various environments of this beautiful country sitting across the equator. Since then, though, the country has gone through several stages of development and its human and physical environment have changed too, under the influence of local and global processes such as land tenure and land use, and climate changes.

Inspired by the book by Ojany and Ogendo and, at the same time, aware that we could not replicate their intent, we therefore focused on the natural aspects of Kenya's environment. In doing so, we wanted to harvest from as much as possible the local scientific knowledge of the country, presenting the readers with the state of the art of Kenyan knowledge on Kenya itself. At times, this proved difficult as the still too few scientists in the country are in great demand for many initiatives and they made an exceptional effort to contribute to this book. At the same time, we also wanted to expose to the Kenyan and international readers the work of some researchers from other countries who are concentrating their studies on Kenya. Needless to say that this volume does not pretend to be a complete outlook of Kenya's environment, but it wants to present an updated overview of its physical environment, its hazards, and part of its natural environment management needs. Being a synthetic overview of Kenya's natural environment, we tried to be accessible for a variety of readers: from the academic audience, to the university students, to the ones who look for a summary of Kenyan environments, who hopefully will find an up-to-date reference.

This volume would ideally need to be complemented by a companion text on the Human Geography of Kenya in the present times, presenting its complex and quickly changing human and social developments.

This book is divided in two sections: the first one addresses the Natural Resources of Kenya, with 18 chapters, and the second section its Natural Hazards, in 5 chapters. The book has a combination of main chapters, usually larger in length and scope, and some smaller chapters which are either geographically specific or topic specific.

In Chapter 1, Opiyo et al. describe the general geology of the country that is an important backbone for most of the other natural aspects such as climate, vegetation, soil, and waters, let alone human population distribution and its economic activities. Kenya is a paradise for geologists, offering rising tropical coastline, active continental rifting, active and extinct volcanism, proterozoic complexes, and an almost complete suite of rocks and structural patterns.

Chapter 2, by Opiyo et al., follows naturally the first one and presents the mineral, oil, and gas resources of Kenya. It is an insight of special interest in these times when new deposits of rare earths, gold, gemstones, and other minerals as well as gas and oil are being assessed and/or exploited, and more are to come. This chapter addresses the occurrence of minerals, from the traditional soda ashes from rift lakes to the oil and gas development always in the frame of Kenyan legislation and the history of production.

Chapter 3, by Gichaba, reviews the main physiographic regions and drainage basins of Kenya, highlighting the great variety of morphology that the country shows, with desert, large endoreic regions, and an almost 500-km-long monoclinal slope from the eastern edge of the rift to the coastal belt.

Chapter 4, by Olago, describes the Quaternary geomorphological evolution of Kenya that has been shaped by global, hemispheric, and regional climate changes interacting with tectonism and volcanism to generate the present-day landscapes and environments. This has had an enormous influence on hominid evolution, that from the Rift valley spread across the whole globe, largely conditioned by geographic and geomorphologic constrains.

In Chapter 5, Obiero and Onyando provide a summary of the factors influencing Kenyan climate, which is dominated most notably by the ITCZ fluctuation across the year, and the topographic barriers and corridors of the country, resulting in a great variability of climates across the whole country: from deserts in the north to the lush tropical rainforest environment just few hundreds of kilometres to the west, to the wet coastal climates on the eastern margin.

Chapter 6, by Kiema, illustrates the new methods and data available for remote sensing analysis of climate in Kenya. It focuses particularly on the potential and real applications of remote sensing for integrated water management in ASAL areas (making up to almost two third of Kenya) and also on lake management with two examples from the two large freshwater lakes in Kenya.

In Chapter 7, McClain and Subalusky present the first case study focused on the environmental flows in the Mara River Basin, the world renowned safari destination for rich game viewing and the thrilling spectacle of the annual wildebeest migration. They explore the minimum flow requirement for ecosystem services and conclude with recommendations for an improved sustainable environmental flow management in this delicate ecosystem.

In Chapter 8, Kuria presents one of the most important and conditioning resources of Kenya with two thirds of its territory classified as ASAL: groundwater. Here, the reader will find a very welcome update on the groundwater quantity and quality of Kenya, largely influenced by the geological setting in the highlands as well as in the lowlands. The chapter was built from the analysis of the more than 10,000 borehole databases held by the Ministry of Water, and it highlights its potential and constraints for use in a modern scientific way. Finally, it presents for the first time a regional groundwater potential map of Kenya.

In Chapter 9, by Omuombo et al., the coastal waters are addressed with special regard to currents and tides, but also pollution and coastal water management practices. It represents an updated summary of the potential and threats of coastal waters in Kenya.

Chapter 10, by Oludhe, presents a new emerging field of interest in the energy sector: renewable energy resources of Kenya. This contribution touches on solar, wind, hydro, geothermal, and biogas resources of Kenya, presenting several maps of the energy potential of the country.

Chapter 11, by Omuto, showcases the major soil and data types in Kenya. This chapter shows where the gaps are in soil data in the country, and nevertheless, it presents a new soil map for Kenya that shows how the country is dominated by soil types that have strong crop production limitations under rain-fed agriculture but are good for the development of pastoral resources. These insights are very relevant for future development policies.

Chapter 12, by Brink et al., shows years of research conducted on vegetation changes in Kenya and the region by this group of researchers. The chapter highlights how the predominant clearing of natural vegetation in favour of cropland may have long-term impacts on food production and food security, freshwater reserves, and ecosystems, especially in view of the impacts of climate change and climate variability that are already observed in Kenya. The authors make extensive use of freely available remote sensing data for regional and national analysis.

Chapter 13, by Pellika et al., shows a case study with a focus on the agriculture expansion and its consequences in the Taita Hills, in line with the previous chapter. The authors stress the fact that, despite increasing cropland, the clearing of natural vegetation has also some negative impacts on very important ecosystem services, with negative feedback on the expanding agriculture. The authors make extensive use of remote sensing data at different scales combined with repeated field verification campaigns over the past 10 years in the Taita Hills of southwestern Kenya.

Chapter 14, by Palumbo et al., shows the importance of fire ecology in protected areas management. The authors introduce the ecological role of fire on the tropical ecosystems and biodiversity and then they assess the importance of fire in Kenyan ecosystems and protected areas. The authors based their analysis on 10 years of remote sensing observation of fire from space so are able to relate fire occurrence with other biophysical changes such as land cover.

Chapter 15, by Kirui, describes the distribution, types, and commercial uses and also shows the threats and conservation strategies for mangroves in

Kenya. Despite their importance for ecosystem services and sea-level rise protection worldwide, and in Kenya too, these unique coastal forests are among the most threatened habitats in the world and in Kenya: recent overexploitation, port developments, increased sediment supply, and biophysical degradation of waters are posing a real threat to this ecosystem. An opportunity for their protection is legislated in the Forest Act that includes also the sustainable management of mangroves.

Chapter 16, by Nyingi et al., exposes the importance of freshwater ecology of Kenyan highlands and lowlands. They clearly show how these ecological systems are on the verge of collapse that would pose a dramatic threat to the economy of so many communities and of the ecosystems in general. Still, they explain the gaps in knowledge on these precious resources and continue in presenting a much needed updated overview of the existing knowledge.

Chapter 17, by Ogada and Nyingi, shows the management of wildlife and fisheries resources in Kenya: origins, present challenges, and future perspectives. The authors provide a very clear and important historical timeline of what was until recent years the common practice of wildlife management, beautifully illustrated. They then continue presenting the conflicts in human–wildlife dynamics and illustrate this with two case studies from Amboseli and Laikipia, including poaching issues that so sadly are increasing in the country. Finally, they explain the regulatory framework in place on wildlife and fishery.

In Chapter 18, Biamah et al. explore in depth the Environmental Impact Assessment (EIA) process in the Kenyan context. In doing so, they always frame the EIA general principles into the Kenyan environment and laws and regulations that apply in the country.

Chapter 19, by Mulwa et al., opens the section on Natural Hazards. This chapter provides an up-to-date overview of Kenyan seismic hazard, with a review and explanation of Kenya's seismicity in the 1900s to date, combining macroseismic data and instrumental observations, in the frame of the regional geodynamic set-up.

Chapter 20, by Maina-Gichaba et al., focuses on landslide hazard in Kenya. This is an often under-reported hazard in this country despite the high toll of lives and livelihoods, and posing serious challenges to a sustainable development and local population's resilience. This chapter provides an up-to-date description of the factors contributing to the slope processes in Kenya.

In Chapter 21, by Opere, the issue of floods in Kenya is explored in detail. Floods, both seasonal and flash floods, are more recurrent—or at least more reported—in Kenya in recent years. After a review of the historical flood events in Kenya, this chapter addresses also the vulnerability linked to floods, and finally it presents a flood risk map for Kenya.

Chapter 22, by Omuombo et al., addresses the hazard of coastal erosion along the almost 600 km of Kenyan coast, starting from causes and extreme events and concluding with management issues.

In Chapter 23, the concluding chapter, Balint et al., showcase a new index created for monitoring droughts, the Combined Drought Index (CDI), tested so far in Kenya and Somalia. The CDI combines three main meteorological indicators in one, has already been successfully deployed in Somalia, and will hopefully be tested soon in other geo-ecological environments. The CDI makes use of simple, almost everywhere available, data like Temperature, Rainfall and NDVI thus being a very 'low data consuming indicator' quite suitable for developing countries where data availability is low.

Contributors

Numbers in Parentheses indicate the pages on which the author's contributions begin.

Norbert Opiyo Akech (3, 11), Department of Geology, College of Biological and Physical Sciences (CBPS), University of Nairobi, PO Box 30197-00100, Nairobi, Kenya

Zoltan Balint (341), FAO-SWALIM, Nairobi, Kenya (retired)

Elijah K. Biamah (237), Professor of Environmental and Water Systems Engineering, University of Nairobi, Kenya

Catherine Bodart (133), Joint Research Centre—European Commission, Via E. Fermi, 2749 Ispra, Italy

Andreas B. Brink (133), Joint Research Centre—European Commission, Via E. Fermi, 2749 Ispra, Italy

Graeme Buchanan (133), Joint Research Centre—European Commission, Via E. Fermi, 2749 Ispra, Italy

Barnaby J.F. Clark (165), Department of Geosciences and Geography, University of Helsinki, Helsinki, Finland

Marco Clerici (133), Joint Research Centre—European Commission, Via E. Fermi, 2749 Ispra, Italy

François Donnay (133), Joint Research Centre—European Commission, Via E. Fermi, 2749 Ispra, Italy

Nguyen Anh Duong (267), Institute of Geophysics, Vietnamese Academy of Science and Technology, Hanoi, Vietnam

George Eshiamwata (133), Joint Research Centre—European Commission, Via E. Fermi, 2749 Ispra, Italy

Nathan Gichuki (199), University of Nairobi, P.O. Box 30197, Nairobi, Kenya

Alemu Gonsamo Gosa (165), Department of Geography and Planning, University of Toronto, Toronto, Canada

Jean-Marie Gregoire (133), Joint Research Centre—European Commission, Via E. Fermi, 2749 Ispra, Italy

Nina Himberg (165), Faculty of Agriculture and Forestry, University of Helsinki, Helsinki, Finland

Pekka Hurskainen (165), Department of Geosciences and Geography, University of Helsinki, Helsinki, Finland

Jacqueline Kiio (237), Agricultural Economist, Ministry of Agriculture, Kenya

Fumiaki Kimata (267), Research Center for Seismology, Volcanology and Disaster Mitigation, Nagoya University, Nagoya, Japan

Enoch K. Kipseba (293), Mines and Geology Department, Ministry of Environment and Mineral Resources, Nairobi, Kenya

Bernard K. Kirui (133, 193), Department of Natural Resources, Egerton University. P.O. Box 536, 20115 Egerton, Kenya & Joint Research Centre—European Commission, Via E. Fermi, 2749 Ispra, Italy

Benjamin Kogo (237), Environmental and Biosystems Engineer, Norken Consult, Kenya

Zachariah Kuria (83), Department of Geology, College of Biological and Physical Sciences, University of Nairobi, P.O. Box 30197, Nairobi, Kenya

John B. Kyalo Kiema (51), Department of Geospatial and Space Technology, University of Nairobi, Nairobi, Kenya, and Institut d'Enseignement Supérieur (INES) - Ruhengeri, Musanze, Rwanda

Andrea Lupi (133), Joint Research Centre—European Commission, Via E. Fermi, 2749 Ispra, Italy

Eduardo Maeda (165), Department of Geosciences and Geography, University of Helsinki, Helsinki, Finland

Charles Maina-Gichaba (23, 293), Department of Geology, University of Nairobi, Nairobi, Kenya

Moses Masibo (3, 11, 293), Commissioner of Mines and Geology, Ministry of Mining, PO Box 30009-00100, Nairobi, Kenya

Michael E. McClain (67), Department of Water Science and Engineering, UNESCO-IHE, Delft, The Netherlands

Peris Muchiri (341), FAO-SWALIM, Nairobi, Kenya

Josphat K. Mulwa (267), Department of Geology, University of Nairobi, Nairobi, Kenya

Francis Mutua (341), Department of Meteorology, University of Nairobi, Nairobi, Kenya

James Mwang'ombe (165), Kenya Forest Service, Kenya

Dorothy Wanja Nyingi (199, 219), Ichthyology Section and KENWEB, National Museums of Kenya, P.O. Box 40658, Nairobi, Kenya

John P.O. Obiero (39), Department of Environmental and Biosystems Engineering, University of Nairobi, P.O. Box 30197, Nairobi, Kenya

Eric O. Odada (109, 331), Department of Geology, University of Nairobi, P.O. Box 30197–00100, Nairobi, Kenya

Mordecai O. Ogada (199, 219), Laikipia Wildlife Forum P.O. Box 764-10400, Nanyuki, Kenya

Daniel O. Olago (11, 31, 109, 331), Department of Geology, Institute for Climate Change and Adaptation, College of Biological and Physical Sciences (CBPS), University of Nairobi, P.O. Box 30197, Nairobi, Kenya

Christopher Oludhe (115), Department of Meteorology, University of Nairobi, Nairobi, Kenya

Loice M.A. Omoro (165), Faculty of Agriculture and Forestry, University of Helsinki, Helsinki, Finland

Christine A. Omuombo (3, 109, 331), Department of Geology, College of Biological and Physical Sciences (CBPS), University of Nairobi, PO Box 30197-00100, Nairobi, Kenya

Christian T. Omuto (123, 341), Department of Environmental and Biosystems Engineering, University of Nairobi, P.O. Box 30197-00100, Nairobi, Kenya

Japheth O. Onyando (39), Department of Agricultural Engineering, Egerton University, P.O. Box 536, Njoro, Kenya

Alfred Opere (315), Department of Meteorology, University of Nairobi, Nairobi, Kenya

Ilaria Palumbo (133, 181), Joint Research Centre—European Commission, Via E. Fermi, 2749 Ispra, Italy

Paolo Paron, UNESCO-IHE, Institute for Water Education, Department of Water Science & Engineering, P.O. Box 3015 - 2601DA, Delft, The Netherlands

Jean-François Pekel (133), Joint Research Centre—European Commission, Via E. Fermi, 2749 Ispra, Italy

Petri K.E. Pellikka (165), Department of Geosciences and Geography, University of Helsinki, Helsinki, Finland

Mika Siljander (165), Department of Geosciences and Geography, University of Helsinki, Helsinki, Finland

Amanda L. Subalusky (67), Department of Ecology and Evolutionary Biology, Yale University, New Haven, USA

Section I

Natural Resources

Chapter 1

General Geology of Kenya

Norbert Opiyo Akech*, Christine A. Omuombo* and Moses Masibo†
*Department of Geology, College of Biological and Physical Sciences (CBPS), University of Nairobi, PO Box 30197-00100, Nairobi, Kenya
†Commissioner of Mines and Geology, Ministry of Mining, PO Box 30009-00100, Nairobi, Kenya

1 INTRODUCTION

The geology of Kenya can be divided into five major geological successions: (1) the Archaean (Nyanzian and Kavirondian), (2) Proterozoic (Mozambique Belt and Bukoban), (3) Palaeozoic/Mesozoic sediments, (4) Tertiary/Quaternary volcanics and sediments, and (5) Pleistocene to Recent soils, alluvial beach sands, evaporites, fossil coral reefs and sandstones at the coast: alluvial and lacustrine sediments of the Rift Valley. There are also volcanic rocks of the Rift Valley from the younger volcanoes (Figure 1).

2 THE NYANZIAN SHIELD

The rocks forming the Nyanza Craton (Nyanzian and Kavirondian systems) are Archaean in age and are the oldest rocks in the country. Archaean greenstone belts are on a global scale broadly subdivided into two major groups, the older greenstones with supracrustal rocks that formed between 3.5 and 3.3 Ga and the younger, more widespread greenstone belts formed between 3.0 and 2.5 to 2.2 Ga (Windley, 1984). These greenstones in western Kenya fall in the younger group, with Rb/Sr ages being less than 3.0 Ga (Cahen et al., 1984), and are part of the granite–greenstone belt of the Tanzania Craton. This belt (Figure 1) covers much of northern Tanzania and extends through western Kenya into southeastern Uganda. The Kenyan part of the craton is predominantly composed of a greenstone–trondhjemite–tonalite–granite (TTG) sequence, divided into the Nyanzian and Kavirondian sequences. Within the two sequences are small remnants of the older basement commonly found in the southern half in Tanzania of the craton upon which the TTG sequences were emplaced and are also recognised in the Kenyan sector (Opiyo-Akech, 1988).

The volcanic rocks in the Kenyan sector of the Tanzania Craton constitute a bimodal mafic–felsic suite consistent with sequences found elsewhere in the

Developments in Earth Surface Processes, Vol. 16. http://dx.doi.org/10.1016/B978-0-444-59559-1.00001-3

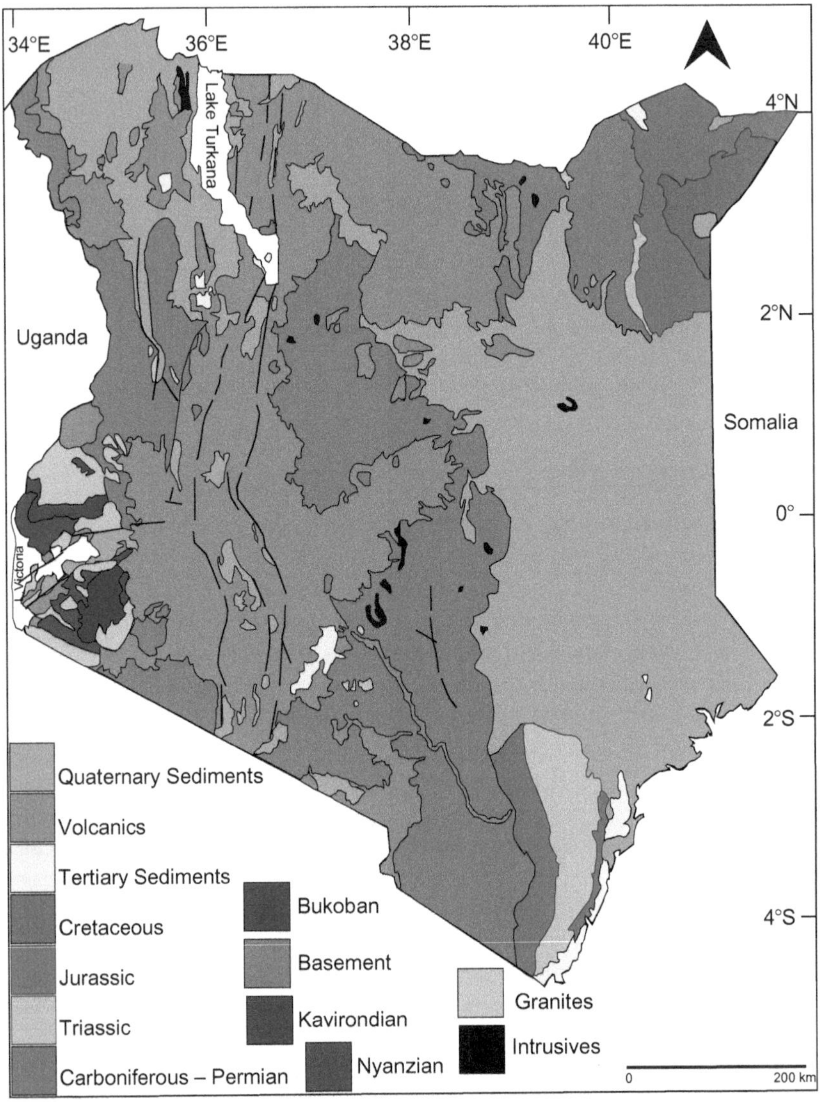

FIGURE 1 Map showing the general geology of Kenya. *Simplified from Survey of Kenya (1962).* (For colour version of this figure, the reader is referred to the online version of this chapter.)

world (Windley, 1984). The Nyanzian consists mainly of felsic volcanics with lesser amounts of mafic lavas, cherts and banded iron formations. In the southern terrain to the south of the Nyanza Rift, the mafic volcanics are tholeiitic while the felsic volcanics are calc-alkaline and high-K (Ichang'i, 1990; Ichang'i and MacLean, 1991). High-K dacites form a predominant volcanic rock type in the Migori segment of this belt and have a positive chemical correlation with most of the granites. The tholeiitic mafic and calc-alkaline felsic volcanics were erupted into a submarine environment, while the younger high-K dacites were erupted subaerially (Ichang'i, 1990). To the north of the Nyanza Rift in the northern terrain, the volcanic succession is in most parts composed of intermediate to acidic sequences of andesites, dacites and rhyolites of apparent calc-alkaline affinity (Opiyo-Akech, 1988; Ichang'i, 1990), together with related agglomerates and tuffs. There are in some localities relatively large exposures of tholeiitic basalts (lchang'i and MacLean, 1991; Opiyo-Akech, 1992).

The Kavirondian Group rests unconformably on the Nyanzian Group. This group consists of rudites typified by polymictic conglomerates, arenites represented by greywackes and grits, sandstones and argillites characterised by shaly rocks. The rocks have well-developed graded bedding typical of turbidites especially within the arenaceous to argillaceous horizons. Both the Nyanzian and Kavirondian systems are isoclinally folded about axes that have an east-westerly trend.

Intruding through the sequences are plutonic rocks that are typical of what are ordinarily found in other granite–greenstone belts of the world. The granitoids are typical of many 'TTG' (trondhjemite–tonalite–granite) plutonic suites that dominate the Archaean but are more potash-rich than most such suites (Ichang'i and MacLean, 1991; Opiyo-Akech et al., 1999a). In the northern terrain belt, the Got Ramogi syenitic intrusion has been encountered (Opiyo-Akech et al., 1999b). Got Ramogi is a relatively small hill with several episodes of syenitic intrusions, the successfully younger phases becoming more silica-deficient. The latest intrusions that occur as pegmatites intruding through older syenites contain large crystals of sodalite as a major mineral phase.

3 THE KISII GROUP

This is an outlier of volcanic rocks that consists of detrital and shallow-water sediments of the Kisii Group that unconformably overlies the Neo-Archaean greenstones and associated granitoid intrusions. The Kisii Group was initially correlated with rocks of the 'Bukoban System' and dated at circa 900 Ma (Huddleston, 1951). From the work by Pinna et al. (2000), it is noted that no comprehensive supracrustal sequence in eastern Africa matches the lithology, petrology and geochronology to the Kisii Group. Single-zircon evaporation Pb/Pb age dating suggests that the Kisii Group has similar ages to the greenstone sequences (2.75–2.58 Ga) in the northern part of the Tanzania Craton, negating the idea that the Kisii Group can be correlated to the Bukoban (Pinna et al., 2000). The volcanics of the Kisii Group are described by Pinna et al. (2000) as being geochemically similar to Phanerozoic continental-arc tholeiitic suites derived from a contaminated mantle. The relationships between these rocks with the underlying Neo-Archaean greenstones have, however, not been fully determined.

4 MOZAMBIQUE BELT

The Mozambique Belt is a structural unit within which a wide variety of meta-sedimentary and meta-igneous rocks are found showing a broad concordance of structural style and metamorphic history. The most characteristic feature of the Mozambique Belt is its structural trend, which is more or less north–south in its entire belt. In most of these rocks, the degree of deformation is intense and the rocks are of high metamorphic grades. They were thus referred to in the earlier literature as the basement system rocks, due to their high degree of metamorphism and deformation. They have a complex history of superimposed deformation and metamorphism consisting of high-grade reworked or reactivated basement (Muhongo, 1991). It is estimated that the important tectothermal events in this belt took place between 845 and 715 Ma (Cahen et al., 1984; Key et al., 1989). Subsequent reworking of the basement has been traced to ages ranging from 528 to 479 (Shibata and Suwa, 1979; Cahen et al., 1984; Frisch and Pohl, 1986). Recent work on the Mozambique Belt has shown that rocks can be subdivided into groups of contrasting lithology, structure and composition of igneous rocks content. These groups are being studied in greater detail in order to come up with proper chronostratigraphic terminology.

There are basic and ultrabasic intrusions within the Mozambique Belt. These basic to ultrabasic igneous complexes range in size from minor intrusives of bosses and dykes to relatively large bodies. They occur both east and west of the Rift Valley. Some of the older basic intrusions have undergone deformation and metamorphism to give ortho-amphibolites and charnockitic gneisses.

5 PALAEOZOIC AND MESOZOIC SEDIMENTS

The geological history of the Palaeozoic and Mesozoic sediments in Kenya is linked to the evolution, faulting and subsequent rifting during the break-up of Gondwanaland in the Mesozoic era (Pepper and Everhart, 1963). This break-up was followed by the formation of an intracratonic basin due to downwarping at the end of the Paleozoic, which was later filled by the easterly dipping Permo-Triassic clastic deposits of the Duruma Group, a sandstone series equivalent to the Karoo system of Southern Africa (Caswell, 1953, 1956; Thompson, 1954; Oosterom, 1988). These Permo-Triassic sediments are exposed in northeastern and southeastern regions of Kenya. Extensive exposures are present in the southeastern region where the sediments comprise of the *Taru*, *Maji-ya-Chumvi*, *Mariakani* and *Mazeras* Formations that consist of grits, shales, sandstones and fluvio-deltaic/Aeolian sequences, respectively (Caswell, 1953, 1956; Ase, 1978, 1981; Cannon et al., 1981; Oosterom, 1988), while in the northeastern region, isolated exposures of these sediments in the *Mansa Guda* Formation of Triassic age consist of conglomerates and quartzites of deltaic origin.

The Jurassic limestones and shales representing the Mesozoic sediments are associated with shallow to deep marine environments. In the northeastern region, they are represented by the *Daua Limestone Series* and the *Mandera Series* (Saggerson and Miller, 1957; Thompson and Dodson, 1958, 1960; Joubert, 1960, 1963), while in the southeastern region, the *Kambe*, *Kibiongoni* and *Mtomkuu* Formations represent these sediments. The Jurassic deposits are intercalated with shales and are fossiliferous in both regions and have yielded fossil ammonites, crinoids, bryozoans, corals, gastropods, nautiloids and brachiopods similar to the Tethyan–Mediterranean assemblages that portray the shallow to deep marine deposition (Caswell, 1953, 1956; Saggerson and Miller, 1957; Thompson and Dodson, 1958, 1960; Joubert, 1960, 1963; Pepper and Everhart, 1963; Ase, 1978, 1981; Cannon et al., 1981; Oosterom, 1988).

The tectonic history of the Cenozoic is characterised by extensive erosion and faulting, eliminating the older Cretaceous deposits. Caswell (1953), Joubert (1960) and Saggerson and Miller (1957) give an account of Freretown limestones in the south and the Marehan Sandstone Series in the north that appear as isolated blocks and hills, respectively. In most cases, the Tertiary and Quaternary volcanics and sediments overlie these deposits unconformably.

6 TERTIARY AND QUATERNARY VOLCANICS

The Tertiary and Quaternary volcanics are predominantly distributed along the East African Rift System (EARS). Trends in age relationships show that the older rocks are found to the north of the rift and generally get younger to the south. There is added complexity in their spatial occurrence associated with the eastward shift of volcanism and tectonics, thus resulting in the variations in occurrences in the floor of the EARS and on the peneplains to the west and east of the Rift Valley (Schluter, 1997; Chorowicz, 2005). The oldest volcanics are basalts and trachytes of the Oligocene age (33–25 Ka) found west of Lake Turkana and linked to the volcanic cycles of the Ethiopian rift (Bellieni et al., 1987; Schluter, 1997). Basalts account for more than half the volume of volcanic rocks in the Kenyan rift volcanic rocks of the Miocene to Plio-Pleistocene age. Most of the basalts occur as Miocene flows that flooded a proto-rift depression in northern Kenya. The oldest Miocene basalts comprise the Samburu, Lodwar, Mt. Elgon and Rusinga basalts. There are post-Miocene basalts of the rift floor that show affinities with lavas on the flanks of rift and resemble rocks of multicentre fields east of the rift. The formation of the Laikipia plateau marks the mid-Miocene volcanic evolution followed by the formation of the Yatta and Kapiti phonolitic flows (Schluter, 1997). Late Miocene and Pliocene volcanics are associated with the Aberdares and the development of Mt. Kenya east of the EARS and the Kamasia and Mau ranges to the west (Baker, 1967; Schluter, 1997). Nephelinites are frequently accompanied by highly undersaturated phonolites (phonolitic nephelinites), late Miocene phonolites, trachytes, rhyolites and ignimbrites that occur on the rift floor in southern Kenya and are associated phases of domal uplift in Plio-Pleistocene times. Quaternary volcanism within the floor of the valley has given rise to varied craters and cinder cones such as Longonot, Menengai and Suswa. Volcanism in the eastern flank of the Kenya rift has been addressed in the section on Quaternary evolution.

7 TERTIARY AND QUATERNARY SEDIMENTS

The Tertiary sedimentary deposits are closely linked to lacustrine, fluvial and Aeolian systems. Along the coast, the Miocene deposits comprise yellow sand marls, mottled clays, conglomerates, limestones and sands of the *Baratamu Formation*. These deposits reflect shallow marine regressive deposition with the basal deposit of marginal marine fluvial environment known as the 'Mombasa crag' and 'Kipevu beds'. The Pliocene deposits are composed of red sands, clays, gravels and conglomerate deposited under fluvial environment and are known as the *Marafa Formation*. The Plio-Pleistocene deposits are a mixed sequence of red beds deposited under Aeolian and beach environments represented by red hills known as the *Magarini Formation* (Caswell, 1953, 1956; Ase, 1978, 1981; Cannon et al., 1981; Oosterom, 1988).

In the rift system, sediments usually occur at the base of volcanic successions, intercalated with them, or occurring in tectonic troughs. The repeated faulting of the Rift Valley floor and the numerous volcanic eruptions created many short-lived basins of internal drainage in which lacustrine and fluviatile sediments accumulated. Most of these sediments are unfossiliferous, but a few are of interest as they contain deposits that bear artefacts and interesting fossils that have been studied extensively. There are many sediment deposits in various parts of Kenya. The more important sediments of middle Pleistocene are the Olorgesailie lake beds, a lacustrine series with much diatomite, mammalian fossils and artefacts. This is also comparable to the Kariandusi sediments near Gilgil and the Kanjera Beds in the Kavirondo Gulf off Lake Victoria. Olorgesailie beds and Kariandusi sediments are in the Rift Valley.

The Quaternary sediments are widely distributed in the country with extensive deposits in the eastern region and some in the northwestern region. Along the coast, these sediments represent marine transgression and regression leading to the deposition of sands and formation of Pleistocene to Recent coral reef complexes of variable onshore and offshore extent (Caswell, 1953, 1956; Ase, 1978, 1981; Cannon et al., 1981; Oosterom, 1988).

REFERENCES

Ase, L.E., 1978. Preliminary report on studies of shore displacement at the southern coast of Kenya. Geogr. Ann. A 60A (3–4), 209–221.

Ase, L.E., 1981. Studies of shores and shore displacement on the southern coast of Kenya—especially in Kilifi District. Geogr. Ann. A 63A (3–4), 303–310.

Baker, B.H., 1967. Geology of the Mount Kenya Area, Geological Survey of Kenya. Report No. 79, 78 pp.

Bellieni, G., Visentin, E.J., Piccirillo, E.M., Zanettin, B., 1987. Volcanic cycles and magmatic evolution in northern Turkana (Kenya). Tectonophysics 143 (1–3), 161–168. http://dx.doi.org/10.1016/0040-1951(87)90085-0.

Cahen, L., Snelling, N.J., Delhal, T., Vail, J.R., 1984. The Geochronology and Evolution of Africa. Clarendon Press, Oxford, 512 pp.

Cannon, R.T., Simiyu Siambi, W.M.N., Karanja, E.M., 1981. The Proto-Indian Ocean and a probable Paleozoic/Mesozoic Triradial rift system in East Africa. Earth Planet. Sci. Lett. 52, 419–426.

Caswell, P.V., 1953. Geology of the Mombasa—Kwale Area. Report No. 24. Geological Survey of Kenya.

Caswell, P.V., 1956. Geology of the Kilifi—Mazeras Area. Report No. 34. Geological Survey of Kenya.

Chorowicz, J., 2005. The East African rift System. J. Afr. Earth Sci. 43 (1–3), 379–410.

Frisch, W., Pohl, W., 1986. Petrochemistry of some mafic and ultramafic rocks from the Mozambique Belt, SE Kenya. Mitt. Österr. Geol. Ges. 78, 97–114.

Huddleston, A., 1951. Geology of the Kisii District. Degree sheet 41, S.E. Quadrant. Rep. Geological Survey of Kenya, 18, 64 pp.

Ichang'i, D.W., 1990. The Migori Segment of the Archean Nyanza Greenstone Belt, Kenya: Geology, Geochemistry and Economic Mineral Potential. Ph.D. ThesisMcGill University, Montreal, 144 pp.

Ichang'i, D.W., MacLean, W.H., 1991. The Archean volcanic facies in the Migori segment, Nyanza greenstone belt, Kenya: stratigraphy, geochemistry and mineralization. J. Afr. Earth Sci. 13, 277–290.

Joubert, P., 1960. Geology of the Mandera-Damassa Area. Report No. 48. Geological Survey of Kenya, 65 pp.

Joubert, P., 1963. Geology of the Wajor-Wajir Bor Area. Report No. 57. Geological Survey of Kenya, 34 pp.

Key, R.M., Charsley, J.J., Hackman, B.D., Wilkinson, A.F., Rundie, C., 1989. Superimposed Upper Proterozoic collision-controlled orogenesis in the Mozambique belt of Kenya. Precambrian Res. 44, 197–225.

Muhongo, S., 1991. The Mozambique Belt: a polyorogenic mobile belt. In: Muhongo, S. (Ed.), Geology for Economic Development, UNESCO Newsletters, vol. 8, pp. 5–14.

Oosterom, A.P., 1988. The Geomorphology of Southeast Kenya. Ph.D. ThesisWageningen Agricultural University, Netherlands, 227 pp.

Opiyo-Akech, N., 1988. Geology and Geochemistry of the Late Archaean Greenstone Associations, Maseno Area, Kenya. Ph.D Thesis, University of Leicester, England, 149 pp.

Opiyo-Akech, N., 1992. Tectono-magmatic emplacement of the Kenyan granite greenstones: geochemical evidences from the Maseno area, western Kenya. Afr. J. Sci. Technol. B 6 (2), 27–33.

Opiyo-Akech, N., Tarney, J., Hoshino, M., 1999a. Petrology and geochemistry of granites from the Archaean terrain north of Lake Victoria, western Kenya. J. Afr. Earth Sci. 29, 283–300.

Opiyo-Akech, N., Ndege, M.M., Okoth, P., Omenge, J., 1999b. Geomorphology, Geology and water resources of the Got Ramogi area. In: Odhiambo, T.R. (Ed.), GOT RAMOGI SACRED FOREST: Principles of Forest Conservation and Utilization in Tropical Africa.

Pepper, J.F., Everhart, G.M., 1963. The Indian Ocean: The Geology of Its Bordering Lands and the Configuration of Its Floor, Department of Interior United States, Geological Survey, 33 pp.

Pinna, P., Cocherie, A., Thiéblemont, D., Jezequel, P., 2000. The Kisii Group of western Kenya: an end-Archæan (2.53 Ga) late orogenic volcano sedimentary sequence. J. Afr. Earth Sci. 30 (1), 79–97.

Saggerson, E.P., Miller, J.M., 1957. Geology of the Takabba-Wergudud Area, Mandera District. Report No. 40. Geological Survey of Kenya, 42 pp.

Schluter, T., 1997. Geology of East Africa. Gebruder Borntraeger, Berlin-Stuttgart, 484 pp.

Shibata, K., Suwa, K., 1979. A geochronological study on granitoid gneiss from the Mbooni Hills, Machakos Area, Kenya. In: 4th Preliminary Report of African Studies. Nagoya University, Japan, pp. 163–167.

Survey of Kenya, 1962. Geological Map of Kenya. Geological Survey of Kenya, Nairobi, Kenya.

Thompson, A.O., 1954. The Geology of the Malindi Area. Report No. 36. Geological Survey of Kenya.

Thompson, A.O., Dodson, R.G., 1958. Geology of the Derkali-Melka Murri Area. Report No. 43. Geological Survey of Kenya, 35 pp.

Thompson, A.O., Dodson, R.G., 1960. Geology of the Bur Mayo—Tarabaj Area. Report No. 47. Geological Survey of Kenya, 49 pp.

Windley, B.F., 1984. The Evolving Continents, second ed. Wiley, Chichester.

Chapter 2

Mineral, Oil and Gas Resources

Norbert Opiyo Akech*, Moses Masibo† and Daniel O. Olago‡

*Department of Geology, College of Biological and Physical Sciences (CBPS), University of Nairobi, PO Box 30197, Nairobi, Kenya

†Commissioner of Mines and Geology, Ministry of Mining, P.O. Box 30009,00100, Nairobi, Kenya

‡Department of Geology, and Institute for Climate Change and Adaptation, College of Biological and Physical Sciences (CBPS), University of Nairobi, P.O. Box 30197, Nairobi, Kenya

1 MINERAL RESOURCES

1.1 Introduction

The mining sector in Kenya is still largely underexplored, and there is a possibility of discovering more minerals through systematic exploration in the country. Investment in the sector has been hindered by lack of proper legislation and regulation. The Mining Act Chapter 306 of the Laws of Kenya has governed the sector. The Act was promulgated in 1940 and was last reviewed in 1987 to match the prevailing contemporary international practice. A comprehensive review of the Mining Act started in 1993 and is now complete and is soon to be published into law. The reviewed act provides for less discretionary powers to the licencing authorities and greater security of tenure. A mineral policy, which did not exist in the current act, has also been formulated.

The law will factor and regularise the activities of small-scale miners. It will also take cognisance of the requirements of the new Environmental Management and Coordination Act of 1999. The Environmental Act has now been passed into law. The Government of Kenya has also embarked on the process of formulating a mineral policy, which does not currently exist in the current statutes.

In spite of the challenges, there has been a general improvement and steady growth in the performance of Kenya's mineral sector in recent years. This is poised to improve with the enactment of the new law and the formation of a full-fledged Mining Ministry.

1.2 Mineral Occurrence in Kenya

Minerals in Kenya include soda ash, fluorite (fluorspar), titanium, niobium, rare earth elements (REEs), diatomite, carbon dioxide, gold, coal, iron ore,

Developments in Earth Surface Processes, Vol. 16. http://dx.doi.org/10.1016/B978-0-444-59559-1.00002-5

vermiculite, kyanite, manganese, chromite, silica sand, gemstones, gypsum and limestone.

In summary, mineralization in Kenya falls into the following major geologic settings:

- **i.** Gold, base metals and banded iron formations occur in the Archaean Greenstone Belts in the southwestern part of Kenya.
- **ii.** Gold, base metal (including iron ore, copper ore and manganese ore deposits), industrial mineral (including limestones, kyanite, vermiculite and wollastonite) and coloured gemstone occurrences in the Neoproterozoic Mozambiquian system in various parts of Kenya.
- **iii.** Carbonatites with phosphorus, niobium and REE mineralization as well as soda ash, diatomite, bentonite, carbon dioxide and fluorspar deposits associated with the Tertiary rift system.
- **iv.** Heavy mineral sands are associated with Cenozoic detrital sediments at the Coast.
- **v.** Gypsum deposits associated with recent sediments in various parts of the country.
- **vi.** Coal deposits in Pleistocene basins (Mui) and Karoo sedimentary series (Coast).
- **vii.** Manganese, lead/zinc, uranium, REEs and barite.

The minerals that are found and exploited in Kenya in significant quantities are soda ash (trona) in Lake Magadi area, fluorspar at Kimwarer in Kerio Valley, niobium and REEs on Mrima Hill in Kwale County and titanium in Kwale, Malindi and Lamu areas. There is also reasonable potential for gold in Kakamega, Vihiga, Migori, Transmara, Bondo, Siaya, Pokot and Turkana as well as the Nandi area. Coal occurs in Mwingi and Mutitu areas and is the subject of current investigations to establish its viability for exploitation. Iron ore occurs in parts of Taita, Meru, Kitui, Kilifi and Samia Hills areas where there are efforts by the Ministry and private prospecting companies to undertake its evaluation. Manganese ore occurs in Kilifi, Taita and Samburu areas where a number of exploration companies have shown interest. Other mineral occurrences include diatomite at Kariandusi near Gilgil, vermiculite on Kinyiki Hill, gypsum in El Wak, in Garissa, Tana River, Kajiado and Turkana areas, natural carbon dioxide at Kereita in Kiambu, and a variety of gemstones in Taita, Kwale, Kitui, Mwingi, Kajiado, Isiolo, Pokot, Turkana and Marsabit areas.

1.3 Mineral Production in Kenya

Mining activities have taken place in Kenya for many years up to the present time, but most of the medium- to large-scale mining of gold and base metals took place during pre-independence days. The largest gold in the country was the Rosterman mine near Kakamega town in western region that operated from the 1930s and closed down in 1952. A medium-scale copper and gold mine was in operation at Macalder in Migori County from 1956 until it closed down in the late 1960s. There is however increased exploration activities in the region and possibilities of reworking the huge tailings left behind from the earlier activities for gold are likely.

Lead ore mines were operated in the Kinagoni and Vitengeni areas of Coast region till the 1970s.

In the post-independence years, mining operations have been dominated by the production of a variety of industrial minerals, among which are soda ash, fluorspar, diatomite and limestone. Gold and gemstone production became the main activity of small-scale miners who have operated continuously in different parts of the country until the late 2000s when medium-scale mines were opened in western Kenya. The charts in Figures 1 and 2 show Kenya's mineral production and export for the period 2007–2011 and Table 1 shows the value of export (Mines and Geology Department).

Soda ash is produced from the mineral trona that occurs in Lake Magadi, which is situated within the Great Rift Valley. The commodity is produced by Tata Chemicals, formerly known as Magadi Soda Company Limited, in one of the biggest known natural sources of trona in the world. The trona is also mined together with common salt and for the country Lake Magadi is also a major source of crude salt.

Fluorspar is mined to the east of the town of Eldoret within the Rift Valley. The Kenya Fluorspar Company Limited operates the mine. The mine produces acid-grade fluorspar of which the bulk is for export. Production of fluorspar averages about 100,100 tonnes per year.

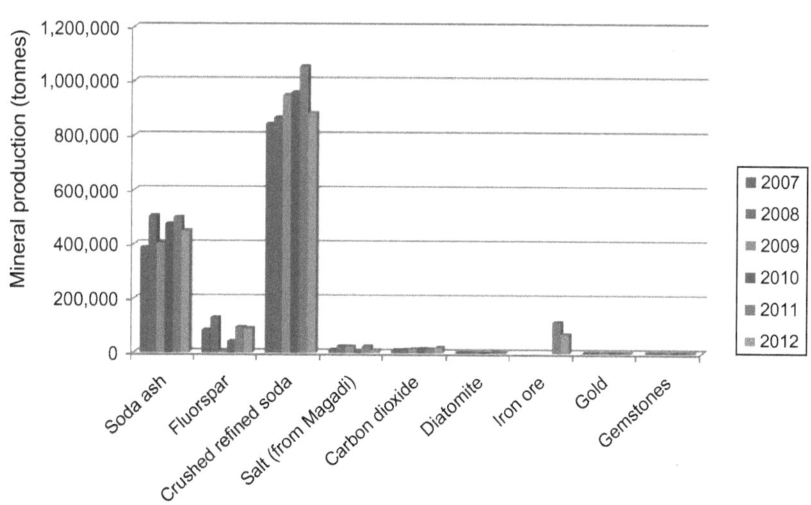

FIGURE 1 Kenya mineral production (2007–2012). (For colour version of this figure, the reader is referred to the online version of this chapter.)

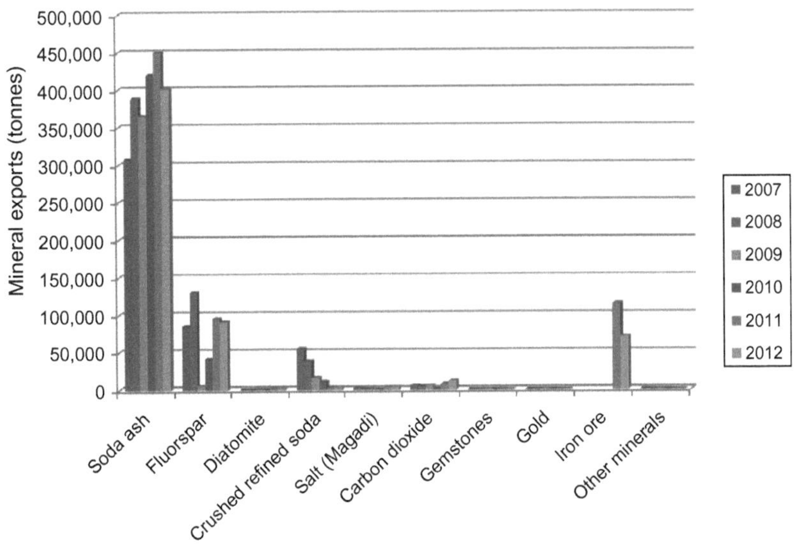

FIGURE 2 Kenya mineral exports (2007–2012). (For colour version of this figure, the reader is referred to the online version of this chapter.)

Diatomite is produced about 130 km to the northwest of Nairobi, but it is known to occur in a number of localities within the Rift Valley. The commodity is produced for both export and local markets. Its potential is high.

Vermiculite is mined in the northern central part of the Rift Valley Province within the pre-Cambrian rocks of Mozambique Belt.

Limestone and lime products, marbles and dolomites widely occur in the country and huge deposits of the commodities are known to exist in the form of sedimentary limestones and its metamorphic form 'marble'.

Natural carbon dioxide associated with the rift system is currently being extracted from Kereita, about 50 km northwest of Nairobi and Sosian near Eldoret by Carbacid (CO_2) Limited and marketed for industrial purposes within the East African region.

Gypsum is mined in Tana River in the coastal belt, Kajiado near Nairobi and in Turkana in northwestern Kenya. The gypsum mined is supplied to local cement plants and some is exported to Uganda.

As regards *gemstones*, Kenya boasts of a wide range of coloured and ornamental stones, which are mined in the country. These include, mainly in order of importance, *ruby*, *tsavorite*, *sapphire*, *corundum*, various types of *garnet*, *peridot*, *tourmaline*, *aquamarine*, *zoisite* and *rhodolite*.

Iron ore deposits are found in the form of oxides: magnetite (Fe_3O_4), hematite (Fe_2O_3), goethite, limonite and carbonates—siderite and anchorites. Some of iron ore deposits are found at Marmante in the greater Tharaka area, Tseikuru in Mwingi, Ikutha in Kitui County, Migori County, Kishushe in

TABLE 1 Kenya's Mineral Production and Value of Export for the Period 2007–2012

		2007		2008		2009		2010		2011		2012	
Mineral	Unit	*Quantity*	*Value (×1000)*	*Quantity*	*Value (×1000)*	*Quantity*	*Value (×1000)*	*Quantity*	*Value (×1000)*	*Quantity*	*Value (×1000)*	*Quantity*	*Value (×1000)*
Soda ash	MT	386,578	4,769,457	502,846	8,881,714	404,904	6,085,430	473,689	6,980,230	499,052	7,353,976	449,269	9,388,035,457
Fluorspar	MT	85,115	995,867	130,100	1,949,194	5500	122,540	40,750	789,421	95,051	3,984,413	91,000	2,941,615,258
Crushed refined soda	MT	843,043	5,448,046	865,788	6,974,924	948,076	13,202,957	959,160	7,156,271	1,054,236	7,865,631	882,801	7,657,487,131
Salt (from Magadi)	MT	11,596	58,068	24,345	139,171	24,125	154,730	6194	35,276	24,638.	140,323	9980	65,664,808
Carbon dioxide	MT	11,028	78,630	12,317	296,324	15,097	333,145	16,345	466,653	15,770.	450,239	19,919	639,987,028
Diatomite	MT	1842	81,184	1776	85,937	1697	102,723	224	8913	712.50	14,667	1746	85,573,165
Iron ore	MT									114,800.00	649,345	70,500	397,707,729
Gold	GM	3,023,325	3923	339,923	592,891	1,055,330	2,283,998	2,355,176	6,216,996	1,635,597	5,650,820	3,642,122	13,919,684,925
Gemstones	GM	8,927,000	110,975	20,934	178,606	39,409	141,324	167,625	226,397	310,096	230,875	120,901,528	157,467,488
Grand total			15,465,119		19,098,761		22,426,847		21,880,159		26,340,292		35,253,222,990

Taita-Taveta, Ishiara in Mbeere, Bukura southwest of Kakamega, Baragoi in Samburu, Uyoma in Bondo area, Maturu in Lugari District, Malakisi in Bungoma District, Ruri in Suba District, Homa Hills at Homa Bay, Mfangano Island in Suba District and God Nyango and Rachuonyo in Nyanza.

Lead was being produced from galena, which was being mined in small quantities in the coastal belt, where it was smelted together with scrap lead to manufacturer lead batteries.

Silica sand deposits are found in Ramisi area of Kwale District where they are exploited for glass manufacture.

1.4 Other Minerals

Manganese deposit occurrences are within Kilifi district along the coast, Samburu and Moyale Districts in the Rift Valley and in parts of Isiolo in Eastern Province. Production of this mineral has commenced albeit in a small scale.

Copper occurrences have been reported within various parts of the country including Mwingi, Taita-Taveta and West Pokot where some operators have obtained exploration licences to determine their economic viability.

Chromite deposits are found in Sigor, West Pokot in Rift Valley Province and at Dabel, Moyale District in Eastern Province. Further exploration is required to prove their economic viability.

Niobium and other rare earth elements occur at Mrima Hill in Kwale where they are being explored for possible exploitation. Other occurrences are in Ruri and Homa Hill areas in Homa Bay.

Titanium is known to occur in large quantities in the coastal area. Mining of the mineral is expected to commence soon. Base titanium is in the process of commencing production.

Coal deposits of considerable quantities have been discovered within the Mui Basin of Mwingi and Kitui Districts. Other occurrences are reported within the Karoo Basin in Kwale and Kilifi Districts.

2 OIL AND GAS RESOURCES

Petroleum exploration in Kenya[1] begun in the 1950s with the first well being drilled in 1960. In 1986, the petroleum exploration and production legislation in Kenya was revised to provide suitable incentives and flexibility to attract international exploration interest in the country. Several companies such as Amoco and Total drilled a number of wells between 1985 and 1990, which turned out to be dry but with some indications of oil and gas. Following the recent discovery in March 2012 (Ngamia 1 well) and November 2012 (Twiga 1 well) of hydrocarbon deposits in the Lokichar Basin by Tullow Oil, exploration efforts have been intensified in the related Mesozoic and

1. This historical text is summarised from the NOCK website (www.nock.go.ke).

Early Tertiary rift basins of Kenya with a view to meeting the global, regional and local demand for energy. If the reserves prove commercially viable, Kenya will be well positioned to advance towards production. Here, we examine some of the key basins with hydrocarbon potential. Other basins such as the Lotikipi, Kavirondo and Magadi Basins are not as well studied in terms of hydrocarbon potential, but surveys are ongoing in these regions as well.

2.1 Lokichar Basin

The Lokichar Basin, in which the first oil discovery has been made, has a 7-km-thick infill of which the upper half had been calibrated by the approximately 2950 m deep Loperot-1 well (Morley et al., 1999). This approximately 3-km-thick sediment pile includes alternating packages of coarse to fine sandstones tens to hundreds of metres thick, which from bottom to top are called the Lokhone and Auwerwer Formations, respectively. The Lokhone Formation includes two major shale intervals named the Loperot (lower) and Lokhone (upper) Shale Members (Morley et al., 1999; Talbot et al., 2004). The Loperot-1 well penetrated the Lokhone Shale Member between 925 and 1385 m, and the Loperot Shale Member was found between 2325 and 2950 m. The Loperot and Lokhone Shale Members are proposed to be Eocene–Oligocene and Oligocene–Early Miocene in age, respectively (Morley et al., 1999). The Lokhone Shale Member is demonstrated to contain excellent source rocks (1–17% TOC), while the Loperot Shale Member presented TOC values of 0.2–3.3% (Morley et al., 1999; Talbot et al., 2004). The upper part of the Lokichar Basin fill, formed by the Auwerwer Sandstone Formation, is overlain by basaltic lava flows (Auwerwer Basalts), which are of Middle Miocene age (Morley et al., 1992). The two Loperot and Lokhone Shale Members correspond to major phases of lake environment occupying the Lokichar Basin during Eocene to Oligocene–Early Miocene times. Organic geochemistry and palynofacies analyses have demonstrated a phytoplankton-dominated organic matter, accumulating in a large and deep freshwater body, with characteristics similar to the offshore waters of the modern Lake Tanganyika or Malawi (Talbot et al., 2004).

2.2 Lake Turkana Basin

The present-day Lake Turkana Basin is part of a string of major N–S-oriented half-grabens that developed from Cretaceous through Eocene–Oligocene to Plio-Pleistocene times (Mugisha et al., 1997; Morley et al., 1999; Hautot et al., 2000; Tiercelin and Lezzar, 2002; Tiercelin et al., 2004). The northern end of this basin is an N–S-trending half-graben that formed during Middle to Late Miocene times, following a major phase of volcanic activity dated Late Eocene to Middle Miocene that resulted in the deposition of the 2.5-km-thick Turkana Volcanics (Bellieni et al., 1987; Zanettin et al., 1983). The structure of the northern half of the Turkana Basin has been partly imaged using offshore reflection seismic data of the PROBE Project (Rosendahl et al., 1986; Dunkelman et al., 1988, 1989) and few seismic lines acquired by Amoco Kenya Petroleum Company (Wescott et al., 1999). The major N–S-oriented border faults that delineate (to the west) the Turkana Rift Basin are imaged on the onshore seismic line. Multifold seismic reflection and geologic mapping indicate that at least 3 km of Neogene sediments fills beneath Lake Turkana, including a series of half-graben basins, with centrally located Quaternary volcanic centres, which are linked end to end by structural accommodation zones (Dunkelman et al., 1988, 1989). The deep basin configuration below the Turkana Volcanics is almost unknown because of the considerably degraded quality of seismic data below the subsurface volcanics (Morley et al., 1992; Wescott et al., 1999).

2.3 Kerio and Baringo Basins

Between the equator and 1° latitude north, extension mechanisms occurred during Palaeogene time (Mugisha et al., 1997; Hautot et al., 2000), contemporaneous with rifting in Northern Kenya, Southern Sudan and Ethiopia (Morley et al., 1992; Hendrie et al., 1994). This resulted in the initiation of the two oldest deep rift basins of the Central Kenya Rift, the Kerio Basin to the west and the Baringo Basin to the east (Tiercelin and Lezzar, 2002). Exposed lake- and fluvial-type sediments of possible Palaeogene age are green laminated shales and sandstones forming the Kimwarer and Kamego Formations that outcrop in the Kerio and Baringo Basins, respectively (Renaut et al., 1999; Hautot et al., 2000). These two formations represent the upper part of a several-kilometre-thick sediment pile of possible Palaeogene age that is only illustrated by geophysical methods (Mugisha et al., 1997; Hautot et al., 2000). They lie at the base of a nearly entire rift-fill sequence (known as the Tugen Hills sequence) that is considered as one of the best exposed successions of Neogene sediments in the East African Rift System (e.g. Andrews and Banham, 1999). Thick basaltic and phonolitic series more or less extensively capped the Kerio and Baringo Basins (Samburu, Sidekh and Elgeyo Formations) between 23 and 10 Ma (Chapman and Brook, 1978; Hill et al., 1986). During this period, fluviolacustrine sedimentation continued to develop in the Kerio Basin of Central Kenya, with the deposition of the 400-m-thick Tambach Formation dated between 16 and 14 Ma (Renaut et al., 1999). The Tambach Formation was in turn capped by the Uasin Gishu phonolites by 12 Ma (Lippard, 1973; Chapman and Brook, 1978) and was followed by a second phase of rift tectonics that affected the Kerio Basin, initiating the deposition of the Ngorora fluviolacustrine formation in an almost 100-km-long and 40-km-wide faulted basin (Tiercelin and Lezzar, 2002). The lower half of this thicker Ngorora Formation is only 'geophysically' represented, while the upper 400 m of sediments outcrops largely in the Tugen Hills region between the Kerio and Baringo Basins (Tiercelin and Lezzar, 2002).

2.4 Anza Basin

The Anza graben/rift is a large multiphase continental rift basin that links Lamu Embayment of southeastern Kenya with the South Sudan rifts. The graben trends approximately NW–SE, with its axis passing through Marsabit and Habaswein in Eastern Kenya. The graben can be divided into two major sectors—the northern and southern sectors (Bosworth and Morley, 1994; Dindi, 1994). The Anza graben is the southeastern continuation of the Cretaceous oil- and gas-prolific Central African Rift System, which extends into Kenya from the Melut and Muglad Basins of Southern Sudan where Cretaceous working petroleum systems have been proved and oil discovered. Geophysical and exploratory drilling data demonstrate that the basement depth varies from 8 to 10 km. The graben is generally concealed beneath Late Tertiary/Recent volcanic rocks and the sub-Miocene basin is infilled by various superficial unconsolidated sediments. Quaternary sediments that variably mask the older sedimentary sequence are common within the Anza graben. The major sediments in the graben were deposited mainly in a continental, mainly fluvial and lacustrine setting. Continental sedimentation dominated in most of the Anza Basin during the Cretaceous. At least one marine incursion occurred as demonstrated by the Cenomanian section in the N'dovu-1 well (Bosworth and Morley, 1994). Only Late Cretaceous sediments have been reached by drilling in the central and southern Anza rift and this is attributed to the thickening of the Upper Cretaceous and Tertiary section towards the southeast. Rifting continued in the Anza rift during the Tertiary and a thick sequence (locally in excess of 3000 m) of Palaeocene–Miocene fluvial and lacustrine sediments was deposited (Bosworth and Morley, 1994). The geology of the southern Anza graben is comprised of sediments, sands and volcanics of Quaternary age underlain by Permo-Triassic and Mesozoic rocks that are rarely exposed. In addition, Tertiary and Mesozoic sequences also occur at depth and these have been encountered in oil exploration boreholes (Dindi, 1994; Mariita, 2003). The tectonic structures associated with the graben are not readily observable because of the sediment and/or volcanic mask.

2.5 Lamu Basin

The Lamu Embayment occupies the coastal onshore and offshore areas of southeast Kenya and was initiated in the Karoo (Permo-Triassic) times. This fault-bounded basin formed as a result of the Palaeozoic–Early Mesozoic phase of rifting that developed at the onset of Gondwana dismemberment. The resultant graben was filled by Karoo (Permian–Early Jurassic) continental siliciclastic sediments. Carbonate deposits associated with the Tethyan sea invasion dominated the Middle to Late Jurassic Basin fill (Simiyu, 1989). Cessation of the relative motion between Madagascar and Africa in the Early Cretaceous heralded passive margin development and deltaic sediment progradation until the Palaeogene. Shallow seas transgressed the basin in the Miocene when another carbonate regime prevailed. In offshore depo-centres, the 12,000–13,000 m sedimentary column in the embayment consists mostly of marine sandstones, shales and carbonates, which range from the Karoo through the Tertiary times (cf. Walters and Linton, 1973). In contrast, onshore sediment thicknesses range from 3250 m in the area of Garissa High to a probable maximum of 10,000 m in the coastal areas (NOCK, 2008). This variation in sediment thickness results, in part, from abundant hiatuses caused by periodic erosion or non-deposition. Thus, the basin depositional history is characterised by pulses of transgressive and regressive cycles bounded by tectonically enhanced unconformities dividing the total sedimentary succession into discrete megasequences: source-rock strata occur within the Palaeogene depositional cycle and were lowered into the oil window in Miocene time, when the coastal parts of the basin experienced the greatest amount of subsidence (NOCK, 1987). The tectono-eustatic pulses of the Tertiary brought about source and reservoir strata into a spatial relationship in which hydrocarbons could be entrapped. A basement high on the continental shelf has potential for Karoo sandstone and Jurassic limestone reservoirs.

REFERENCES

Andrews, P., Banham, P. (Eds.), 1999. Late Cenozoic Environments and Hominid Evolution: A Tribute to Bill Bishop. Geological Society, London.

Bellieni, G., Justin Visentin, E., Piccirillo, E.M., Zanettin, B., 1987. Volcanic cycles and magmatic evolution in northern Turkana (Kenya). Tectonophysics 143, 161–168.

Bosworth, W., Morley, C.K., 1994. Structural and stratigraphic evolution of the Anza rift, Kenya. Tectonophysics 236, 93–115.

Chapman, G.R., Brook, M., 1978. Chronostratigraphy of the Baringo Basin, Kenya Rift Valley. In: Bishop, W.W. (Ed.), Geological Background to Fossil Man. Geological Society of London, Scottish Academic, London, pp. 207–223.

Dindi, E.W., 1994. Crustal structure of the Anza-graben from gravity and magnetic investigations. Tectonophysics 236, 359–371.

Dunkelman, T.J., Karson, J.A., Rosendahl, B.R., 1988. Structural style of the Turkana Rift, Kenya. Geology 16, 258–261.

Dunkelman, T.J., Rosendahl, B.R., Karson, J.A., 1989. Structure and stratigraphy of the Turkana Rift from seismic reflection data. J. Afr. Earth Sci. 8, 489–510.

Hautot, S., Tarits, P., Whaler, K., Le Gall, B., Tiercelin, J.-J., Le Turdu, C., 2000. Deep structure of the Baringo Rift Basin (Central Kenya) from three-dimensional magnetotelluric imaging: implications for rift evolution. J. Geophys. Res. 105 (B10), 23,493–23,518.

Hendrie, D.B., Kusznir, N.J., Morley, C.K., Ebinger, C.J., 1994. Cenozoic extension in northern Kenya: a quantitative model of rift basin development in the Turkana region. In: Prodehl, C., Keller, G.R., Khan, M.A. (Eds.), Crustal and Upper Mantle Structure of the Kenya Rift. Tectonophysics, vol. 236. Elsevier B.V., Amsterdam, The Netherlands, pp. 409–438.

Hill, A., Curtis, G., Drake, R., 1986. Sedimentary stratigraphy of the Tugen Hills, Baringo, Kenya. In: Frostick, L.E., Renaut, R.W., Reid, I., Tiercelin, J.-J. (Eds.), Sedimentation in the African Rifts. Geol. Soc. London Spec. Publ., vol. 25. Blackwell Scientific Publications, Oxford, UK, pp. 285–295.

Lippard, S.J., 1973. Plateau phonolite lava flows, Kenya. Geol. Mag. 110, 543–549.

Mariita, N.O., 2003. An integrated geophysical study of the northern Kenya rift crustal structure: implications for geothermal energy prospecting for Menengai area. PhD Dissertation, University of Texas at El Paso, USA.

Morley, C.K., Wescott, W.A., Stone, D.M., Harper, R.M., Wigger, S.T., Karanja, F.M., 1992. Tectonic evolution of the northern Kenyan Rift. J. Geol. Soc. Lond. 149, 333–348.

Morley, C.K., Wescott, W.A., Harper, R.M., Cunningham, S.M., 1999. Geology and geophysics of the Rukwa Rift. In: Morley, C.K. (Ed.), Geoscience of Rift systems—Evolution of East Africa. American Association of Petroleum Geologists Studies in Geology, vol. 44. Cambridge University Press, Cambridge, UK, pp. 91–110.

Mugisha, F., Ebinger, C.J., Strecker, M., Pope, D., 1997. Two-stage rifting in the Kenya rift: implications for half-graben models. Tectonophysics 278, 63–81.

NOCK, 1987. Aeromagnetic data interpretation of the Winam Gulf and the East African Rift areas in Kenya. National Oil Corporation of Kenya, Unpubl. Rep. No. 26/2 16D.

Renaut, R.W., Ego, J., Tiercelin, J.-J., Le Turdu, C., Owen, R.B., 1999. Saline, alkaline palaeo-lakes of the Tugen Hills-Kerio Valley region, Kenya Rift Valley. In: Andrews, P., Banham, P. (Eds.), Late Cenozoic Environments and Hominid Evolution: A Tribute to Bill Bishop. Geological Society, London, pp. 41–58.

Rosendahl, B.R., Reynolds, D.J., Lorber, P.M., Burgess, C.F., McGill, J., Scott, D., Lambiase, J.-J., Derksen, S.J., 1986. Structural expressions of rifting: lessons from Lake Tanganyika, Africa. In: Frostick, L.E., Renaut, R.W., Reid, I., Tiercelin, J.-J. (Eds.), Sedimentation in the African Rifts. Geol. Soc. London Spec. Publ., vol. 25. Blackwell Scientific Publications, Oxford, UK, pp. 29–43.

Simiyu, S.M., 1989. Geophysical studies in Lamu Embayment to determine its structure and stratigraphy. M.Sc. Thesis, University of Nairobi (unpubl).

Talbot, M.R., Morley, C.K., Tiercelin, J.J., Le Hérisse, A., Potdevin, J.L., Le Gall, B., 2004. Hydrocarbon potential of the Meso-Cenozoic Turkana Depression, northern Kenya. II. Source rocks: quality, maturation, depositional environments and structural control. Mar. Pet. Geol. 21, 63–78.

Tiercelin, J.-J., Lezzar, K.-E., 2002. Geological and structural setting of the East African Lakes. In: Odada, E.O., Olago, D.O. (Eds.), The East African Great Lakes: Limnology, Palaeolimnology and Biodiversity. Advances in Global Change Research, vol. 12. Kluwer Academic Publishers, Amsterdam, The Netherlands, pp. 3–60.

Tiercelin, J.J., Potdevin, J.L., Morley, C.K., Talbot, M.R., Bellon, H., Rio, A., Le Gall, B., Vétel, W., 2004. Hydrocarbon potential of the Meso-Cenozoic Turkana Depression, northern Kenya. I. Reservoirs: depositional environments, diagenetic characteristics, and source rock-reservoir relationships. Mar. Pet. Geol. 21, 41–62.

Walters, R., Linton, R.E., 1973. The sedimentary basins of coastal Kenya. In: Sedimentary Basins of the African Coasts, Part 2. Association of African Geological Surveys, pp. 133–158.

Wescott, W.A., Wigger, S.T., Stone, D.M., Morley, C.K., 1999. Geology and geophysics of the Lotikipi Plain. In: Morley, C.K. (Ed.), Geoscience of Rift Systems—Evolution of East Africa. American Association of Petroleum Geologists Studies in Geology, vol. 44. AAPG, Tulsa, OK, USA, pp. 55–65.

Zanettin, B., Justin Visentin, E., Bellieni, G., Piccirillo, E.M., Rita, F., 1983. Le volcanisme du Bassin du Nord-Turkana (Kenya): age, succession et évolution structurale. In: Popoff, M., Tiercelin, J.J. (Eds.), Rifts et Fossés anciens. Bulletin des Centres de Recherches Exploration—Production Elf—Aquitaine, vol. 7. Faculté des sciences et techniques de Saint-Jérôme, Marseilles, France, pp. 249–255.

Chapter 3

Relief, Physiography and Drainage

Charles Maina-Gichaba
Department of Geology, University of Nairobi, Nairobi, Kenya

1 INTRODUCTION

Kenya's relief stretches from sea level to about 5,200 m a.s.l. m at the peak of Mt. Kenya. Its description is contained in some of the early work by Ojany (1966), O'Connor (1966) and Ojany and Ogendo (1988). However, detailed studies of the regional relief have been carried out by various authors (Kadomura, 1970; Oosterom, 1988; Touber, 1990; Dijkshoorn et al., 2010). Combined with its tropical latitudinal location, this relief range creates varied physical environment with characteristics that are almost equatorial sharply contrasting with semiarid and arid environments (Ojany and Ogendo, 1988). Topography is described as both simple and diverse. Its simplistic form is shown by the fact that the relief can easily be separated into lowlands and uplands, while diversity is exemplified by the presence of varied landform types, which according to Sayre et al. (2013), can be divided into smooth and irregular plains, escarpments, hills and low and high mountains with breaks.

The widespread plains and plateau typical of large parts of the African continent are remnants of old erosion surfaces formed during the Tertiary period (Ojany and Ogendo, 1988). These are now warped and broken by faults in many areas, while elsewhere volcanic activity has produced further modifications. Earth movements, particularly in late Tertiary, and Quaternary times, have resulted in the formation of the major mountain blocks and Rift Valley systems. These were accompanied by extensive volcanic lava emissions, which cover a significant percentage of the country's land surface. As a consequence of volcanism and earth movements, the drainage has been dislocated, interrupted and modified, and there is hardly a river that has not been affected. Many lakes have been formed in downwarped or downfaulted areas. In coastal regions, the history has been further complicated by Pleistocene changes of sea level (Abuodha, 1992). Major physiographic regions are associated with the drainage patterns of the country. Figure 1 shows the digital

Developments in Earth Surface Processes, Vol. 16. http://dx.doi.org/10.1016/B978-0-444-59559-1.00003-7

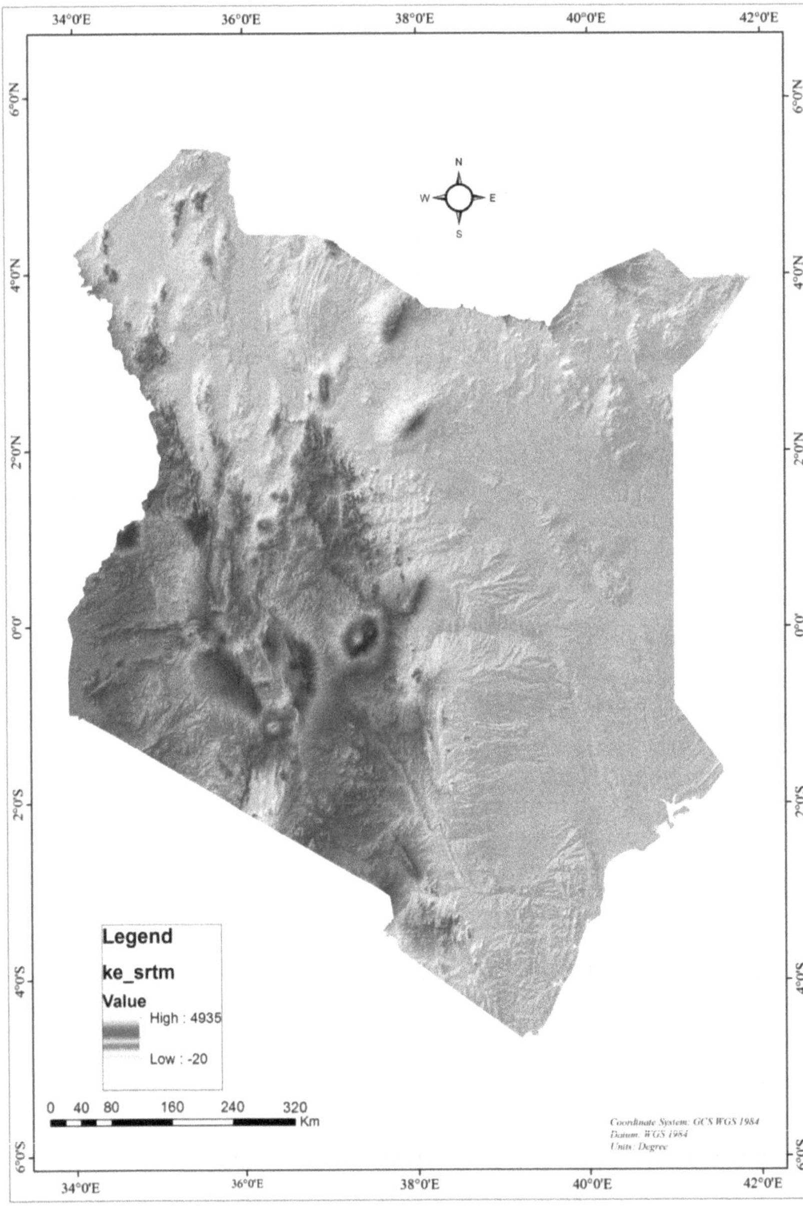

FIGURE 1 Digital elevation model showing the general relief of Kenya. Compiled at the GIS Lab, Department of Geology, University of Nairobi. (For colour version of this figure, the reader is referred to the online version of this chapter.)

elevation model of the landscape in Kenya, showing the various landscape features as a function of elevation.

2 MAJOR PHYSICAL REGIONS OF KENYA

Landforms refer to elements in the landscape characterised by a distinctive surface expression, internal structure or both and sufficiently conspicuous to be included in a physiographic description. They may be defined as a particular association that can be seen at a single view. The major landforms recognised in Kenya can be described using the following characteristics: (a) relief intensity, vertical difference between the higher and lower parts of the landscape, and (b) slope (class), the dominant overall slope class of a landform (FAO, 2003).

Using these two criteria the following landforms are recognised in Kenya:

1. Mountains and major scarps
2. Volcanic footridges
3. Footslopes and pediments
4. Piedmont plains
5. Pediplains
6. Plateaus and high-level structural plains
7. Uplands and dissected peneplains
8. Plains
9. Floodplains
10. Lowland

2.1 Mountains and Major Scarps

A mountain is a natural elevation of the earth's surface rising more or less abruptly from the surrounding levels and attaining an altitude that, relative to adjacent elevation, is with a relief intensity of more than 300 m.

2.2 Volcanic Footridges

The term 'volcanic footridges' was coined to describe a landform that consists of broad parallel, rather convex interfluves alternating with deeply incised valleys, often with convex slopes and with a narrow valley bottom.

2.3 Footslopes and Piedmonts

A footslope is a general term for the lower portion of a hillslope.

Piedmonts are broad, flat gently sloping surface of low relief flanking an upland area. Generally they are erosion surface, often cut by stream erosion exposing the older stratigraphy and pedogenetic processes.

2.4 Piedmont Plains

A Piedmont plain is a plain situated at the foot of mountains or hills. The equivalent of the Piedmont alluvium or Piedmont plain in semiarid lands is bajada.

Generally, these plains are gently sloping, formed by the lateral coalescence of a series of separate but confluent alluvial fans.

2.5 Plateaus and High-Level Structural Plains

A plateau is an elevated flat or levelled ground, commonly bound, on at least one side, by an abrupt scarp.

High-level structural plains are included in this physiographic unit. They are relatively flat areas underlain by horizontal or slightly inclined strata of unequal resistance to weathering.

2.6 Uplands and Dissected Peneplains

These landforms are surfaces of erosion and former accumulation, which are undergoing processes of land degradation of slight to moderate intensity.

A plain is an extensive region of comparatively flat, smooth and level to gently undulating land having no or few prominent surface irregularities.

2.7 Floodplains

These are strips of relatively low smooth plains or flat land adjacent to a river channel, formed or in process of being formed by the present river and covered with river water when flooded. They are made up of alluvium (sand, silt and clay) deposited during floods.

2.8 Lowlands

These are landforms of depressed topography without visible drainage outlets.

3 THE DRAINAGE PATTERNS

The drainage system in Kenya is shown in Figure 2. It is apparent that there is a dense drainage network in the country although many streams are ephemeral, while only the major rivers are perennial.

The country can be divided into five major drainage basins (Figure 3) (NEMA, 2003): are Lake Victoria, Rift Valley, Athi/Sabaki, Tana, and Ewaso Nyiro/North-Eastern basin. Drainage is influenced by the country's topography and underlain geology. The main rivers drain radially from the central highlands into the Rift Valley, eastward into the Indian Ocean and westward into Lake Victoria, while those north of Mount Elgon and from the highlands along the Sudan–Ethiopian border drain mainly into Lake Turkana.

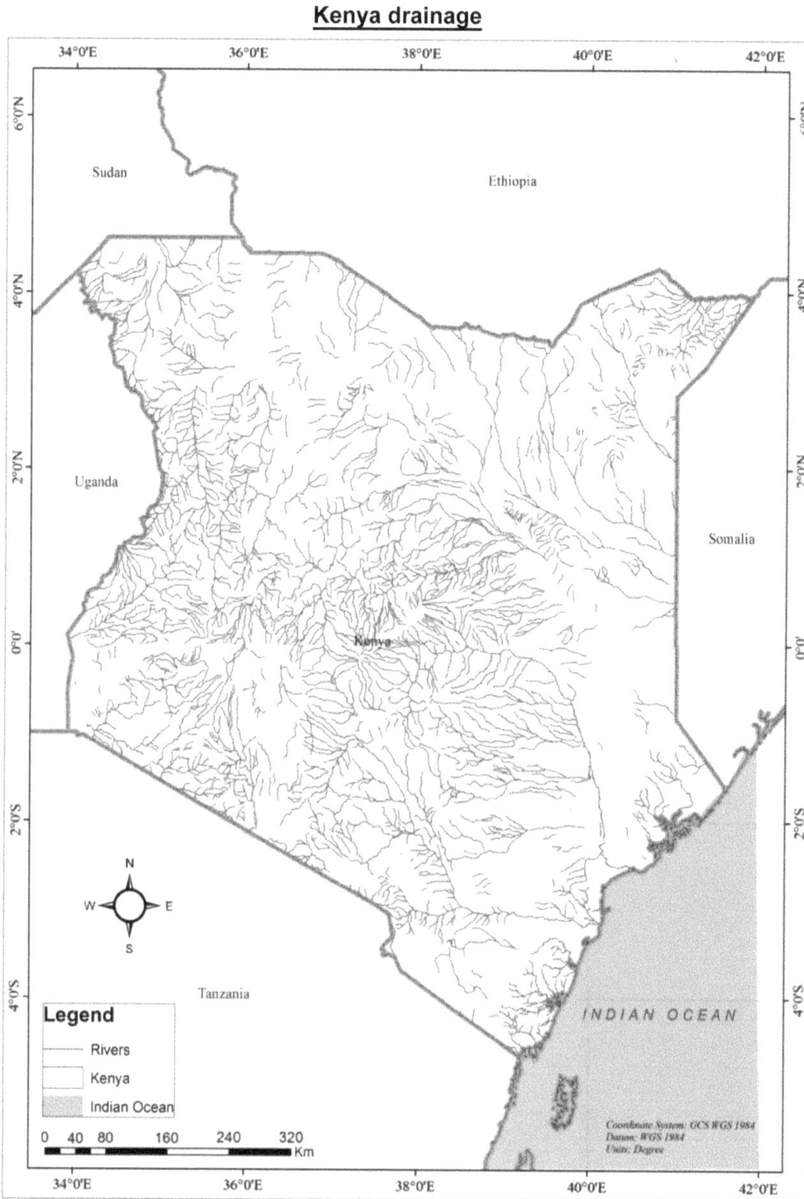

FIGURE 2 Drainage pattern in Kenya. Compiled at the GIS lab, Department of Geology, University of Nairobi. (For colour version of this figure, the reader is referred to the online version of this chapter.)

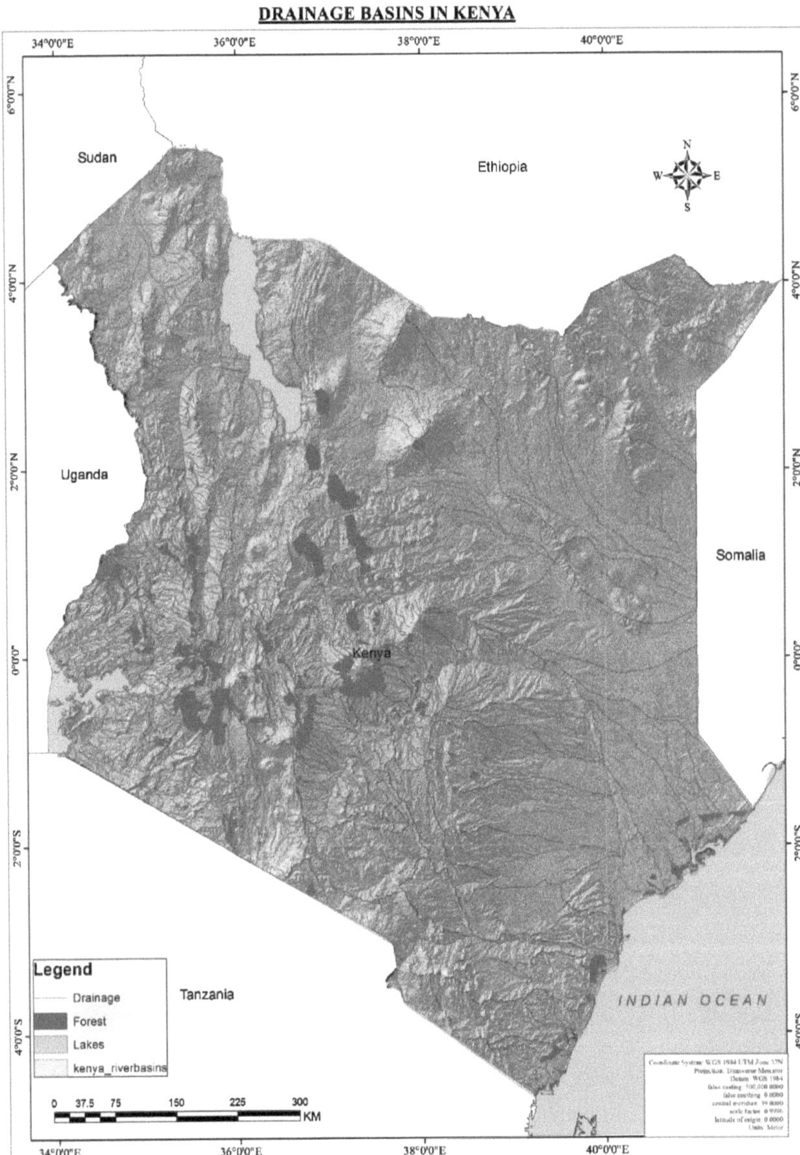

FIGURE 3 Major drainage basins of Kenya. *http://www.apipnm.org/swlwpnr/reports/y_sf/z_ke/ke.htm*. (For colour version of this figure, the reader is referred to the online version of this chapter.)

TABLE 1 Percentage Area Drained by the Respective Drainage Area Basin as a Percentage of Total Land Area of Kenya

Basin	Percentage of Total Land Area (%)
Lake Victoria	8
Rift Valley Inland Drainage	22.4
Athi River and Coast	11.55
River Tana	21.7
Ewaso Nyiro North	36.3

1. Lake Victoria basin (46,229 km^2) comprises the area west of the Rift Valley that drains into Lake Victoria.
2. The Rift Valley basin (130,452 km^2) is an endorheic basin, an area of internal drainage, discharging into Lake Turkana in the north and Lake Natron to the south.
3. The Athi/Sabaki River basin (66,837 km^2) comprises the southern part of the country east of the Rift Valley. It drains the southern slopes of the Aberdare Ranges and flanks of the Rift Valley south of the Athi River.
4. The Tana River basin (126,026 km^2) drains the eastern slopes of the Aberdare Ranges, the southern slopes of Mt. Kenya and the Nyambene Ranges and discharges into the Indian Ocean. The Tana is the largest river in Kenya.
5. The Ewaso Nyiro/North-Eastern basin (210,226 km^2) comprises the northern part of Kenya and drains the north-eastern slopes of the Aberdare Ranges and Mt. Kenya into the Indian Ocean.

Although Kenya has numerous rivers, a small number are permanent, they are: the Tana, Athi, Nzoia, Yala, Sondu, Nyando and Mara (Nyadawa and Mwangi, 2010). Several of the rivers have been dammed upstream to provide hydroelectric power, irrigation water and water for domestic use.

Table 1 shows the percentage area drained by the respective drainage area basin as a percentage of total land area of Kenya. Thus, the largest two drainage basins in Kenya are the River Tana basin and the endorheic Rift Valley basin constituting about 45% of the total land area.

It is important to note that Kenya has five water towers (Government of Kenya, 1992, 1994), which form the sources of all the rivers draining the basins mentioned earlier. These are: (a) Mt. Kenya, (b) the Aberdare Ranges, (c) the Mau Complex, (d) Mt. Elgon and (e) the Cherangany Hills. These valuable ecosystems are, however, under continuous threat from human encroachment, the result of which is manifested in the drying up of some rivers and streams.

REFERENCES

Abuodha, P.A.W., 1992. Geomorphology and sedimentology of the Mombasa-Diani area: implications for coastal zone management. Unpublished M.Sc. thesis, University of Nairobi, Nairobi, p. 155.

Dijkshoorn, J.A., Macharia, P.N., Huting, J.R.M., Maingi, P.M., Njoroge, C.R.K., et al., 2010. Soil and terrain conditions for the upper Tana River catchment, Kenya: Green water credits report 11/ISRIC report 2010/09. ISRIC—World Soil Information, Wageningen. http://www.isric.org/isric/webdocs/Docs/ISRIC_Report_2010_09.pdf, 47 pp. with data set.

Food and Agriculture Organization (FAO), 2003. Gate Way to Land and Water Information; Kenya National Report. http://www.apipnm.org/swlwpnr/reports/y_sf/z_ke/ke.htm (accessed 15.05.13.).

Government of Kenya, 1992. National Water Master Plan, 1992. The study on the National Water Master Plan. Department of Water, Nairobi, Government of Kenya.

Government of Kenya, 1994. The Kenya National Environment Action Plan (NEAP). Government of Kenya, Nairobi.

Kadomura, H., 1970. The landforms in the Tsavo-Voi area, Southern Kenya: an interim report on the geomorphology in the Basement System rocks in a tropical semi-arid region of East Africa. Geogr. Rept. Tokyo Metrop. Univ. (5), 1–23.

National Environmental Management Agency, NEMA, 2003. State of the environment report, Kenya. Government Printer, Nairobi.

Nyadawa, M.O., Mwangi, J.K., 2010. Geomorphic characteristics of Nzoia River Basin. JAGST 12 (2), 2010 Geomorphologic Characteristics.

O'connor, A.M., 1966. East African topographic mapping. East Afr. Geogr. Rev. (4), April, 1966, 69–76.

Ojany, F.F., 1966. The physique of Kenya: a contribution in landscape analysis. Ann. Assoc. Am. Geogr. 56 (2), 183–196, June 1966. UN-HABITAT; 1966.

Ojany, F.F., Ogendo, R.B., 1988. KENYA: A Study in Physical and Human Geography. Longman, Kenya.

Oosterom, A.P., 1988. The Geomorphology of Southeast Kenya. Ph.D. thesis, Agricultural University, Wageningen, p. 227.

Sayre, R., Comer, P., Hak, J., Josse, C., Bow, J., Warner, H., Larwanou, M., Kelbessa, E., Bekele, T., Kahl, H., Amena, R., Andriamasimanana, R., Ba, T., Benson, L., Boucher, T., Brown, M., Cress, J., Dassering, O., Friesen, B., Gachathi, F., Houcine, S., Keita, M., Khamala, E., Marangu, D., Mokua, F., Morou, B., Mucina, L., Mugisha, S., Mwavu, E., Rutherford, M., Sanou, P., Syampungani, S., Tomor, B., Vall, A., Vande Weghe, J., Wangui, E., Waruingi, L., 2013. A New Map of Standardized Terrestrial Ecosystems of Africa. Association of American Geographers, Washington, DC, p. 24.

Touber, L., 1990. Landforms and soils of Isiolo District, Kenya: A site evaluation for rangeland use. Report no 29. Wageningen, The Netherlands.

Chapter 4

Quaternary Evolution

Daniel O. Olago
Department of Geology, Institute for Climate Change and Adaptation, College of Biological and Physical Sciences (CBPS), University of Nairobi, P.O. Box 30197, Nairobi, Kenya

1 INTRODUCTION

The basal age of the Quaternary Period, which is characterised by periodic widespread glacier activity (glacials), with intervening warm episodes (interglacials), is placed at 2.58 Ma (Gibbard et al., 2010). Over long timescales, the changes in precipitation in eastern Africa are driven by the precession cycle of 23,000 years (Trauth et al., 2001) and its higher-precession harmonics (Olago et al., 2000). Indeed, the precession signal has been observed in many proxy records across Africa (Olago et al., 2007). The Quaternary evolution of Kenya has thus been shaped by global, hemispheric and regional climate changes that have interacted with tectonism and volcanism to generate the present-day landscapes and environments (Figure 1). These changes have had major impacts on and implications for hominin, animal and plant evolution within the country. This chapter outlines key aspects or features of Kenya's geomorphological evolution in the Quaternary.

2 THE WESTERN KENYA REGION

Lake Victoria occupies a position between the uplifted border plateaus of the eastern and western branches of East African Rift System (EARS). It resulted from river reversal and ponding as a consequence of rift margin uplift (Bishop and Posnansky, 1960; Ebinger, 1989; Scholz et al., 1990, 1998) and thus can also be described as a 'tectonically induced lacustrine system' (Tiercelin and Lezzar, 2002). Precisely when this occurred in the Pleistocene is not known: (1) possibly during Early to Middle Pleistocene age (1.6–0.8 Ma), as suggested by lacustrine sequences 100 m above the present lake near the Kavirondo Gulf on the Kenyan side of the lake or (2) during Middle to Late Pleistocene (younger than 0.8 Ma), on the basis of lake series exposed 130 m above the present lake, in the Kagera River Valley (Tiercelin and Lezzar, 2002). The younger age is favoured by seismic reflection profiles of the lake that indicate a basal age of about 400,000 years based on the

Developments in Earth Surface Processes, Vol. 16. http://dx.doi.org/10.1016/B978-0-444-59559-1.00004-9

Age (Ma)	Global climate transitions	East African climate and environment			Western Kenya region	Kenya rift system			Eastern Kenya region	Coastal Kenya region	
						Volcanism	Tectonics	Lake basin formation			
Present		General increasing trend of aridity and hydrological variabiliyt with precession-linked wet phases and lake level changes	Change in lowland vegetation from woodland to savanna types	Marine transgressions and regressions	Short-lived period of desiccation of LVB about 15 ka; Nile outflow ca. 7.5 Ka	Hot springs and fumaroles-geothermal systems	Minor faulting	Mostly small, shallow rift lakes; desiccation/lowstands during LGM, highstands at ca.10 to 9 Ka	No volcanic activity	Retreating shore line and sand dune building in some areas	Shoreline terraces related to glacial–interglacial cycles and isostatic readjustments
0.1							Several phases of rift deformation and migration		Mount Kenya and Nyambeni Range volcanism declining in frequency and intensity	Sand dune ridges	
0.2											
0.3										Onshore–offshore coral reef complex	
0.4						End of flood volcanism	Formation of axial volcanoes				
0.5											
0.6					Formation of LVB?			Basins formed during Mio-Pliocene times achieve final morphology; Creation of new lake basins			
0.7											
0.8											
0.9	The Mid-pleistocene revolution									Data scarcity	
1.0											
1.1											
1.2											
1.3											
1.4											
1.5											
1.6						Widespread flood volcanism					
1.7											
1.8					Formation of LVB(?)						
1.9	Development of walker circulation										
2.0									Formation of Chalbi basin		
2.1											
2.2											
2.3											
2.4											
2.5											
2.6	Onset of Northern hemisphere glaciation								Volcanism of Mount Kenya, Nyambeni Range, Kulal and other NNE to NE trending hills to the north.		
2.7											
2.8											
2.9											
3.0											

FIGURE 1 Summary of key aspects of the Quaternary evolution of Kenya (LVB, Lake Victoria basin).

assumption of modern sedimentation rates and the effects of compaction (Johnson et al., 1996). The presence of three erosional boundaries (lowstands) in the seismic sections implies that the lake may be responding to the 100,000 year glacial–interglacial cycles (Johnson et al., 2000).

3 THE KENYA RIFT SYSTEM

The Kenya Rift Valley (130,452 km^2) is a topographically closed area that drains into Lake Turkana in the north and Lake Natron to the south. The present-day general architecture of the EARS is made of a series of relatively narrow (40–70 km wide) faulted troughs that are distributed in two distinct branches, eastern and western, all over a length of more than 4000 km (Tiercelin and Lezzar, 2002). The major fault scarps are the Elgeyo and Nguruman escarpments that reach 1500 m and the Aberdares that reach over 2000 m above the rift floor (Schlüter, 1997), but it is estimated that the fault throws are of the order of 3000–4000 m (Baker and Wohlenberg, 1971). The rift floor is characterised by grid/box faulting: here, the faults are short, parallel and close together, have throws of less than 100 m (Schlüter, 1997) and are estimated to have an average spacing of 1.5 km (Baker, 1986).

Major tectonic and/or volcanic events are known to have occurred in the EARS between 2 and 0.5 Ma, creating new lake basins or bringing basins formed during Mio-Pliocene times to their final morphology (Tiercelin and Lezzar, 2002). Flood volcanism persisted for the whole rift from 2 to 1.5 Ma and ended at about 0.8–0.3 Ma (Baker et al., 1988), followed by several close phases of rift deformation and migration (Chapman et al., 1978; Le Gall et al., 2000). Such tectonic phases were associated with the formation of a suite of large, axial trachytic caldera volcanoes (Suswa, Longonot, Menengai, etc.) and several other smaller volcanic complexes (Tiercelin and Lezzar, 2002) and resulted in the development on the rift floor of inner troughs of widths ranging from 15 to 35 km (Chapman et al., 1978; Dunkley et al., 1993). Present-day lakes such as Lake Baringo or Lakes Nakuru, Elmenteita and Naivasha are the remnants of largest lake domains developed during Lower–Middle Pleistocene times and are represented by rare sedimentary deposits sometimes uplifted or downfaulted along the inner trough margins (Richardson, 1966; Richardson and Richardson, 1972; Kamau, 1974; Renaut, 1982; Tiercelin and Lezzar, 2002). The lakes in the Kenya Rift are generally small (maximum 30×20 km) and shallow (from 5 to 50 m water depth) and have more or less saline characteristics, mainly because they lie in areas under semiarid climatic conditions and/or have no outlet; Lake Turkana, at the northern end of the Kenya Rift, is the exception in terms of dimensions, with a length of 250 km and a maximum water depth of 125 m (Tiercelin and Lezzar, 2002).

Minor faulting has continued in the rift to the present day. There are also a number of hot springs and fumaroles that indicate the presence of remnant volcanic activity within the Kenya Rift (e.g. Clarke et al., 1990). Hot springs occur in places such as Olkaria south of Lake Naivasha, along the fringes of Lakes Bogoria, Baringo and Magadi, at Homa Bay in the Kavirondo Rift and the Barrier Volcanic Complex in Lake Turkana. Mounts Longonot, Eburru and Suswa are known for fumarolic activity (Schlüter, 1997).

4 THE EASTERN KENYA REGION

The peak area of Mt. Kenya represents the plug of a long-extinct volcano, the present summit of which must be very much below the level of the original crater wall (Coe, 1967). The mountain is now in the youthful stages of dissection: the upper part has suffered the greatest erosion, having been subjected to at least two periods of glaciation (Baker, 1967). The volcanic activity of Mt. Kenya dates back to the late Pliocene but continued into the Quaternary; its first eruptions took place from 3.5 to 3 Ma (Baker et al., 1971). Basalts and phonolites, together with pyroclastics of similar composition, constitute most of its bulk, but late basalt flows and ashes were erupted from parasitic vents (Baker, 1967; King, 1970). The Quaternary geology of Mt. Kenya is characterised by a volcanic rock suite (the Mount Kenya Volcanic Series), which shows a general decline in the frequency and intensity of volcanic eruptions with time (Baker, 1967). The updoming of the eastern flank of the rift system resulted in some marked fluvial incision by the large Tana River during the Quaternary (Veldkamp et al., 2007). Mt. Kenya marks the southwestern extremity of a large field of basaltic cones and vents, the Nyambeni Volcanic Series (Baker, 1967). The Nyambeni Mountain Range erupted partly at the same time as Mount Kenya, and its active phase continued into Pleistocene times (4.5–0.5 Ma) (Brotzu et al., 1984). The Plio-Quaternary central volcanism of the region, extending all the way to the Ethiopian border and producing the Nyambeni range, Marsabit, Demo Dera, Kulal, Hurri Hills and Dida Gola areas, is a product of linear multicentre activity that is probably related to NNE to NE-trending features (Brotzu et al., 1984).

The Chalbi basin is an inland drainage basin, with a playa in the middle, which is located 60 km east of the rift in the northern part of Kenya. A lake or swamp environment existed there from 2.5 to 2 Ma (Nyamweru and Bowman, 1989), and it probably took its present form during the early Pleistocene through a combination of volcanic and tectonic events (Nyamweru, 1986). Conditions in the Chalbi were influenced by the general climatic fluctuations of the terminal Pleistocene–Holocene that affected East Africa (Nyamweru, 1986).

5 THE COASTAL KENYA REGION

Several factors such as the tectonic evolution, climate, wave and tidal regime, sedimentation and river discharge control the geomorphology of the Kenyan coast (Abuodha, 2003). During the Quaternary, coastal Kenya has been characterised by marine transgressions and regressions causing the deposition of dune sands and the growth of coral limestone intercalated with continentally derived deposits (Schlüter, 1997). The coral reef is best developed in the central part of coastal Kenya: its extent offshore is variable, up to a maximum of 2 km, and onshore, it may extend to a maximum of 6 km (Schlüter, 1997). The reef complex has been assigned a Middle Pleistocene age (Karanja et al., 1993). Four post-Pliocene marine terraces were recognised by Caswell (1956) at 61, 30.5, 9 and 4.5 m above present sea level (Caswell, 1956), while eight were recognised by Oosterom (1988). These have, however, not been well constrained agewise. The Pleistocene sand dune ridges formed from reworked medium- to fine-grained sands aligned in the direction of the southeast monsoon winds dot the coastline. The distribution of these dunes is in such a way that the recent dunes are prominent seaward, whereas the older dunes are further inland (Karanja et al., 1993). Wave action and tides from the sea have led to the retreating shoreline in some sections of the coast: this is represented by the presence of white sandy beaches and sheltered bays generally composed of quartz, carbonates and shell fragments in areas such as Mombasa, Watamu and Malindi (Caswell, 1953, 1956; Ase, 1978, 1981; Cannon et al., 1981; Oosterom, 1988).

REFERENCES

Abuodha, P.A.W., 2003. Geomorphology of the Kenyan coast: Not as a Result of Sea-Level Change alone. Buletim Geologico, no. 43 Mozambique, 2-7. Paper presented during the 'Impact of sea level rise: Past Present–Future' Inhaca Island, Mozambique 4-8 November 2002. Organized and sponsored by the Commission of the Holocene of the International Union for Quaternary Research (INQUA). http://www.oceandocs.org/bitstream/1834/407/1/1Abuodha.pdf, cited 15th April 2013.

Ase, L.E., 1978. Preliminary report on studies of shore displacement at the southern coast of Kenya. Geogr. Ann. 60A (3–4), 209–221.

Ase, L.E., 1981. Studies of shores and shore displacement on the southern coast of Kenya—especially in Kilifi District. Geogr. Ann. 63A (3–4), 303–310.

Baker, B.H., 1967. Geology of the Mount Kenya Area. Geol. Surv. Kenya, Rep. 79, 78, Nairobi.

Baker, B.H., 1986. Tectonism and volcanism of the southern Kenya Rift Valley and its influence on sedimentation. In: Frostick, L.E. et al., (Ed.), Sedimentation in the African Rifts, Geol. Soc. London Spec. Publ., vol. 25, pp. 267–284, London.

Baker, B.H., Wohlenberg, J., 1971. Structure and evolution of the Kenya Rift Valley. Nature 229, 538–542.

Baker, B.H., Williams, L.A.J., Miller, J.A., Fitch, F.J., 1971. Sequence and geochronology of the Kenya rift volcanics. Tectonophysics 11, 191–215.

Baker, B.H., Mitchell, J.G., Williams, L.A.J., 1988. Stratigraphy, geochronology and volcano-tectonic evolution of the Kedong-Naivasha-Kinangop region, Gregory Rift Valley, Kenya. J. Geol. Soc. Lond. 145, 107–116.

Bishop, W.W., Posnansky, M., 1960. Pleistocene environments and early man in Uganda. Uganda J. 24, 41–61.

Brotzu, P., Morbidelli, L., Nicoletti, M., Piccirillo, E.M., Traversa, G., 1984. Miocene to Quaternary volcanism in eastern Kenya: sequence and geochronology. Tectonophysics 101 (1–2), 75–86.

Cannon, R.T., Simiyu Siambi, W.M.N., Karanja, E.M., 1981. The Proto-Indian Ocean and a probable Paleozoic/Mesozoic Tri-radial rift system in East Africa. Earth Planet. Sci. Lett. 52 (2), 419–526.

Caswell, P.V., 1953. The Geology of the Mombasa—Kwale area. Geological Survey of Kenya, Report No. 24.

Caswell, P.V., 1956. The Geology of the Kilifi—Mazeras area. Geological Survey of Kenya, Report No. 34.

Chapman, G.R., Lippard, S.J., Martyn, J.E., 1978. The stratigraphy and structure of the Kamasia Range, Kenya Rift Valley. J. Geol. Soc. Lond. 135, 265–281.

Clarke, M.C.G., Woodhall, D.G., Allen, D., Darling, G., 1990. Geological, Volcanological and Hydrogeological Controls on the Occurrence of Geothermal Activity in the Area Surrounding Lake Naivasha Kenya. Ministry of Energy, Nairobi, Kenya and British Geological Survey, UK, pp. 1–348.

Coe, M.J., 1967. The Ecology of the Alpine Zone of Mount Kenya. Dr. W. Junk Publishers, Dordrecht, The Netherlands, p. 136.

Dunkley, P.N., Smith, M., Allen, D.J., Darling, W.G., 1993. The geothermal activity and geology of the northern sector of the Kenya Rift Valley. p. 185, British Geological Survey Research Report SC/93/1.

Ebinger, C.J., 1989. Tectonic development of the western branch of the East African rift system. Bull. Geol. Soc. Am. 101, 885–903.

Gibbard, P.L., Head, M.J., Walker, M.J.C., 2010. Formal ratification of the Quaternary System/ Period and the Pleistocene Series/Epoch with a base at 2.58 Ma. J. Quat. Sci. 25 (2), 96–102.

Johnson, T.C., Scholz, C.A., Talbot, M.R., Kelts, K., Ricketts, R.D., Ngobi, G., Beuning, K.R.M., Ssemmanda, I., McGill, J.W., 1996. Late Pleistocene desiccation of Lake Victoria and rapid evolution of cichlid fishes. Science 273, 1091–1093.

Johnson, T.C., Kelts, K., Odada, E., 2000. The Holocene history of Lake Victoria. Ambio 29 (1), 2–11.

Kamau, C., 1974. The Lake Naivasha Basin. Nairobi University, p. 185, Report, Geography Department Research Report SC/93/1.

Karanja, F.M., Siambi, W.S., Sewe, C., 1993. Excursion Guide to the Kenya Coastal Geology and Marine Terraces. International Workshop Past Global Changes, Mombasa, pp. 1–12, Mombasa.

King, B.C., 1970. Vulcanicity and rift tectonics in East Africa. In: Clifford, T., Gass, I.G. (Eds.), African Magmatism and Tectonics. Oliver and Boyd, Edinburgh, pp. 263–283.

Le Gall, B., Tiercelin, J.-J., Richert, J.P., Gente, P., Sturchio, N.C., Stead, D., Le Turdu, C., 2000. A morphotectonic study of an extensional fault zone in a magma-rich rift: the Baringo Trachyte fault system, central Kenya Rift. Tectonophysics 320, 87–106.

Nyamweru, C.K., 1986. Quaternary environments of the Chalbi basin, Kenya: sedimentary and geomorphological evidence. In: Frostick, L.E., Renaut, R.W., Reid, I., Tiercelin, J.-J. (Eds.), Sedimentation in the African Rifts. Geol. Soc. Spec. Publ. No. 25: pp. 297–310.

Nyamweru, C.K., Bowman, D., 1989. Climatic changes in the Chalbi Desert, North Kenya. J. Quat. Sci. 4, 131–139.

Olago, D.O., Street-Perrott, F.A., Perrott, R.A., Ivanovich, M., Harkness, D.D., Odada, E.O., 2000. Long-term temporal characteristics of palaeomonsoon dynamics in equatorial Africa. Glob. Planet. Change 26, 159–171.

Olago, D.O., Umer, M., Ringrose, S., Huntsman-Mapila, P., Sow, E.H., Damnati, D., 2007. Palaeoclimate of Africa: an overview since the last glacial maximum. In: Otter, L., Olago, D.O., Niang, I. (Eds.), Global Change Processes and Impacts in Africa: A Synthesis. East African Educational Publishers, Nairobi, pp. 1–32.

Oosterom, A.P., 1988. The geomorphology of southeast Kenya. Wageningen Agricultural University, p. 227, Ph.D. thesis.

Renaut, R.W., 1982. Late Quaternary geology of the Lake Bogoria fault-trough, Kenya Rift Valley. University of London, p. 498, Unpublished Ph.D. thesis.

Richardson, J.L., 1966. Changes in level of Lake Naivasha, Kenya, during postglacial times. Nature 209, 290–291.

Richardson, J.L., Richardson, A.E., 1972. History of an African rift lake and its climatic implications. Ecol. Monogr. 42, 499–534.

Schlüter, T., 1997. Geology of East Africa. Gebruder Borntraeger, Berlin-Stuttgart, p. 484.

Scholz, C.A., Rosendahl, B.R., Versfelt, J.W., Rach, N., 1990. Results of high-resolution echo-sounding of Lake Victoria. J. Afr. Earth Sci. 11, 25–32.

Scholz, C.A., Johnson, T.C., Cattaneo, P., Malinga, H., Shana, S., 1998. Initial results of 1995 IDEAL seismic reflection survey of Lake Victoria, Uganda and Tanzania. In: Lehman, J.T. (Ed.), Environmental Change and Response in East African Lakes. Kluwer, Dordrecht, pp. 47–58.

Tiercelin, J.-J., Lezzar, K.E., 2002. A 300 million years history of rift lakes in central and east Africa: an updated broad review. In: Odada, E.O., Olago, D.O. (Eds.), The East African Great Lakes: Limnology, Palaeolimnology and Biodiversity. Advances in Global Change Research, vol. 12. Kluwer Academic Publishers, Dordrecht, The Netherlands, p. 586.

Trauth, M.H., Deino, A., Strecker, M.R., 2001. Response of the East African climate to orbital forcing during the last interglacial (130-117 ka) and the early last glacial (117-60 ka). Geology 29 (6), 499–502.

Veldkamp, A., Buis, E., Wijbrans, J.R., Olago, D.O., Boshoven, E.H., Marée, M., van den Berg van Saparoea, R.M., 2007. Late Cenozoic fluvial dynamics of the Tana River, Kenya, an uplift dominated record. Quat. Sci. Rev. 26, 2897–2912.

Chapter 5

Climate

John P.O. Obiero* and Japheth O. Onyando†
*Department of Environmental and Biosystems Engineering, University of Nairobi, P.O. Box 30197, Nairobi, Kenya
†Department of Agricultural Engineering, Egerton University, P.O. Box 536, Njoro, Kenya

1 THE CLIMATE OF KENYA

1.1 Introduction

A thorough understanding of climate characteristics and variation is important in agriculture including pastoralism, which plays a dominant role in the Kenyan economy. The main factors that control climate include latitude, altitude, characteristics of prevailing winds, distance from the sea or any form of sizeable water body and topography. Topography becomes significant especially when the area under study has diversity of relief, which can form the barrier to prevailing winds. Vegetal cover is also important but normally is a result of the previously mentioned factors. The pressure belts are also significant in controlling climate as the belts shift with the movement of the overhead sun resulting into seasons. The greatest insolation directly usually below the overhead sun creates a region of lowest pressure. The low pressure is also known as the heat trough or equatorial trough or the intertropical convergence zone. It is a point of convergence of air masses. An air mass is a large body of air covering hundreds of kilometres and that has a definite source region and marked uniformity in temperature or humidity throughout its entire length and width. As a body of air moves, it will be known by its source region, and as it moves across a section of the country, it transports and introduces its own particular climatic characteristics throughout the country over which it passes. Convergence of air masses normally results into an upward movement of air over the area of low pressure, thereby causing cooling, condensation and precipitation. Air masses, also referred to as air streams, are therefore important as carriers of moisture.

Developments in Earth Surface Processes, Vol. 16. http://dx.doi.org/10.1016/B978-0-444-59559-1.00005-0

2 AIR MASSES IN KENYA

Climate in the East African Region is influenced by three major air streams (Nicholson, 1996) which therefore determine the Kenyan climate. These air-streams are Congo air with westerly and south westerly flow, north east and south east trade winds. In the Eastern part of the country, the airstream which has Arabian source is known as the Arabian northeast trade winds. The northeasterly wind is the northeast monsoon that has crossed a stretch of the northern part of the Indian Ocean and brings some rain to the coastal regions during the period November to March. By about the month of April, the wind system has diverged and many of the central, southern and eastern parts of the country begin to gain the influence of the southeast trade winds from the Indian Ocean. Since these winds have crossed a large body of water over the southern Indian Ocean, they are the source of the major rains in Kenya. This air mass persists more or less with the same consistency until about August. From about the month of July, high winds penetrate from the west as the Congo airstream. This Congo airstream is completely unstable and storms easily form and develop. Its influence is felt mostly in the western parts of Kenya. The air masses are illustrated in Figure 1. Pressure patterns also have a major influence on the Kenyan Climate. Pressure belts shift with the movement of the overhead sun thereby causing seasons. The greatest insolation obtained

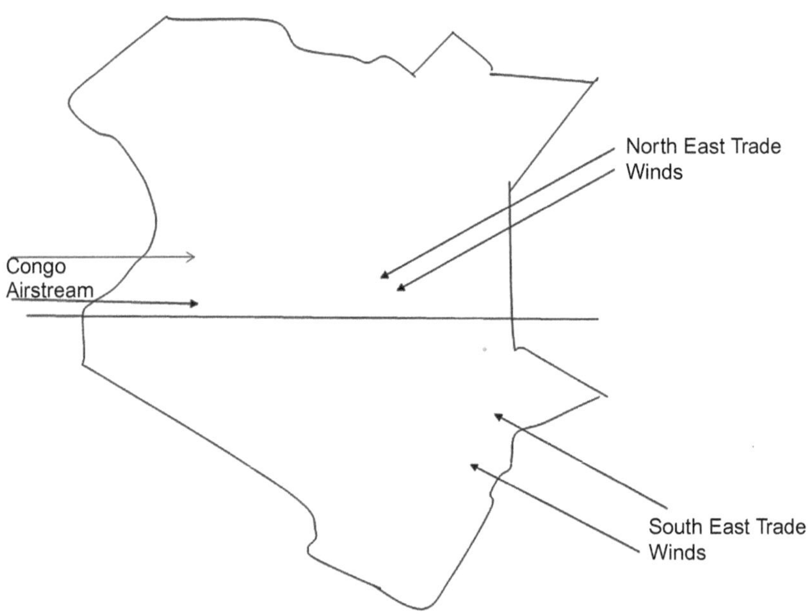

FIGURE 1 Air masses in Kenya. Harmattan—dry winds—November–March; Northeast trade winds—moist—November-March; Southeast trade winds—moist—April–August; Congo airstream—moist—July. (For colour version of this figure, the reader is referred to the online version of this chapter.)

directly below the overhead sun creates lowest pressure which is known as a convergence zone. The Inter Tropical Convergence Zone (ITCZ) is an area of the atmosphere above the earth's surface where the two airstreams (The North East and South East trade winds) converge. It is characterised by warm air rising and the formation of clouds resulting into rainfall. It oscillates north and south across the equator as the earth revolves around the sun. The ITCZ intensifies the weather patterns for an area so that when an area expects rainfall when ITCZ is overhead, the rainfall is expected to be greater and more intense than when ITCZ is not passing over.

3 RAINFALL AND TEMPERATURE

3.1 Rainfall in Kenya

The climatic factor of greatest economic significance is rainfall. The average annual rainfall in Kenya ranges from 250 to 2500 mm. The distribution of rainfall in Kenya is irregular in time and space, that is, not uniform. The climate is characterized by alternating wet and dry seasons. Because of the great variability of rainfall received from one place to another, the mean annual rainfall is mostly considered. The percentages of land area receiving selected amounts of annual rainfall is illustrated in Table 1.

From Table 1 earlier, it is observed that only 15% of Kenya area receives reliable rainfall of over 760 mm (30 in.) per annum (Griffiths, 1972). The areas with over 1270 mm (50 in.) mean annual rainfall include most of the catchment areas and are relatively small. This emphasizes the vital need for strict acceptable hydrological, agricultural and forestry conservation practices if the rivers are to have dry season flow. In Kenya, the regions with good reliable rainfall are confined to the narrow coastal belt and the highlands. The transition from humid to dry areas is particularly sharp and is related to the sharp changes in topography especially east of the Rift Valley where there are two distinct wet seasons. The longer rainy season of the highlands west of the Rift Valley offers a more reliable growing season for annual crops and this has long been the nucleus of agricultural production in Kenya. A summary of annual rainfall from various rainfall stations in major towns in Kenya is tabulated in Table 2.

TABLE 1 Areal Proportions of Rainfall Received in Kenya

Rainfall Received	Percentage Area
≤508 mm (20 in.)	72%
508–760 mm	13%
760–1270 mm	12%
≥1270 mm	3%

TABLE 2 Summary of Mean Annual Rainfall and Periods for Various Locations on Kenya

Location	Mean Annual Rainfall	Average No. of Rainy Days Annually
Nairobi	879	120
Mombasa	1204	145
Nakuru	871	137
Eldoret	973	134
Nyeri	922	100
Wajir	249	29
Kisumu	1278	139
Magadi	378	67
Lodwar	165	23

The reason for the large rainfall variation (Table 2) is partly because of the tremendous topographical contrasts including the great altitude range and partly the distribution and presence of large water bodies such as Lake Victoria and Indian Ocean.

3.2 Temperature

The average annual temperatures in Kenya range from <10 to 30 °C. Elevation and relief are the two factors that have the greatest influence on the temperatures in Kenya; however, at the coastal areas and around Lake Victoria, the influence of land and sea breezes is more important. The highland areas in the country are the regions with low temperature, while the low-lying areas stand out as areas with higher temperatures. As is expected, there is a large range from below freezing point in the snow-capped mount Kenya to over 40 °C in some parts of north and northeastern Kenya. The low-lying northern plains are the hottest and generally the maximum temperatures commonly exceed 35 °C with March as the hottest month. Night temperatures in these areas are generally above 25 °C. Nyanza and western province mostly show great variations from hot low-lying plains to cool temperatures in the hills surrounding Kakamega, Kericho and Kisii areas. The rift valley, including the adjoining highlands, shows a large diurnal range in temperature of over 20 °C. The hottest parts are the low-lying rift valley floor with Magadi frequently recording over 40 °C. The central areas of Kenya including Nairobi generally enjoy pleasant temperatures with a daily range of about 15 °C. The coastal belt is an area of fairly high temperatures ranging 26 °C to

TABLE 3 Temperature and Sunshine Data for Selected Locations in Kenya

Station	Elevation	Mean Max.	Mean Min.	Ave. Sunshine Hours
Low-lying areas				
Lodwar	506	34.9	23.7	3582
Lamu	30	29.2	24.1	3242
Mombasa	16	30.1	23.4	2987
Kisumu	1146	29.4	17.1	2842
Highlands				
Nairobi	1798	23.6	11.6	2503
Nanyuki	1946	23.7	8.5	2172
Nakuru	1836	26.4	10.1	2584

32 °C on average with a very small diurnal range of between 7 and 9 °C. The coolest months are July and August where afternoon temperatures may average 28 °C and night temperatures of 20 °C. The mean maximum and minimum temperatures recorded in the various locations are tabulated in Table 3.

Table 3 shows that the low-lying areas and coastal regions have higher temperatures and more hours of sunshine than the highlands.

4 CLIMATIC REGIONS IN KENYA

There are seven climatic regions in Kenya.

4.1 Modified Equatorial Climate of the Coast

As the other equatorial climate in this region, this region has no dry season but has a well-developed double maximum rainfall in May and October. There is high average temperatures and humidity throughout the year. The Kenyan coast differs from the equatorial climate being well known for its high annual rainfall totals in that it receives much lower totals.

4.2 Modified Tropical Climate of the Kenya Highlands

The modification is caused by the high varied relief. The resulting climate is therefore cooler than a tropical continental climate. The amount of rainfall received at different places depends very much on the position of the station in relation to the rain-bearing winds. For example, Meru on the northeastern slopes of Mount Kenya receives about 1320 mm of rain per year, while Nyeri on southwestern Kenya only receives 737 mm a year. This region has the best climate in the country. Day temperatures are cool and nights may be chilly.

4.3 Modified Equatorial Climate of Lake Victoria Basin

In this region, modification results from relief and influence of the lake. The rainfall totals in the region show considerable variations and are in any case much lower than a typical equatorial climate. The enclosed basin of Lake Victoria tends to increase temperatures, although the lake itself exerts a cooling influence on the surrounding areas.

4.4 Modified Equatorial Climate of Northwestern Kenya

This is a continuation of the climate of Eastern and North Uganda. High relief lowers the temperature in parts of the region.

4.5 Tropical Climate of the Narok and Southern Taita/Kwale Areas

This is part of the tropical climate that dominates the central mainland of Tanzania. The high relief of some localities exerts a cooling effect so that taken altogether, the climate here is more tolerable than the semi-desert climate of eastern Kenya.

4.6 Tropical Continental Semi-Desert Climate of Eastern Kenya

This vast region receives <500 mm of annual rainfall. The mean annual temperature is between 22 and 27 °C with a very wide daily range of about 11 °C. The skies are generally clear and great variations in the mean annual rainfall can be expected.

4.7 Desert Climate of Central Northern Kenya

A considerable stretch of northern Kenya receives <250 mm of rainfall and its climatology may be regarded as desert. The aridity of this area is a continuation of that of Arabia. In this region, the skies are cloudless and visibility is only hindered by sandstorms, which are of common occurrence.

5 CLIMATE AND AGRICULTURE

The relation between climate and agricultural potential in Kenya has been recognized. Crop growth is influenced by the balance between rainfall, evaporation and temperature. The length and intensity of rainy and dry seasons are more important than the annual variations.

5.1 Climatic Zones in Relation to Crop Production

The major climates in the world (FAO Bulletin 52, 1984) show broad climatic divisions defined in terms of monthly mean temperatures, seasonality of rainfall and temperature regimes during the growing period. The suitability for each of the major climates in consideration for particular crop adaptability group is also discussed. The agro-climatic zones consist of two major components, the moisture availability zones and temperature zones. The ratio of average annual rainfall and average annual evaporation determines the moisture availability zones. The temperature zones are based on the average annual temperature. The agro-climatic zones are tabulated in Table 4.

The agro-climatic zones have a direct bearing on potential for land use. The ratio of average annual rainfall to average annual potential evaporation has been used in delineating the agro-climatic zones in Kenya. The agro-climatic zones provide a tool for assessing which areas are suitable for various land-use alternatives, with emphasis on suitability for crop production. Table 5 illustrates the classification with respect to potential for crop production (Biamah and Cherogony, 1995).

The boundary criteria for the temperature zones are illustrated in Table 6. The moisture ranges and corresponding suitability for various crops and animal production as well as forestry are illustrated in Table 7. The crops that can be produced in the indicated temperature ranges for the various zones are shown in Table 8. Agro-climatological crop list details are well documented in Jaetzold and Schmidt (1982), detailing crop growing periods, crop varieties, average number of days to maturity and rainfall required.

TABLE 4 Criteria for Establishing Boundaries for Moisture Availability Zones

Zone	R/E_o Ratio	R/E_o (%)	Climate Designation
I	>0.8	>80	Humid
II	0.65–0.80	65–80	Sub-humid
III	0.5–0.65	50–65	Semi-humid
IV	0.4–0.5	40–50	Semi-humid to Semi-arid
V	0.25–0.40	25–40	Semi-arid
VI	0.15–0.25	15–25	Arid
VII	<0.15	<15	Very arid

R, average annual rainfall.
E_o, average annual potential evaporation.

TABLE 5 Agro-Climatic Zones and Potential for Crop Production in Kenya

Zone	R/E_o (%)	Classification	Risk of Maize Crop Failure
I	>80	Humid	Extremely low (0–1%)
II	65–80	Sub-humid	Very low (1–5%)
III	50–65	Semi-humid	Fairly low (5–10%)
IV	40–50	Semi-humid to semi-arid	Low (10–25%)
V	25–40	Semi-arid	High (25–75%)
VI	15–25	Arid	Very high (75–95%)
VII	<15	Very arid	Extremely high (95–100%)

TABLE 6 Boundary Criteria for Temperature Zones

Zone	Altitude (ft)	Mean Annual Temperature	Climate Designation
1	0–3000	24–30	Fairly hot to very hot
2	3000–4000	22–24	Warm
3	4000–5000	20–22	Fairly warm
4	5000–6000	18–20	Warm temperate
5	6000–7000	16–18	Cool temperate
6	7000–8000	14–16	Fairly cool
7	8000–9000	12–14	Cool
8	9000–10,000	10–12	Very cool
9	>10,000	<10	Cool to very cold

6 CLIMATE EXTREMES

6.1 Overview

Kenya is affected by a range of hazards emanating from extremes of climate that often translate into natural disasters arising from drought, severe storms, tropical cyclones and floods. The most catastrophic events in recent times have been the droughts of the early 1980s, the late 1980s, early 1990 and 2002/2003. The 1997/1998 El Niño, the strongest in the century, caused a lot of damages in East Africa, with heavy floods washing away roads, bridges and railway tracks and destroying many hectares of crops. The National Meteorological Services (NMSs) in the region have the responsibility of the protection of life and property of the community. Each NMS contributes to disaster

TABLE 7 Range of Crops, Types of Animal Production Systems and Forest Species Suited for Various Moisture Ranges

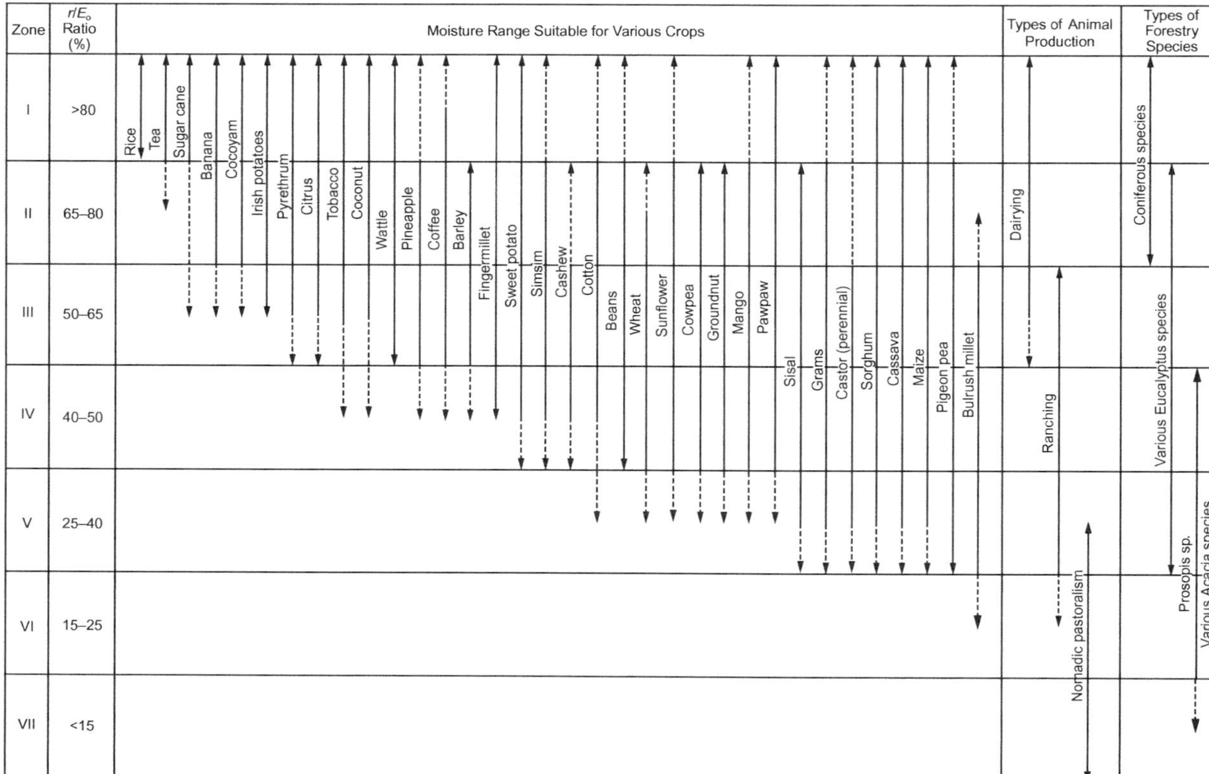

TABLE 8 Ranges of Crops Suitable Under Various Altitude and Temperature Ranges

Temperature Zone	Altitude (ft)	Average Annual Temperature (°C)
9	> 10,000	< 10
8	9000 – 10,000	10 – 12
7	8000 – 9000	12 – 14
6	7000 – 8000	14 – 16
5	6000 – 7000	16 – 18
4	5000 – 6000	18 – 20
3	4000 – 5000	20 – 22
2	3000 – 4000	22 – 24
1	<3000	24 – 30

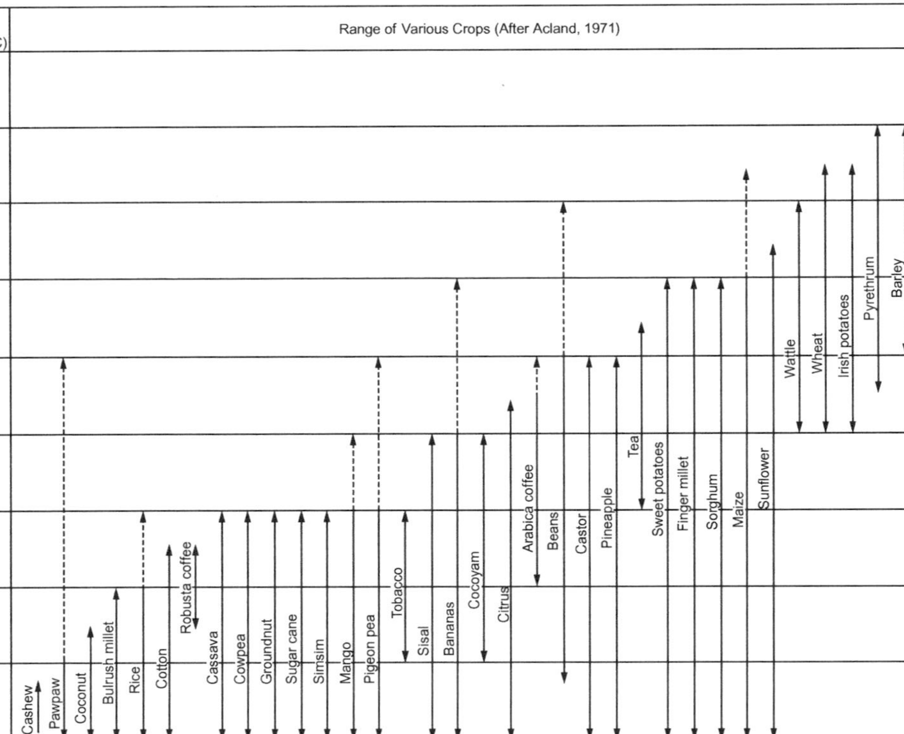

management through its forecasts, advisories and warnings of severe weather and the monitoring of climate especially rainfall.

6.2 Floods

Flood hazards are caused by natural events, which transform into disasters through complex interactions between social, economic, political and environmental processes. A Strategy for Flood Management for the country therefore not only directly deals with the flood hazard but also has strong links with national social, economic and other development policies. Disaster prevention and mitigation due to floods is therefore a multidisciplinary endeavour wherein development activities in different sectors of the economy help in the prevention of the disasters and reduce the vulnerability of the society.

6.3 Drought

Drought is an extended period of months or years when a region notes a deficiency in its water supply whether surface or underground water. Generally, this occurs when a region receives consistently below average precipitation. It can have a substantial impact on the ecosystem and agriculture of the affected region. Although droughts can persist for several years, even a short, intense drought can cause significant damage and harm the local economy. As a drought persists, the conditions surrounding it gradually worsen and its impact on the local population gradually increases. Occurrence of drought is partly related to climate change as observed by Ongwenyi et al., 2000, who noted in a study that increased drought is anticipated in arid and semi arid parts of Kenya which will experience drier conditions due to reduced rainfall and increased evapotranspiration rates. This global warming related impact will therefore significantly affect water resources. Regions that already a favourable amount of rainfall will be expected to receive even higher amounts of rainfall.

7 A NOTE ON METEOROLOGIC NETWORKS AND OBSERVATIONS IN KENYA

In Kenya, systematic collection of all weather data is carried out by the Kenya Meteorological Department (KMD). Most of the weather-related disasters discussed in Section 6 can be avoided by issuing alerts and warnings. These are only possible if the station network is good. KMD has invested a lot on station network, but there is still a long way to go before optimum monitoring network is achieved. The network of synoptic meteorological stations (operational over 24 hour period) is illustrated in Figure 2. The purpose of weather station network is to provide accurate and representative weather information to an area. Automated, electronic weather stations are the least costly means to accomplish this objective over a large area. When properly sited and maintained and sensors are kept calibrated, weather stations will provide accurate data for computation and forecasting weather.

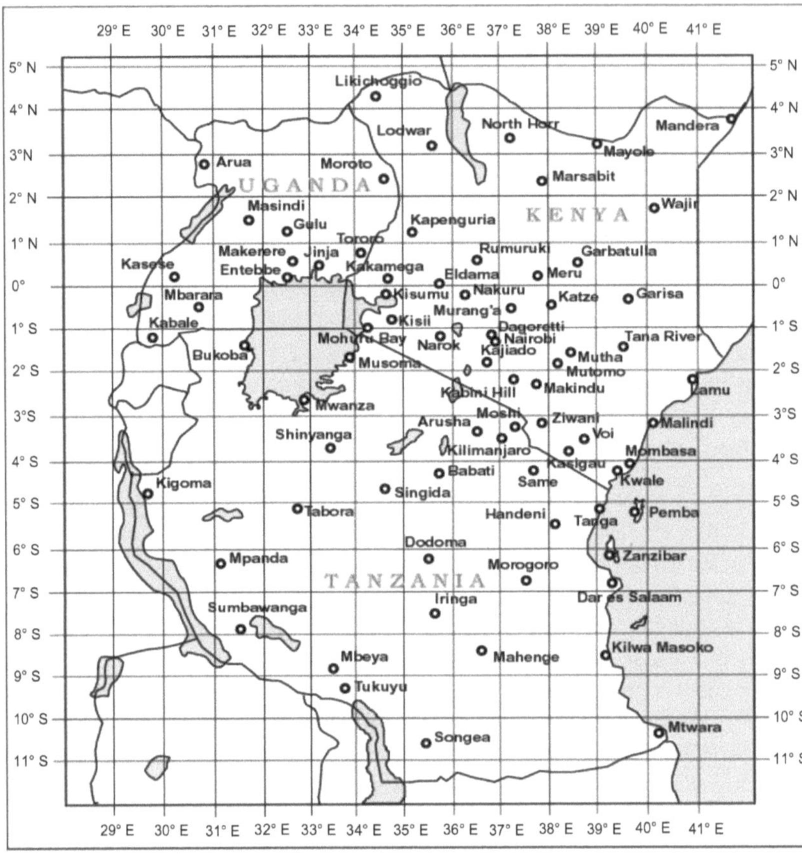

FIGURE 2 Surface synoptic and rainfall station network in East Africa (map not drawn to scale). (For colour version of this figure, the reader is referred to the online version of this chapter.)

REFERENCES

Acland, J., 1971. East African Crops; An Introduction to the Production of Field and Plantation Crops in Kenya, Tanzania and Uganda. Longman, by arrangement with FAO, London.

Biamah, E.K., Cherogony, R.K., 1995. Agricultural Watershed Resource Degradation and Conservation. Lecture Manual, Dept. of Agricultural Engineering, University of Nairobi.

FAO Bulletin 52, 1984. Land qualities and their assessment. In: Guidelines: Land Evaluation for Rainfed Agriculture, Food and Agriculture Organization of the United Nations, Land and Water Development Division, pp. 77–136.

Griffiths, J.F., 1972. Climates. In: Morgan, W.T.W. (Ed.), East Africa: Its Peoples and Resources. Oxford University Press, Ely House, London, pp. 107–118.

Jaetzold, R., Schmidt, H., 1982. Farm Management Handbook of Kenya, vol II/A. Ministry of Agriculture, Kenya, pp. 33–48.

Nicholson, S.E., 1996. A review of climate dynamics and climate variability in Eastern Africa. The limnology, climatology and paleoclimatology of the East African lakes, 25–56.

Ongwenyi, G.S., Kitheka, J.U., Denga, F., 2000. The Impact of Climatic Changes on Land and Water Resources Management in Kenya. Proceedings of the Fourth National Workshop on Land and Water Management in Kenya: Towards Sustainable Land use, Kikuyu, Kenya, 15-19 February, 1993, pp. 219–222.

Chapter 6

Remote Sensing Application Supporting IWRM in Kenya

John B. Kyalo Kiema
Department of Geospatial and Space Technology, University of Nairobi, Nairobi, Kenya
Institut d'Enseignement Supérieur (INES) - Ruhengeri, Musanze, Rwanda

1 CONCEPT OF REMOTE SENSING

Remote sensing is defined as the art, science and technology through which the characteristics of objects/targets either on, above or even below the Earth's surface are identified, measured and analysed without direct contact existing between the sensors and the objects or events being observed (see, e.g. Lillesand et al., 2010; Richards, 2009, etc.). This allows for information about such object features to be obtained by sensing and recording reflected or emitted energy and processing, analyzing and applying that information.

Electromagnetic radiation is normally used as the information carrier in remote sensing. Such electromagnetic radiation, which is either reflected or emitted from objects, normally constitutes the basic remote sensing data. This can be detected by a sensor usually on board airborne (e.g. aircraft or balloon) or spaceborne (e.g. satellites and space shuttles) platforms. As an analogy, our very own eyes serve as our own 'remote sensors' by enabling us to decipher information about surrounding object features without necessarily being in direct physical touch with such objects.

The process of remote sensing is characterized by various stages and interactions summarized in Figure 1 as (a) an energy source or illumination used to provide electromagnetic energy to the object or target of interest; (b) interaction between the electromagnetic radiation and the atmosphere; (c) interaction between the object or target and the electromagnetic radiation; (d) recording of reflected and emitted energy from the object or target by the sensor; (e) transmission, reception and processing of recorded energy into an image; (f) interpretation and analysis of image to extract desired information and (g) application of the information about the object or target in order to better understand it, reveal some new information or assist in solving a particular problem.

Although it is now possible to expand the object feature based in image interpretation beyond the traditional spectral domain to include spatial and

Developments in Earth Surface Processes, Vol. 16. http://dx.doi.org/10.1016/B978-0-444-59559-1.00006-2

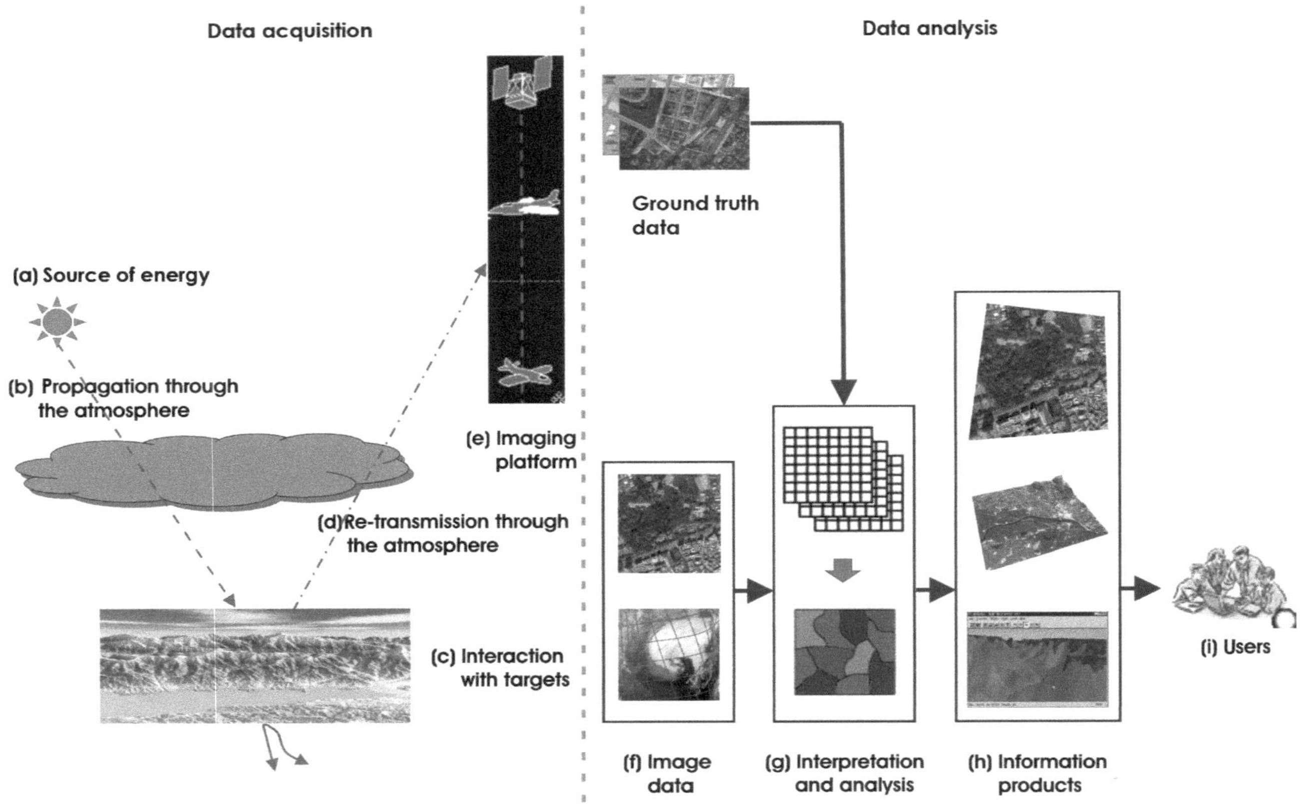

FIGURE 1 Remote sensing process. *Source: Awange and Kiema (2013).* (For colour version of this figure, the reader is referred to the online version of this chapter.)

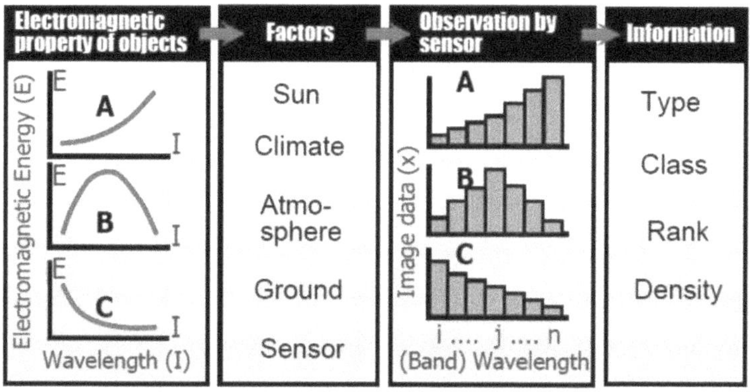

FIGURE 2 Flow of remote sensing. *Source: Murai (2004).* (For colour version of this figure, the reader is referred to the online version of this chapter.)

other dimensions, the practice of remote sensing still relies heavily on the spectral characteristics of objects. Accordingly, each object has a unique spectral signature of reflection or emission dependent on the nature of the object itself, sun, climate, atmosphere, ground condition and sensor, among other factors. This allows the discrimination of the object type, class, rank or density to be made through image interpretation and analysis as illustrated in Figure 2.

Strictly speaking, the technology of remote sensing evolved gradually into a scientific discipline only after the Second World War. Like most other mapping technologies, the early developments in remote sensing were mainly driven by military use, with civilian applications emerging much later. Today, the range of remote sensing applications varies from archaeology, agriculture, cartography, civil engineering, meteorology and climatology, coastal studies, emergency response, forestry, geology, geographic information systems (GISs), hazards, land use and land cover, natural disasters, oceanography, water resources management, etc. Furthermore, the introduction of high-resolution sensors in 1999 has given an impetus to new applications in non-conventional areas like urban mapping, disaster management, location-based services and car and pedestrian navigation.

From its humble beginning, remote sensing technology has today grown in stature to influence virtually all aspects of human endeavour and the environment. Coupled with the availability of historical remote sensing (time-series) data, the overall reduction in data costs and increased spatial resolution in remote sensing data, remote sensing technology appears poised to make an even greater impact in many socioeconomic and political endeavours of mankind. To realize the full potential of this mapping technology, however, it is imperative to integrate remote sensing with other related technologies that deliver geospatial data and information such as global navigation satellite systems (GNSSs), inertial mapping units or other rotation sensors, GISs, wireless sensor networks, global system for mobile communication and the Internet (Awange and Kiema, 2013).

2 STATUS OF STORED WATER IN KENYA

In Kenya, as elsewhere, water has multiple uses, namely, nutritional, domestic, recreational, navigational, waste disposal and ecological as it is a habitat for living and non-living organisms (biodiversity). And because it is indispensable to manufacturing, agriculture, fisheries, wildlife survival, tourism and hydroelectric power generation, it is a vital factor of economic production (NEMA, 2011). Kenya's freshwater endowments cover surface waters, groundwater and wetlands. Surface water bodies include lakes, rivers, swamps, springs, dams and water pans dispersed within the Tana, Athi, Ewaso Nyiro, Rift Valley and Lake Victoria drainage basins as shown in Figure 3. They occupy a combined area of 11,230 km^2, which represents about 1.9% of the country's total area (GoK, 2010a).

In general, people living in arid and semiarid areas of Kenya rely heavily on groundwater. This often constitutes the only reliable source of water with rainfall permeating the soil providing most of Kenya's groundwater resources (Nyaoro, 1999). Groundwater is also an important supplementary source of

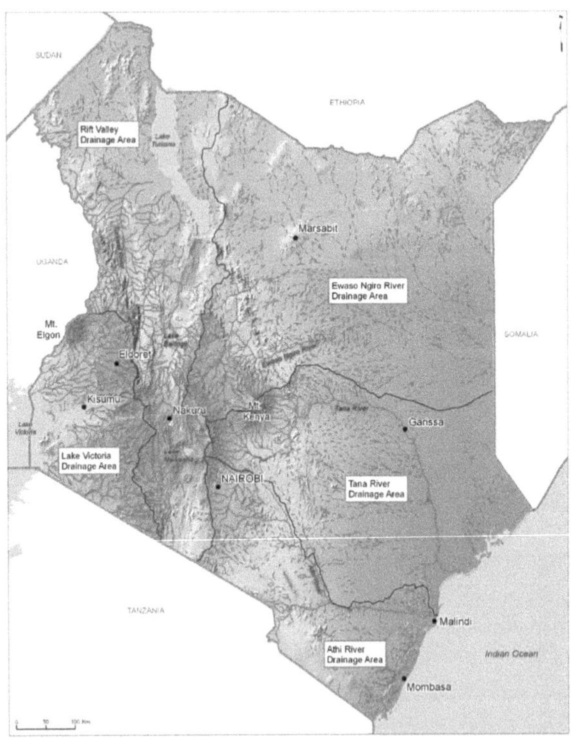

FIGURE 3 Major drainage areas and rivers in Kenya. *Source: WRI et al. (2007).* (For colour version of this figure, the reader is referred to the online version of this chapter.)

water for many urban households, particularly in Nairobi, Mombasa and Nakuru (NEMA, 2011). Details on superficial and groundwater sources in Kenya are discussed in Chapters 4 and 5, respectively.

Quick access to clean and safe water in sufficient quantities still remains a basic challenge to many poor households in the country. This is primarily because demand for water far outstrips the stock of renewable freshwater, with the corollary that Kenya suffers from water scarcity. Factually, Kenya's renewable freshwater per capita endowment is estimated at 21 billion m^3 that works out to about 548 m^3 per capita per year (World Bank, 2010). This figure is well below the widely accepted Falkenmark Water Stress Indicator that places the water scarcity threshold at 1000 m^3 per capita per year (Falkenmark et al., 1989). And the prognosis is bleak as it is projected that the per capita water endowment in Kenya will plummet to about 235 m^3 by 2020 (GoK, 2010b), qualifying the state of water affairs in Kenya as 'acute water scarcity' (NEMA, 2011).

The country's water scarcity has a range of negative ramifications. For instance, it results in higher incidences of waterborne, water-related or sanitation-related diseases such as malaria, diarrhoea and skin infections (NEMA, 2011). Moreover, increasing cases of water conflicts, especially between pastoralist and farming communities along lower and upper river catchments, are also likely to heighten food insecurity. At the gender level, since women are the primary collectors, users and managers of water for domestic use, water scarcity disproportionately affects them because it is they who have to walk long distances, often all day, in search of water (NEMA, 2011).

Against the previously mentioned background is the growing realization today that availability of sufficient, accessible and quality freshwater is not only a matter of great socioeconomic and political importance but also one of fundamental human rights. This means that a blind focus on economic growth without paying requisite attention to the state of salient contributors such as water stored in aquifers is detrimental, since a fall in the amount of the available water would definitely mean reduced yields and overall increased poverty levels (Awange and Kiema, 2013).

In conclusion, the National Environment Management Authority (NEMA) in its 'State of the Environment and Outlook 2010 Report' categorically avers that there is need to ground the concept of Integrated Water Resources Management (IWRM) in Kenya. This would have to take cognizance of the multifaceted nature of water problems and require a comprehensive management of water resources based on an ecosystem approach while appreciating the needs of the diverse users and the broad range of potential impacts of water use. This is an especially useful proposition given that some of the vital sources of the country's freshwater, such as Lake Victoria and the Mara River, are transboundary and the IWRM concept would thus help to foster targeted regional cooperation (NEMA, 2011).

3 CONTRIBUTION OF GEOSPATIAL INFORMATION IN SUPPORTING IWRM

IWRM is defined as that process that promotes the coordinated development and management of water, land and related resources in order to maximize the resultant economic and social welfare in an equitable manner without compromising the sustainability of vital ecosystems (GWP, 2000). In general, IWRM has been globally accepted as a prudent and efficient way of managing water resources in a sustainable manner (see, e.g. Rahaman and Varis, 2005; Rubin et al., 2006; GWP, 2000; GWP, 2004; etc.).

In the long term and to draw tangible benefits, however, IWRM needs to be implemented within a Decision Support System framework. Furthermore, it also needs to integrate hydrometeorological datasets and geospatial information, as well as satellite imagery, besides supporting GIS and modelling functionality for different water resources models (Awange and Kiema, 2013). This section attempts to highlight the increasingly important role played by geospatial information in contributing towards developing a better and sustainable IWRM regime. Thereafter, two case studies extensively discussed in Awange and Kiema (2013) are used to demonstrate the application of remote sensing in the management of water resources.

The availability of techniques that deliver information on the changes in stored water at a more local scale is the first step towards realizing an efficient water society (Awange and Kiema, 2013). However, to be effective, such techniques need to be supported by accurate and timely geospatial and hydrometeorological datasets. With both the techniques and datasets in place, water resources managers are able to make accurate, timely and evidence-based decisions and interventions.

Information about water resources and the environment is inherently geographic. Maps, whether on paper or in digital format, continue to be the medium for presentation of the spatial distribution and character of land and its waters. They also continue to serve as platforms for expression of engineering plans and designs. Johnson (2009) argues that weather patterns, rainfall and other precipitation and resultant water run-off are primary driving forces for land development, water supplies and environmental impacts and pollution. Water resources systems including dams and reservoirs, irrigated lands and canals, water supply collection and distribution systems, sewers and storm-water systems and floodplains are designed in response to a complex mix of topography and drainage patterns, population and land use, sources of water and related environmental factors (Johnson, 2009).

Furthermore, GIS provides an integrating data and modelling environment for planning and design in water resources engineering (Johnson, 2009). A GIS provides a means to collect and archive data on the environment. Measurements of location, distance and flow by GNSSs are typically handled in digital format and quickly integrated into a spatial database. Data processing, synthesis and modelling activities can draw on these data using the GIS, and analysis results can be archived. The GIS database can then be used to generate reports and maps, often interactively, to support decision-making in which design alternatives are best and the impacts of these (Johnson, 2009).

Since maps are a powerful communication medium, this information can be presented in public forums within the context of participatory GIS so that citizens concerned with conservation, planning and design choices can better understand and be more involved (see, e.g. Longley et al., 2005) in the process of decision-making. Moreover, GIS data and analysis results can also be shared with other users to learn and compare experiences from similar projects within the framework of spatial data infrastructure (SDI). The concept and application of SDI is discussed in Groot and McLaughlin (2000), with Kenya's own preparedness in setting up a national SDI articulated in Mulaku et al. (2007).

4 CASE STUDIES

4.1 Lake Victoria Basin

Lake Victoria, the world's third largest lake and the largest in the developing world, is a source of water for irrigation, transport, domestic and livestock uses and supports the livelihoods of more than 30 million people living around it (Awange and Ong'ang'a, 2006). Figure 4 shows Lake Victoria Basin. Nicholson (1998) documents its significance as an indicator of environmental and climate change on long-term scales. Though the lake has continued to attract worldwide attention due to its importance, its recent decline has caused great alarm as to whether the lake is actually drying up! Since the 1960s, the lake level has experienced significant fluctuation (see, e.g. Nicholson, 1998). However, the water level is reported by Riebeek (2006) to have experienced a drastic drop in the last 10 years causing great alarm. According to Kull (2006), the lake levels have dropped by more than 1.1 m below the 10-year average.

Awange et al. (2008) undertook a satellite analysis of the entire Lake Victoria Basin in an attempt to establish the cause of the decline in the lake's water levels. The Gravity Recovery and Climate Experiment (GRACE) and CHAllenging Minisatellite Payload (CHAMP) satellites together with data from the Tropical Rainfall Measuring Mission (TRMM) satellite were employed in the study. Using 45 months of data spanning a period of 4 years (2002–2006), the GRACE satellite data were used to analyse the gravity field variation caused by changes in the stored waters within the lake basin. Figure 5 presents the annual variation of the geoid in the lake's basin during the high rain season months of March, April and

FIGURE 4 Lake Victoria Basin. *Source: Kayombo and Jorgensen (2006).* (For colour version of this figure, the reader is referred to the online version of this chapter.)

May for the period 2002–2006. The GRACE results indicated that the basin's total water storage (TWS) dramatically decreased at a rate of 6.20 mm/month. These changes are expressed in equivalent water thickness (also known as TWS). For the period 2002–2006, the results signalled a general decline in the lake basin's water volume at a rate of 1.83 km^3/month (Awange et al., 2008).

To validate the GRACE results, TRMM data for the same period of time were used to compute mean rainfall at a spatial resolution of 25×25 km from which the rainfall trends were analysed. To assess the effect of evaporation, GNSS remote sensing data (59 CHAMP satellite occultations) for the period 2001–2006 were analysed to define if tropopause warming took place. The results indicated that the tropopause temperature fell in 2002 by about 3.9 K and increased by 2.2 K in 2003 and remained above the 189.5 K value of 2002. The tropopause heights showed a steady increase from a height of 16.72 m in 2001 and remained above that value reaching a maximum of 17.59 km in 2005, an increase in height by 0.87 m. Temperatures did not, therefore, increase drastically to cause massive evaporation.

TRMM results indicated the rainfall over the basin (and directly over the lake) to have been stable during this period. Since rainfall over the period remained stable and temperatures did not increase drastically to cause increased evaporation, the remaining major contributor during the period

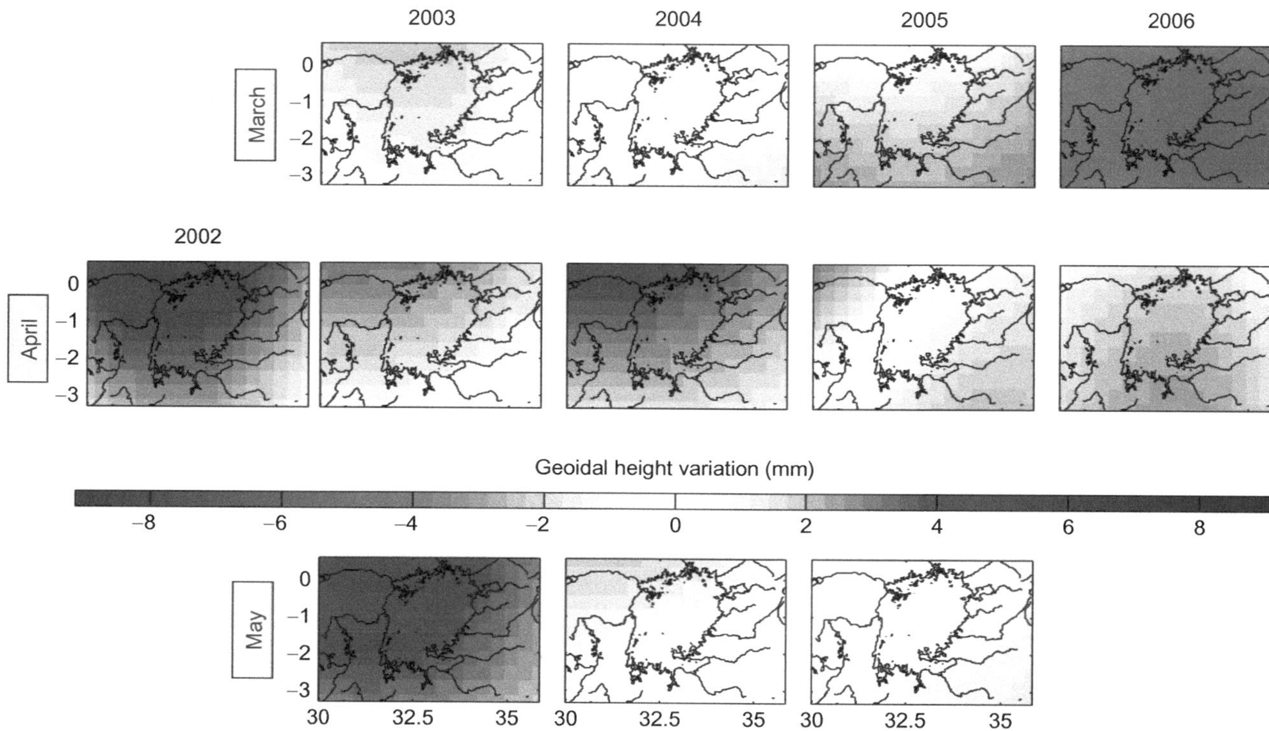

FIGURE 5 Inter-annual comparison in the geoid during the high rainy season of MAM from 2002 to 2006. *Source: Awange et al. (2008).* (For colour version of this figure, the reader is referred to the online version of this chapter.)

2002–2006 was suspected to be discharged from the expanded Owen Falls Dam. Awange et al. (2008) concluded, thanks to the remote sensing data, that the fall in Lake Victoria's water level between 2001 and 2006 was most likely due to human impact on the basin's environment (i.e. expanded dam) as opposed to natural factors.

4.2 Lake Naivasha Basin

Lake Naivasha is the only freshwater lake in the Great Rift Valley of East Africa in an otherwise soda/saline lake series. Figure 6 illustrates the

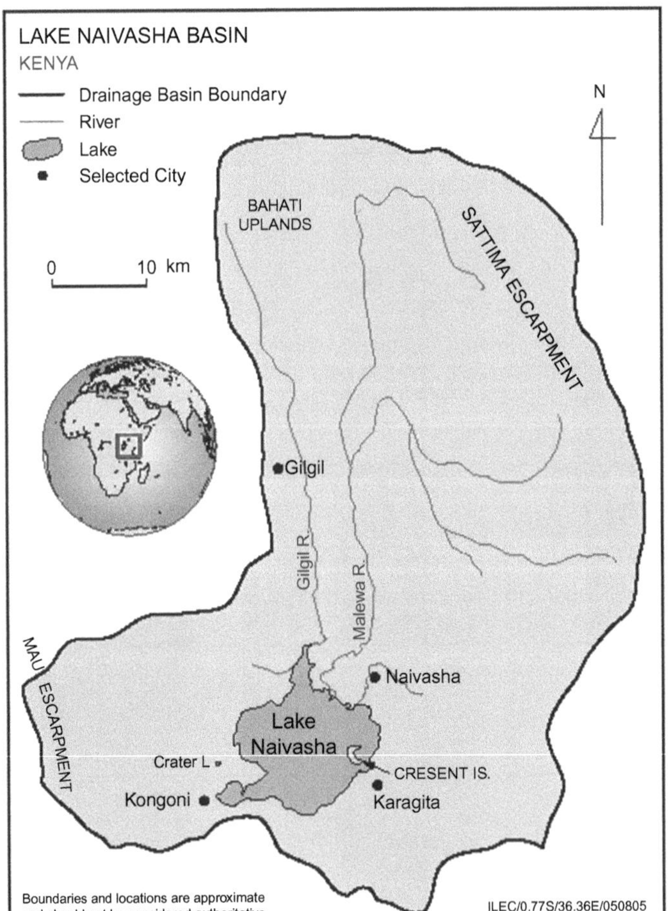

FIGURE 6 Lake Naivasha drainage system. *Source: Becht (2007).* (For colour version of this figure, the reader is referred to the online version of this chapter.)

Naivasha basin drainage system. In fact, it is the freshness of the water of the lake that is the basis for its diverse ecology (Harper et al., 1990). However, recent years have seen a rapid decline in its extent to the point where questions are being raised in the local media as to whether the lake is actually dying. So unique is Lake Naivasha in the chain of East African Great Rift Valley lakes that in 1995 it was declared a Ramsar site due to its importance as a wetland. Lake Naivasha and its basin supports a rich ecosystem with hundreds of bird species, papyrus fringes filled with diverse wildlife species, swampy areas where waterfowl breed and feed and, at the same time, magnificent views of the nearby volcanoes. The lake also supports local fishery and tourism and is used for recreation, water sports, subsistence farming and hunting. The surrounding lands are dominated by the cultivation of flowers, vegetables, fruits and cereals and power generation (Becht and Harper, 2002).

Several studies on the fluctuation in Lake Naivasha's water levels have been carried out during the course of the last half century or so. Richardson and Richardson (1972), for instance, state that the lake was nearly twice as extensive in the 1920s as it was in 1960–1961 with the decrease attributed to a slight trend of decreasing rainfall during this period, averaging 5 mm/year over the basin, between 1920 and 1949 (see also Sansome, 1952) and an increase in human consumption from river influent and boreholes. In the 1980s, the fall of the lake's level continued, with the local Olkaria geothermal power station and subsurface drainage thought to be the main culprits (Darling et al., 1990).

In the last two decades, the lake's level has continued to drop with floriculture being blamed for excessive water extraction from the lake and aquifers and the smallholder farms in the upper catchment being blamed for nutrient loading. For example, Mekonnen and Hoekstra (2010) observed that the total virtual water exported in relation to the cut flower industry from the Lake Naivasha basin was 16 Mm^3/year during the period 1996–2005. Other factors that have also been suggested as influencing Lake Naivasha's water changes include irregular rainfall patterns (Harper et al., 1990) and trade winds (Vincent et al., 1979).

All the earlier discussions point to the need for reliable mapping of the lake and its basin in order to properly understand the dynamics of the lake's basin. The need for accurately monitoring the lake was captured by Becht and Harper (2002), who state that there is an urgent need to accurately measure all abstractions and provide consistent, reliable, hydrologic and meteorologic data from the catchment, so that a 'safe yield' may be agreed upon by all stakeholders and the sustainable use of the lake waters achieved.

However, lack of reliable basin mapping techniques in the past has hampered the proper mapping of changes in the lake basin while also not allowing accurate predictions of the likely future scenario, despite modelling methods being used to calculate its water balance Becht and Harper (2002). The situation is compounded by the fact that Lake Naivasha has no surface outlet that could assist in hydrologic monitoring and the fact that changes in its water level occur rapidly, over the order of several metres within just a few months, shifting the shoreline by several metres (Becht and Harper, 2002).

To better understand the decline of water storage across the Ramser-Lake Naivasha several different types of datasets were used in the study reported in Awange et al., (2013) including (1) Gravity Recovery and Climate Experiment (GRACE) time-variable gravity field products to derive total water storage (TWS) variations within a region covering the Lakes Naivasha and Victoria basins; (2) precipitation records based on Tropical Rainfall Measurement Mission (TRMM) products to evaluate the impact of climate change; (3) satellite remote sensing (Landsat) images to map shoreline changes and to correlate these changes over time with possible causes; and (4) satellite altimetry observations to assess fluctuations in the lake's level. In addition, data from an in situ tide gauge and rainfall stations as well as the output from the African Drought Monitor (ADM) model are used to evaluate the results.

This study confirms that Lake Naivasha has been steadily declining with the situation being exacerbated from around the year 2000 (see Table 1), with water levels declining at a rate of 10.2 cm/year and a shrinkage in area of 1.04 km^2/year. The results show a correlation coefficient value of 0.68 between the quantity of flower production and the lake's level for the period 2002–2010 at 95% confidence level (see Figure 7), indicating the probable impact of anthropogenic activities on the lake's level drop. These results show

TABLE 1 Summary of the Changes in Area of Lake Naivasha and the Surrounding Bare Land and Vegetation

Year	Lake Area (km^2)	Vegetation (km^2)	Bare Land (km^2)	Mean Lake Level (asl)	Lake Area (%)	Vegetation (%)	Bare Land (%)
1989	113.67	95.35	174.18	1885.41	0.00	0.00	0.00
1995	130.47	90.37	162.35	1886.60	14.78	−5.22	−6.89
2000	136.42	26.70	219.77	1887.74	20.01	−72.00	26.17
2006	130.07	48.44	204.12	1885.82	14.43	−49.20	17.19
2010	126.01	48.93	207.69	1884.44	10.85	−48.69	19.24

Source: Awange et al. (2013).

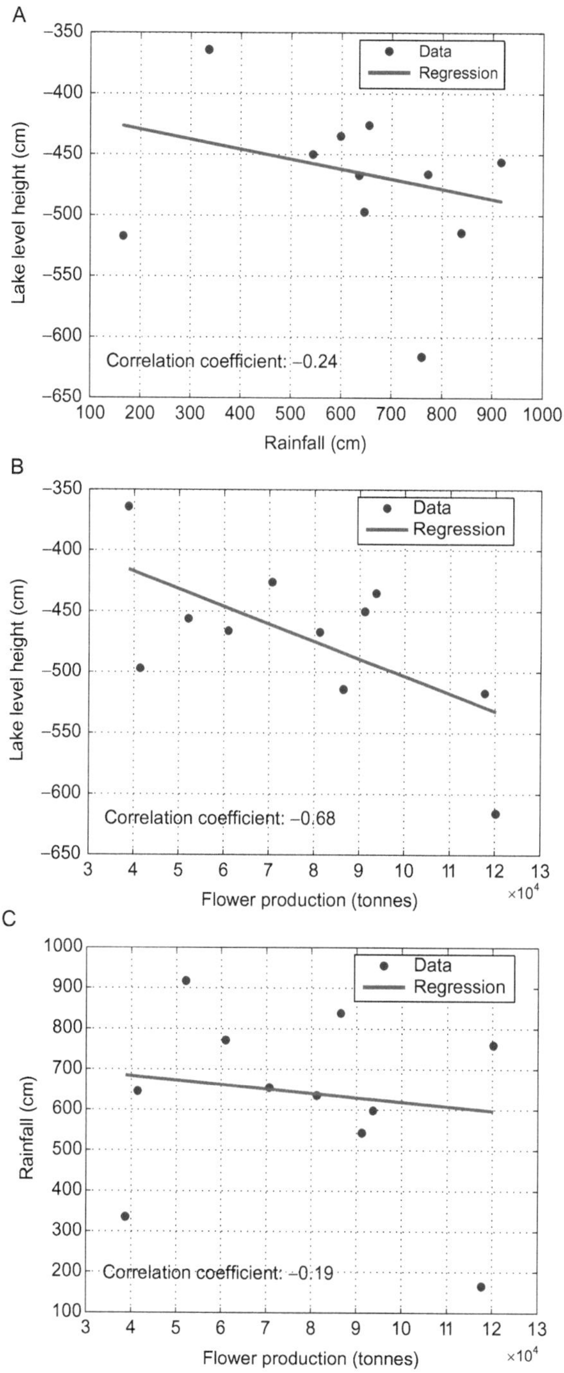

FIGURE 7 Comparing annual average lake levels with (a) rainfall observed at the Naivasha station and (b) flower exports. (c) Comparing annual average rainfall of the Naivasha station and flower exports. The solid lines are fitted linear trends, along with the correlation coefficients. *Source: Awange et al. (2013).* (For colour version of this figure, the reader is referred to the online version of this chapter.)

that there is need for the different communities and interest groups that depend upon Lake Naivasha to better formulate their management plans to ensure sustainable use of this delicate water resource.

ACKNOWLEDGEMENTS

The author wishes to acknowledge the academic support and friendship of Prof. Dr.-Ing. Joseph L. Awange, Curtin University of Technology, Australia. Special regards to also Rev. Fr. Dr. F. Hagenimana, Vice Rector Academics at INES-Ruhengeri, Rwanda.

REFERENCES

Awange, L.J., Kiema, J.B.K., 2013. Environmental Geoinformatics: Monitoring and Management. Environmental Science and Engineering, Springer Verlag, Berlin Heidelberg, 541pp. http://dx.doi.org/10.1007/978-3-642-34085-7_1.

Awange, L.J., Ong'ang'a, O., 2006. Lake Victoria: Ecology, Resources and Environment. Springer, Heidelberg.

Awange, J.L., Sharifi, M.A., Ogonda, G., Wickert, J., Grafarend, E.W., Omulo, M.A., 2008. The falling lake Victoria water level: GRACE, TRIMM and CHAMP satellite analysis of the lake basin. Water Resour. Manag. 22, 775–796. http://dx.doi.org/10.1007/s11269-007-9191-y.

Awange, J.L., Forootan, E., Kusche, J., Kiema, J.B.K., Omondi, P.A., Heck, B., Fleming, K., Ohanya, S.O., Goncalves, R.M., 2013. Understanding the decline of water storage across the Ramser-Lake Naivasha using satellite-based methods. Adv. Water Resour. 60, 7–23, Elsevier. http://dx.doi.org/10.1016/j.advwatres.2013.07.002.

Becht, R., 2007. Environmental Effects of the Floricultural Industry on the Lake Naivasha Basin. ITC. Carolina, B.F. 2002. Competition over water resources: analysis and mapping of water-related conflicts in the catchment of Lake Naivasha (Kenya). MSc Thesis, ITC.

Becht, R., Harper, D.M., 2002. Towards an understanding of human impact upon the hydrology of Lake Naivasha, Kenya. Hydrobiologia 488, 1–12.

Darling, W., Allen, D., Armannsson, H., 1990. Indirect detection of subsurface outflow from a Rift Valley Lake. J. Hydrol. 113 (1–4), 297–306.

Falkenmark, M., Lundquist, J., Widstrand, C., 1989. Macro-scale water scarcity requires micro-scale approaches: aspects of vulnerability in semi-arid development. Nat. Resour. Forum 13 (4), 258–267.

Global Water Partnership (GWP), 2004. Catalyzing Change: Handbook for Developing IWRM and Water Efficiency Strategies. http://www.gwpforum.org/gwp/library.

GoK, 2010a. National Water Harvesting and Storage Management Policy. Ministry of Water and Irrigation, Kenya.

GoK, 2010a. Statistical Abstract. Kenya National Bureau of Statistics (KNBS).

Groot, R., McLaughlin, J. (Eds.), 2000. Geospatial Data Infrastructure: Concepts, Cases and Good Practice. Oxford University Press, New York.

GWP, 2000. Integrated Water Resources Management. Technical Advisory Committee (TAC). TEC Background Papers No. 4, 65pp., www.gwpforum.org/gwp/library/tacno4.pdf.

Harper, D.M., Mavuti, K.M., Muchiri, S.M., 1990. Ecology and management of Lake Naivasha, Kenya, in relation to climatic change, alien species' introductions, and agricultural development. Environ. Conserv. 17 (04), 328–336.

Johnson, L.E., 2009. Geographic Information Systems in Water Resources Engineering. CRC Press Taylor & Francis Group, 978-1-4200-6913-6.

Kayombo, S., Jorgensen, S.E., 2006. Lake Victoria: experience and lessons learned brief. International Lake Environment Committee. Lake Basin Management Initiative, Kusatsu, Japan. Available at: http://www.ilec.or.jp/eg/lbmi/reports/27_Lake_Victoria_27February2006.pdf.

Kull, D., 2006. Connections between Recent Water level drops in Lake Victoria, Dam Operations and Drought. Available at: http://www.irn.org/programs/nile/pdf/060208vic.pdf.

Lillesand, T.M., Kiefer, R.W., Chipman, J.W., 2010. Remote Sensing and Image Interpretation. John Wiley, New York.

Longley, P.A., Goodchild, M.F., Maguire, D.J., Rhind, D.W., 2005. Geographic Information Systems and Science. John Wiley & Sons, West Sussex, England, p. 486.

Mekonnen, M.M., Hoekstra, A.Y., 2010. Mitigating the water footprint of export cut flowers from the Lake Naivasha Basin, Kenya: Value of water research report series no. 45. UNESCO-IHE, Delft, The Netherlands.

Mulaku, G.C., Kiema, J.B.K., Siriba, D.N., 2007. Assessment of Kenya's readiness for geospatial data infrastructure take off. Surv. Rev. 39 (306), 328–337.

Murai, S., 2004. Remote Sensing and GIS Notes. Japan International Cooperation Agency.

National Environment Management Authority (NEMA), 2011. Kenya: State of the Environment and Outlook 2010. Summary for Decision Makers. National Environment Management Authority.

Nicholson, S.E., 1998. Historical Fluctuations of Lake Victoria and other lakes in the Northern Rift Valley of East Africa. In: Lehman, J.T. (Ed.), Environmental Change and Response in East African Lakes. Kluwer, Dordrecht, pp. 7–35.

Nyaoro, J.R., 1999. Case Study on Kenya River Watershed Management and Arising Conflicts. Report on the Conference on Cooperation or Conflict: Ways of Managing Scarce Natural Resources in Africa, 18–23 June 1999. Civil Society Conflict Prevention Network and Citizens' Security Council (KATU), Maputo, Mozambique. Online at, http://www.katunet.fi/suomi/pdf/NATURAL%20RESOURCE%20REPORT.pdf.

Rahaman, M.M., Varis, O., 2005. Integrated water resources management: evolution, prospects and future challenges. Sustainability: Science, Practice, & Policy 1 (1), 15–21. http://ejournal.nbii.org/archives/vol1iss1/0407-03.rahaman.html.

Richards, J.A., 2009. Remote Sensing with Imaging Radar. Springer-Verlag, Berlin Heidelberg, New York.

Richardson, J.L., Richardson, E., 1972. History of an African Rift lake and its climatic implications. Ecol. Monogr. 42 (4), 499–534.

Riebeek, H., 2006. Lake Victoria's Falling Waters. http://earthobservatory.nasa.gov/Study/Victoria/printall.php.

Rubin, H., Rubin, A., Reuter, C., Köngeter, J., 2006. Sustainable integrated water resources management (IWRM) in a semi-arid area. Int. J. Environ. Cult. Econ. Soc. Sustain. 2 (3), 165–180. http://www.Sustainability-Journal.com.

Sansome, H.W., 1952. The Trend of Rainfall in East Africa. E. Afr. Meteorol. Dep., Tech. Mem. 1. 14 pp.

Vincent, C.E., Davies, T.D., Beresford, U.C., 1979. Recent changes in the level of Lake Naivasha, Kenya, as an indicator of equatorial westerlies over East Africa. Clim. Change 2, 175–189.

World Bank, 2010. World Development Indicators: 2010. 1818 H Street NW, Washington, DC 20433 USA.

World Resources Institute (WRI); Department of Resource Surveys and Remote Sensing (DRSRS), Ministry of Environment and Natural Resources, Kenya; Central Bureau of Statistics (CBS), Ministry of Planning and National Development, Kenya; and International Livestock Research Institute (ILRI), 2007. Nature's Benefits in Kenya: An Atlas of Ecosystems and Human Well-Being. Washington, DC and Nairobi: World Resources Institute. Published by: World Resources Institute, 10 G Street NE, Washington, DC 20002, USA.

Chapter 7

Surface Flows for People and Wildlife in the Transboundary Mara River Basin

Michael E. McClain* and Amanda L. Subalusky†
*Department of Water Science and Engineering, UNESCO-IHE, Delft, The Netherlands
†Department of Ecology and Evolutionary Biology, Yale University, New Haven, USA

1 INTRODUCTION

From an international perspective, the Mara is perhaps the best known of Kenya's rivers due to its association with Maasai Mara National Reserve and Serengeti National Park. These conservation areas and the contiguous ecoregion they protect are major tourist attractions for both Kenya and Tanzania and refuges for the region's disappearing natural heritage. The Mara is the only perennial river flowing through the region, making it a critical dry season source of water for wildlife. Its verdant corridor is also an important habitat for many aquatic and terrestrial species, and its open water is a hazardous obstacle to migrating wildlife.

Organisms dependent on the river rely not only on the presence of water but also on the natural variability of flow throughout the year. Base flows meet the minimal habitat requirements of organisms, but seasonally high flows activate new habitats important for different stages in the life histories of riverine plants and animals. High flows also flush stagnant backwaters, bring nutrients and energy sources, and facilitate migration of larger aquatic organisms along the river's length. Larger floods are also important for shaping the river channel and redistributing sediments in the system. Even occasional droughts and very low flows are important for controlling invasive species and resetting certain ecological processes.

Maintaining these flow-related ecological functions of the Mara River is essential for the preservation of Maasai Mara National Reserve and Serengeti National Park, but the river also provides many important resources for humans. The estimated 2010 population in the Kenyan portion of the Mara is just over 550,000 people, most of whom live in rural areas and practice small-scale rain-fed farming. Large-scale irrigated agriculture is currently

Developments in Earth Surface Processes, Vol. 16. http://dx.doi.org/10.1016/B978-0-444-59559-1.00007-4

limited to a few farms in the middle reaches of the river, but water use on these farms is estimated to account for more than half of all water consumed in the basin (Hoffman, 2007). Domestic water use is estimated at about 20% of the total consumption, and livestock consumes another 17%. In addition, people consume fish, harvest riparian vegetation, and include the river in religious and cultural traditions. Each of these uses is growing in the basin as population increases and plans unfold for new water infrastructure. The most significant plans include the construction of two multipurpose reservoirs in the headwaters of the Mara on the Nyangores and Amala Rivers. The planned combined water storage capacity of the dams is 130 million m^3 to be used for irrigation, domestic consumption, and hydropower generation.

Increased use of surface water in the Mara is necessary for development and improved welfare of people in the basin, but water allocations must also consider and protect the water needs of downstream communities and ecosystems. This requirement is written into the 2002 Kenya Water Act and maintained in a new draft water law, which calls for a 'reserve' to be set aside in water allocation plans 'to (a) satisfy basic human needs for all people who are or may be supplied from the water resource and (b) to protect aquatic ecosystems in order to secure ecologically sustainable development and use of the water resource'. The reserve must be protected in all water allocation decisions and is thus the highest priority in allocation planning. The establishment and protection of a reserve flow is included in the water laws of Tanzania and South Africa as well and represents an adaptation (incorporating basic needs of humans) of the concept of environmental flows, which is recognized in water management guidelines around the world (Le Quesne et al., 2010).

This case study presents a first effort to determine reserve flows for the Mara River in reaches upstream of and within Maasai Mara National Reserve (Figure 1). Reserve flow recommendations have also been made for reaches of the Mara within and downstream of Serengeti National Park in Tanzania, but this case study focuses on Kenyan reaches of the river. The effort was co-ordinated by the World Wide Fund for Nature with financing from USAID and involved the Lake Victoria South Catchment Area of the Water Resource Management Authority (WRMA) and a team of international, Kenyan, and Tanzanian scientists. The results were subsequently approved by the Lake Victoria Basin Council of Ministers of the East African Community for implementation by basin water authorities.

2 ENVIRONMENTAL FLOW ASSESSMENT TO SET THE RESERVE

The default approach in the Kenyan water strategy for setting the reserve is to calculate the flow level that is exceeded 95% of the time in a given reach of the river, known as Q95. This statistical calculation is based on available flow data and establishes an absolute minimum flow to be maintained in the river.

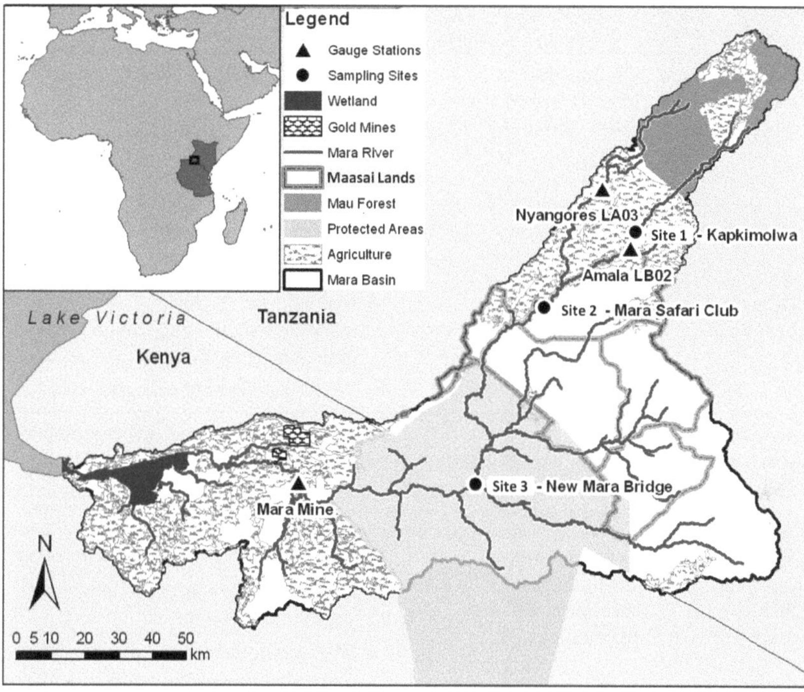

FIGURE 1 The transboundary Mara River Basin of Kenya and Tanzania originates in the Mau Forest and discharges into the Mara wetland and Lake Victoria. Small-scale rain-fed agriculture dominates the land use, along with pastoral lands of the Maasai pastoralist lands. Maasai Mara National Reserve and Serengeti National Park lie at the centre of the basin. Three study sites are located in representative reaches of the river.

It does not consider higher flows or variable flows required to protect ecosystems. Simply prescribing a reserve equal to Q95 will therefore not meet the objectives of the reserve, which include supplying the basic needs of people and protecting ecosystems.

The approach applied in the Mara was based on the building block methodology (BBM) for environmental flow assessment, which was developed in South Africa and has been widely used to determine reserve requirements for more than a decade (King et al., 2000). 'Building blocks' within the BBM refer to specific components of the flow regime considered essential to physical and biological processes in the river (Table 1), and determination of flow recommendations involves implementing several steps in sequence (Figure 2). Scoping activities included collection of existing information on the environmental and ecological condition of the river and present use and management of water resources in the basin. During this phase, the team of scientists was assembled and consultations began with key stakeholders.

Site selection began with a geomorphological survey that classified the river into three uniform macroreaches based on gradient, channel pattern,

TABLE 1 **Definitions of Flow Components (Building Blocks) Targeted in the Recommendations and the Functions They Are Intended to Preserve**

Flow Components	Definitions	Functions
Drought year low flows	The low-flow requirements during the driest month of a drought year	• Maintain hydrological connectivity in the system • Maintain inundation of critical habitats (e.g. riffles) • Sustain flow-sensitive species • Provide natural variability to maintain diverse species assemblage
Drought year high flows	The low-flow requirements during the wettest month of a drought year	• Maintain active channel flows to inundate benches and sustain emergent vegetation • Permit fish passage over obstacles
Maintenance year low flows	The low-flow requirements during the driest month of a maintenance year	• Provide natural variability to maintain diverse species assemblage
Maintenance year high flows	The low-flow requirements during the wettest month of a maintenance year	• Cue migration and spawning in fishes • Inundate macrophytes and emergent vegetation along banks • Displace dominant competitors and allow drift of species into new habitats, promoting increases in species diversity • Maintain groundwater recharge for riparian species
Small annual floods	Small pulses of higher flow that occur in the drier months	• Cue spawning and migration in fishes • Inundate surrounding floodplains to facilitate lateral migration of fauna • Facilitate nutrient transfer between floodplains and the river • Allow germination and seed dispersal of riparian vegetation • Prevent sediment build-up on riverbed, thus increasing habitat variability for invertebrates • Maintain active channel features • Flush out organic matter, thus improving water quality
Major flood events	Major peaks in the river's flow level that occur at a given recurrence interval	• Maintain macrochannel features and provide diversity of physical habitats • Scour bed of sediment deposits • Inundate and recharge larger floodplain, allowing for nutrient transfer

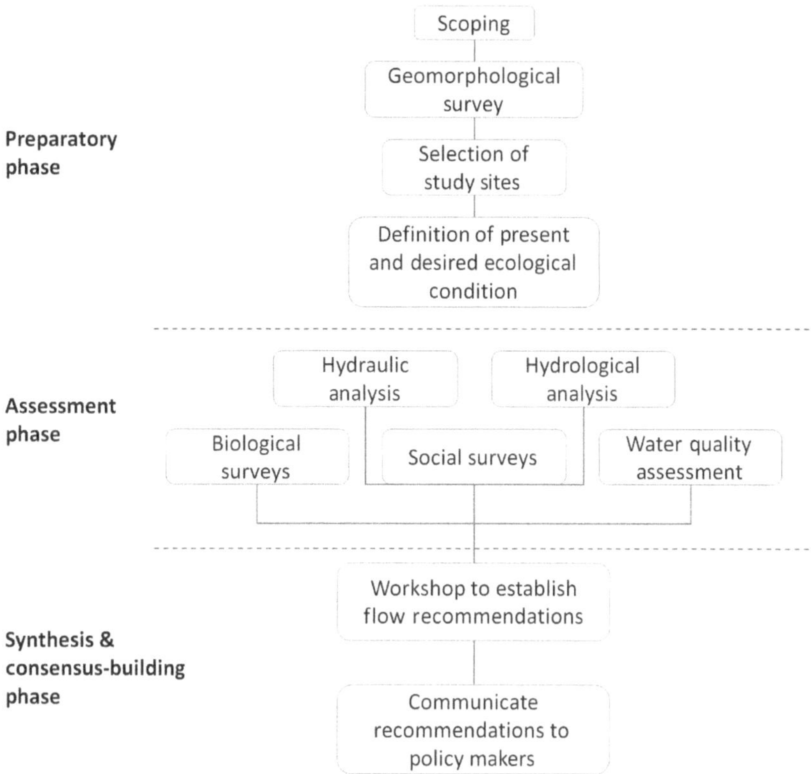

FIGURE 2 Components of the study and progression of activities in the environmental flow assessment framework.

and bed structure (Rowntree and Wadeson, 1999). Suitable sampling sites were chosen in each macroreach with the participation of all team members and based on criteria including ease of accessibility, presence of habitat diversity, sensitivity of habitats to flow changes, suitability for conducting hydraulic analysis, proximity to a flow-gauging site, representation of conditions in the river zone, and presence of critical flow sites (i.e. where flow will stop first if discharges are reduced). WRMA has assigned the highest level of ecological importance to the Mara River in its classification of management units (WRMA, 2009). The team thus adopted this classification and defined the desired level of ecological condition as slightly modified from natural conditions. This allows for use of the river but sets the goal of maintaining a reserve flow regime that will sustain the full ecological integrity and functionality of the system.

The analysis phase consisted of detailed field assessments conducted at each site during March and July 2007 to target medium and low flows, respectively. Unexpected rains caused high flows in July 2007, so a subset of specialists returned in February 2009 to sample the river during critically low flows. Field assessments applied well-accepted methods to describe and assess the condition of the river system. Multiple hydraulic cross sections were surveyed at each site and analysed to determine the relationship between discharge and ecologically relevant variables such as wetted width, depth, and velocity. Detailed observations were made of the geomorphology by mapping the spatial distribution and substrate material of channel features, including pools, riffles, bars, terraces, and benches. Riparian vegetation was characterized along transects extending from the river edge to the furthest extent of the riparian vegetation. Aquatic macroinvertebrates were sampled at all sites and analysed using the South African Scoring System protocol (Dickens and Graham, 2002). Fish were sampled using gillnets placed in all available river habitats (i.e. riffles, runs, and pools) and identified to species level. Gonad state was assessed to gain information about reproductive activities. Water quality data used in the assessment were collected in parallel monitoring activities (GLOWS, 2007). The specialists selected flow-sensitive indicator variables or species to inform them on the river's natural flow regime.

To address the human component of environmental flows, data on population size and growth rate in the Mara River Basin were extracted from the 2006 Kenyan and 2005 Tanzanian census data (Hoffman, 2007). Current and projected water demand were estimated conservatively based on a basic human need of 25 l/person/day (GoK, 2007). Informal surveys and interviews were also conducted in communities dependent on the Mara River to determine the primary resources and services the river provides. Informants were asked to identify riverine resources, resource users, and seasonality of use; prioritize relative importance of resources; compare past and present riverine conditions; and determine the desired state of the river.

To determine patterns of flow in the Mara and its major tributaries, historic and contemporary data were assembled from three gauging stations (Figure 1). Flow data from the Amala River (available from 1955 to 2008) and the Nyangores River (available from 1963 to 2008) were acquired from the Kenya Ministry of Water and Irrigation. Flow data from the Mara River (available from 1970 to 1990) were obtained from the Tanzania Ministry of Water and Irrigation. During periods when data were available for both the Amala and Nyangores, flows were combined to yield an estimate of total discharge in the upper Mara, as these are the river's two primary tributaries. Data were compiled to present average monthly historical flow over the period of record and in wet and dry years and also used to calculate flow duration curves, flood frequency, and low-flow recurrence intervals.

Environmental flow recommendations were determined through a structured and professionally facilitated consensus-building process among participating scientists (King et al., 2000). Specific environmental flow recommendations were determined for each flow regime building block based on the results of field assessments and the experts' knowledge of the flow-related habitat needs of indicator species (Table 1). These needs were translated into flows using relationships developed as part of the hydraulic analysis (King et al., 2000). While the individual flow recommendations of each team member may vary, the facilitated process identified the lowest monthly mean flow rate needed to sustain the entire system in the desired ecological state. In the consultative process, specialists were challenged to provide a sound ecohydrological justification of all recommended flows and to describe potential consequences of not meeting the flow requirement.

Based on the agreed-upon recommendations for average flows during the wettest and driest months of the year, the hydrologist extrapolated recommendations across the entire year to simulate the natural shape of the river's historical hydrograph, as determined from the nearest gauging station. The modified hydrograph, with associated floods, serves as the recommended flow regime and was developed for both *maintenance* and *drought* years. Maintenance refers to years of normal rainfall in which all the ecological, chemical, and geomorphological processes would be expected to operate adequately. Drought refers to years of reduced rainfall in which flows would be required to simply ensure the basic survival of species.

3 RESULTS OF THE ASSESSMENT

The site assessments provided a comprehensive picture of the hydraulic, geomorphological, and biotic characteristics of the upper and mid-reaches of the Mara River and a description of how adjacent human communities make use of riverine resources. The study found little evidence of serious ecological degradation in the Mara River corridor, especially in the midreaches upstream and within Maasai Mara National Reserve (sites 2 and 3). As is typical of medium-sized East African rivers like the Mara, the system contains rather low biodiversity by international standards, but its ecological importance is judged to be high given the role the river plays in the larger ecological and socioeconomic welfare of the Mara–Serengeti Ecoregion. The riparian vegetation observed is typical of the Diospyros riverine forest and Croton riverine thicket that occur throughout the region and rank among the most species-rich and dense woody vegetation in the Mara–Serengeti Ecoregion (Walpole et al., 2004).

Water quality analyses conducted prior to and in parallel with this study identified no systematic water quality problems within the range of normal flows (GLOWS, 2007). Dissolved oxygen concentrations remained near saturation throughout the midreaches of the river and nutrient and organic matter concentrations remained low. Suspended sediment concentrations were generally below 200 mg l^{-1} but varied greatly as a function of flow. There were isolated detections of pesticides and PCBs, but concentrations were low. The generally high quality of water was reflected in the sensitivity of macroinvertebrates collected at the sites. The range of fish age classes and evidence of reproductive activity were also evidence of healthy function in the river.

Communities living near the river identified a number of resources provided by a 'natural' river ecosystem (Table 2). A resource prioritization chart indicated water was the most important resource, followed by riparian vegetation and then the river ecosystem itself, which played a role in cultural and religious practices, such as the Maasai naming ceremony. Soil substrate, wildlife, and fish were also identified as important. Local communities were asked to identify current anthropogenic threats to the river ecosystem. They included riverbank erosion by livestock, pollutants due to human use, and destruction of riparian vegetation by cultivation. All threats were exacerbated in the dry season, when other water sources ran dry and usage was concentrated in the river.

Based on the estimated habitat requirements of biological indicators, hydraulic relationships between habitat and flow, and the outcome of the consensus-building process, the recommended flows during normal rainfall years ranged from average monthly values of 1.3 m^3s^{-1} at site 1 during dry season months to 15.0 m^3s^{-1} at site 3 during wet months. During drought years, the recommended flows declined to 0.3 m^3s^{-1} and 6.0 m^3s^{-1}, respectively (Table 3; Figure 3). These recommendations fall between the 50th percentile of the monthly flow duration curve during wet months of normal (maintenance) years and the 95th percentile of the monthly flow duration curves during drought years (Table 3), relative to the nearest gauging stations. Recommended floods ranged from 2-day events of 12 m^3s^{-1} at site 1 to an annual 3-day flood event of 90 m^3s^{-1} at site 3.

The most important habitat characteristics influencing flow recommendations included (i) dry season base flows that provide sufficient depths for upstream migration of the larger-bodied members of each indicator fish genus (*Labeo* and *Barbus*) and maintain inundation of riffle habitats critical to indicate fish survival; (ii) wet season base flows to inundate lower banks and benches, allowing the input of nutrients to the river, fish passage over larger obstacles, and recharge of groundwater for riparian vegetation; and (iii) wet season high flows (floods) to inundate the flood benches, to facilitate nutrient transfer, to provide access to fish requiring riverbank nursery grounds, and to foster recruitment potential and sustain appropriate density and age structure of riparian vegetation (Figure 4). Flows required to sustain the indicator fish were also judged to be adequate for the indicator insects (Libellulidae and Coenagrionidae). Recommendations for annual floods were influenced by geomorphological considerations. Infrequent large flood events were also required to maintain the structure of macrochannel features, such as terraces and wider banks, but the magnitude of these floods was not quantified explicitly because in the current unregulated state of the Mara River, these large floods occur naturally.

TABLE 2 Summary of the Variety of Ways Communities Utilize the Mara River

River Resources	Resource Use	Desired State of the River
Water	Water for livestock	Sufficient water to provide for livestock, even during droughts, while maintaining acceptable quality for human consumption
	Domestic use	High-enough water quality for human consumption at all times, including low sediment and impurity loads. The need for point of use disinfection is recognized as well
	Irrigation farming	Sufficient water to sustain crops during the dry season when precipitation is low[a]
	Habitat for fish	Dynamic flow regime to cue fish-breeding events
	Recreation, for example, swimming	Sufficient water to allow swimming
	Industrial use, for example, watermills and mines	Sufficient water to maintain industry practices[a]
	Generation of hydroelectric power	Sufficient water levels for hydroelectric power generation[a]
	Cultural/religious practices, for example, baptism	Presence of deep pools where people can carry out cultural practices
Fish	Food	Healthy fish populations
Vegetation	Habitats for wildlife	Intact riparian zone that provides habitat and camouflage for wildlife
	Food	Healthy populations of important food plants
	Medicine	Flow regimes that foster growth of medicinal herbs that are only found in the riparian zone
	Construction material	Intact riparian zones that provide habitat for vines used in construction of the Maasai manyattas
	Cultural/traditional artefacts, for example, rungus	Intact riparian zones that provide habitat for culturally important tree species
	Charcoal	Presence of large tree species that may be used in charcoal production

Continued

TABLE 2 Summary of the Variety of Ways Communities Utilize the Mara River—Cont'd

River Resources	Resource Use	Desired State of the River
Soil sediments	Soil sediments for artworks on houses	Functioning sediment generation process to provide fertile soil
	Sand harvesting	Functioning sediment generation process to provide sands
Wildlife	Tourist attraction, for example, crocodile and hippopotamus	Intact habitat to foster thriving wildlife populations
River ecosystem	Cultural practices (e.g. baptism, circumcision and naming ceremonies)	Sufficient vegetation and deep pools of water to meet cultural needs of the community
	Hotel sites	Adequate water supply and stable riverbanks to allow construction of hotels and restaurants

[a]Some of these uses reflect local people's desire for flows beyond the reserve to meet extractive water needs as well. These additional needs were acknowledged but not included in the final recommended reserve flows. These needs are to be met by flows exceeding the reserve.

Considering human needs, the direct use of the Mara River water for domestic use required a minimal flow, equivalent to a basin total of $0.2\ m^3s^{-1}$ ($\sim$800,000 people $\times$ 25 l/person/day). However, harvesting of riparian plants for food, fibre, and medicine; harvesting of fish; and use of the river ecosystem for cultural practices are highly valued as well, adding additional flow requirements. Use of these ecosystem services highlights the importance of maintaining ecosystem processes that regulate water quality in the river, sustain ecological communities, and maintain an intact riverine ecosystem (Gereta et al., 2003).

4 IMPLICATIONS OF ENVIRONMENTAL FLOW RECOMMENDATIONS

The environmental flow recommendations resulting from this study provide information needed to sustain the Mara River ecosystem for the people and wildlife that depend on it. They also provide the baseline information needed for estimating the amount of water available for additional socioeconomic development. The results indicate that during years of normal rainfall, there is ample flow in the Mara River to meet environmental flow recommendations

TABLE 3 Recommended Environmental Flow Regime for the Mara River, Kenya–Tanzania, Shown as Seasonally Averaged Discharge (m^3/s) and Relative to the Flow Duration Curves (FDCs) of Closest Gauging Stations

Year Type	River Building Blocks	Site 1 (m^3/s) % FDC	Site 2 (m^3/s) % FDC	Site 3 (m^3/s) % FDC
Maintenance (normal rainfall)	Dry season minimum flows	1.3 71%	4.0 80%	6.0 70%
	Wet season minimum flows	2.0 57%	8.0 63%	15.0 45%
	Annual medium floods	12 17% 3/year	16 38% 3/year	25 26% 3/year
	Annual large floods	38 4% 1/year	75 1% 1/year	90 7% 1/year
Drought year (less than normal rainfall)	Dry season minimum flows	0.3 94%	1.0 98%	2.0 92%
	Wet season minimum flows	1.0 76%	4.0 81%	6.0 70%
	Annual floods	4 40% 1/year	25 23% 1/year	20 36% 1/year

The frequency of recommended floods is reported in number of events per year.

and allow for present and even increased extractive uses. The flow available for abstraction, however, is concentrated during wet months and far less water is available for abstraction during dry months (Figure 3). This is a common scenario across Africa and indicates the importance of environmentally appropriate infrastructure to store water in the wet season for use during dry months. The Kenya Water Resource Management Rules (GoK, 2007) require off-channel storage basins for all abstraction permit holders. Full implementation of this rule would go far towards meeting increased demand during the dry season of normal precipitation years.

In contrast, during drought years, environmental flow recommendations likely account for the majority of flows at sites 1 and 2, although both river flows and recommended environmental flows are small at these times. This emphasizes the importance of sustaining very small flows ($<1\ m^3s^{-1}$) during the most severe droughts. The influence of small increments in flow during droughts also highlights the importance of interventions in the upper basin

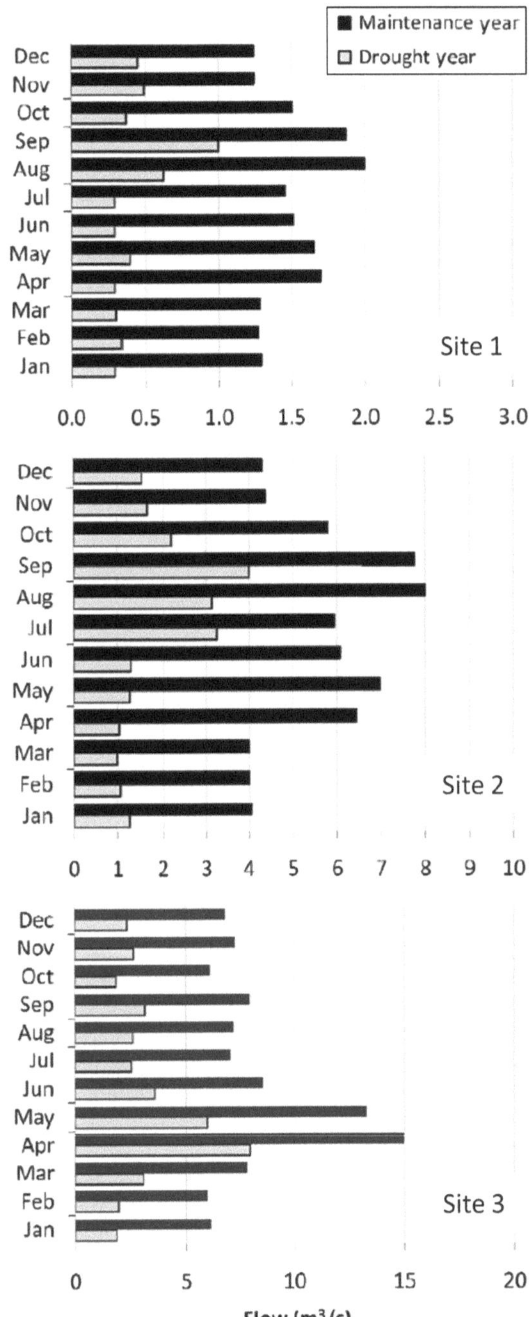

FIGURE 3 Recommended average monthly environmental flows for maintenance years, when all ecological needs of the indicator species can be met, and drought years, when recommended flows are set near minimal, short-term, survival levels.

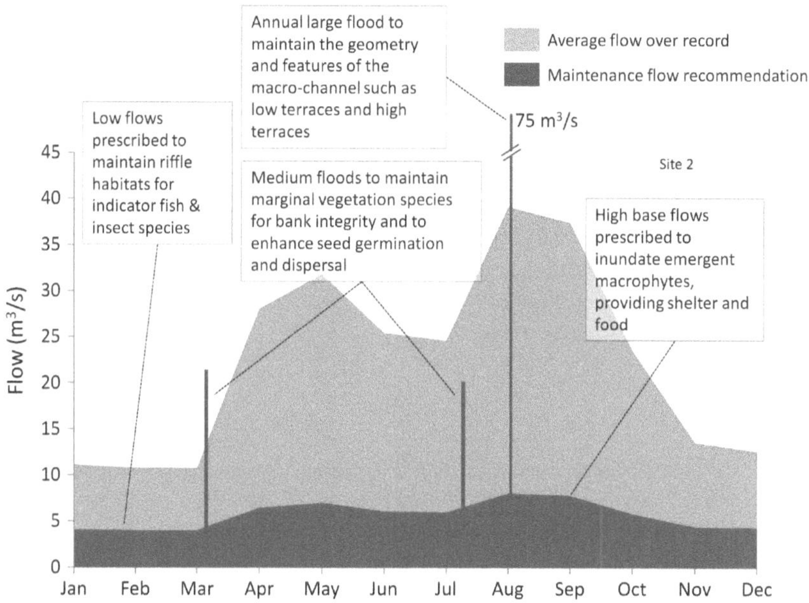

FIGURE 4 Illustration of the average monthly environmental flow recommendations during a maintenance year for site 2 compared to the combined flows of the Amala and Nyangores Rivers upstream. Select motivations for the prescribed flows and floods are indicated.

to protect forests and manage land to maximize infiltration and groundwater recharge. Current patterns of deforestation and small-scale farming practices tend to reduce infiltration and thus reduce dry season river flows (Mango et al., 2011; Defersha and Melesse, 2012). Governmental and non-governmental organizations are promoting improved land use practices in the headwater catchments of the Mara and considering innovative payments for watershed services to finance them (Atisa, 2009; Mendez, 2009). More significant interventions, such as expanding off-channel storage capacity or constructing a minimal number of instream reservoirs, may be necessary to store sufficient water to meet extractive demands during drought periods. The proposed dams partially address the storage needs, but they also pose potential threats to downstream ecosystems. Operating the dams in a way that ensures environmental flow releases will help mediate these threats, but people will have to decide if the gains are worth the risks.

This first assessment of environmental flow requirements for the Mara River was based on the best available data (much of it collected during the study) and the expertise of the scientific team. The recommendations were the result of direct observations of the river and quantification of flow requirements for observed indicator species; thus, the flow recommendations can be implemented by water managers. Planners, however, should apply the precautionary principle when estimating the total available water for development in the basin. This is mainly due to uncertainties in the historical record of flow, which is incomplete in space and time, and uncertainties about future precipitation and run-off patterns (Mango et al., 2011). Another important consideration is that the environmental flow recommendations presented for the protected areas do not consider flow volumes necessary to sustain the Mara wetland on the margin of Lake Victoria or to meet the extractive water needs of Tanzanian communities and industries downstream of Serengeti National Park. Flow volumes crossing the border should therefore be allocated to meet downstream environmental flow recommendations and the extractive needs of Tanzanian water users.

REFERENCES

Atisa, G., 2009. Changing the trend of agricultural expansion through payment of environmental ser-vices to ensure a sustained flow of Mara River in the Mara River Basin (MSc Thesis). Miami, FL: Florida International University.

Defersha, M.B., Melesse, A.M., 2012. Field-scale investigation of the effect of land use on sediment yield and runoff using runoff plot data and models in the Mara River basin, Kenya. CATENA 89, 54–64. http://dx.doi.org/10.1016/j.catena.2011.07.010.

Dickens, C.W.S., Graham, P.M., 2002. The South African scoring system (SASS) version 5 rapid bioassessment method for rivers. Afr. J. Aquat. Sci. 27, 1–10.

Gereta, E. J., Wolanski, E., Chiombola, E. A. T., 2003. Assessment of the environmental, social and economic impacts on the Serengeti ecosystem of the development in the Mara River catchments. Report: Tanzania National Parks, Arusha, Tanzania.

GLOWS (Global Water for Sustainability Program), 2007. Water quality baseline assessment report: Mara River Basin, Kenya-Tanzania. URL: http://glows.fiu.edu/glows/Portals/0/documents/Mara_WaterQuality_AssessmentReport.pdf.

GoK (Government of Kenya), 2007. Water Resource Management Rules. Government Printer, Nairobi, Kenya.

Hoffman, C. M., 2007. Geospatial mapping and analysis of water availability-demand-use within the Mara River Basin (MSc Thesis). Miami, FL: Florida International University.

King, J.M., Tharme, R.E., de Villeers, M.S. (Eds.), 2000. Environmental Flow Assessments for Rivers: Manual for the Building Block Methodology: Water research commission report No.: TT 131/00. Freshwater Research Unit, University of Cape Town, South Africa.

Le Quesne, T., Kendy, E., Weston, D., 2010. The Implementation Challenge: Taking Stock of Government Policies to Protect and Restore Environmental Flows. The Nature Conservancy and WWF.

Mango, L.M., Melesse, A.M., McClain, M.E., Gann, D., Setegn, S.G., 2011. Land use and climate change impacts on the hydrology of the upper Mara River Basin, Kenya: results of a modeling study to support better resource management. Hydrol. Earth Syst. Sci. 15, 2245–2258.

Mendez, S., 2009. An assessment of potential buyers in a payment for ecosystem services scheme in the Mara River Basin, Kenya and Tanzania (MSc Thesis). Delft, Netherlands: UNESCO-IHE Institute for Water Education.

Rowntree, K. M., Wadeson, R. A., 1999. A hierarchical geomorphological model for the classification of selected South African river systems. Water Research Commission Final Report. Water Research Commission, Pretoria.

Walpole, M.J., Nabaala, M., Matankory, C., 2004. Status of the Mara Woodlands in Kenya. Afr. J. Ecol. 42, 180–188.

WRMA (Water Resources Management Authority), 2009. Lake Victoria South Catchment Area Management Strategy. WRMA, Kisumu, Kenya, p. 72.

Chapter 8

Groundwater Distribution and Aquifer Characteristics in Kenya

Zachariah Kuria
Department of Geology, College of Biological and Physical Sciences, University of Nairobi, P.O. Box 30197, Nairobi, Kenya

1 INTRODUCTION

The groundwater occurrence and distribution in Kenya is largely controlled by geology: The Rift Valley System is marked by the presence of intrusive rocks and volcanic flows; the areas underlain by the sedimentary rocks include coastal and northern Kenya; and the aquifers in metamorphic terrains are of very small extent and localised largely in eastern Kenya. Therefore, only a small description is given.

The aspects considered in groundwater include underlying geology of the area and aquifer characteristics, that is, aquifer geometry, groundwater levels in the aquifer, groundwater flow and recharge and specific capacity and transmissivity of the aquifer.

For the purpose of understanding groundwater potential of Kenya, the borehole database of the Ministry of Water and Irrigation was used. The database has over 13,000 boreholes; however, it suffers from gross incompleteness, that is, some boreholes lack location coordinates, elevations and drawdown data. The boreholes with complete data were used to construct the transmissivity map, which shows the groundwater potential of the entire country and groundwater flow directions that mark the recharge and discharge areas. The results accrued from this study present for the first time regional groundwater potential in Kenya. The location of the boreholes used in this study is shown in Figure 1.

2 GROUNDWATER DISTRIBUTION AND AQUIFER CHARACTERISTICS OF SEDIMENTARY ROCKS OF NORTHERN KENYA

2.1 Geologic Setting of Northern Kenya

The geology of northern Kenya has been mapped by Joubert (1963). Earlier information was gathered and concepts developed by various investigators

Developments in Earth Surface Processes, Vol. 16. http://dx.doi.org/10.1016/B978-0-444-59559-1.00008-6

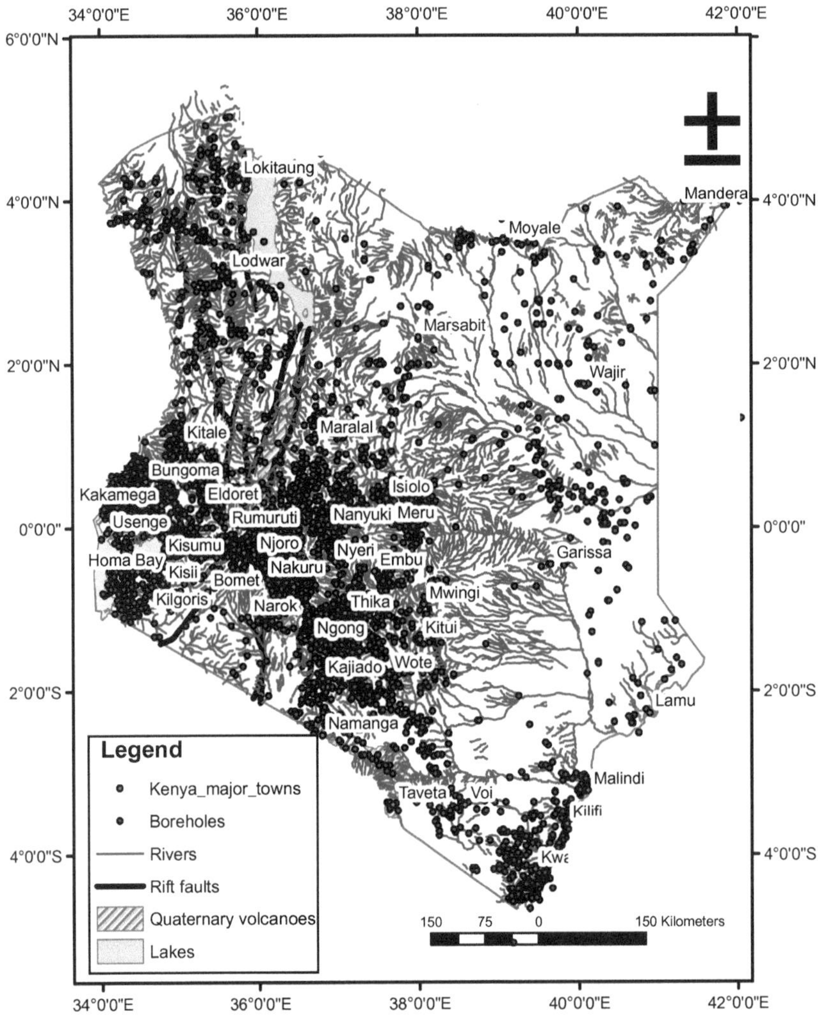

FIGURE 1 Location of the boreholes draped on a drainage system of Kenya. (For colour version of this figure, the reader is referred to the online version of this chapter.)

(e.g. Bestow, 1960). The geologic map of the area bordering north (Thompson and Dodson, 1960) and a review by Swarzenski and Mundorff (1977) make the regional context more clear and provide supplementary clues for interpretation.

Relevant information on the regional geologic units is summarised in Table 1.

As far as is known, Paleozoic rocks are absent in the area, which cause basement rocks to be overlain directly by Mesozoic or even younger rocks. Downwarping of the coastal area at the end of the Paleozoic resulted in a transgression of the sea, in which the Triassic Mansa Guda Formation was

TABLE 1 Stratigraphic Table

Period	Formation	Lithology
Recent		Sandy soils
Pleistocene	Wajir Beds	Gritty soils, laminated limestones Impure limestones, sandstones (?)
Jurassic	Merti Beds	Sandstones Clays and sandy clays
Triassic	Bur Mayo Formation	Grey and brown limestones
Precambrian	Mansa Guda Formation	Conglomerates Sandstones

After Joubert (1963).

deposited as a delta. Recurrence of downwarping in the Lower Jurassic caused another transgression and the deposition of the Jurassic Limestone Series. After regression from the Middle Jurassic times, the Cretaceous continental Marehan Series was deposited. During the Tertiary, erosion of Mesozoic and Precambrian rocks took place; Miocene sediments were deposited but probably largely removed again as a result of Pliocene uplift.

The Pliocene Merti Beds were laid down in at least some parts of the northeastern province. Alternating pluvial and interpluvial periods during the Pleistocene explain the variation in lithology observed in the Wajir Beds and younger formations (Joubert, 1963; Swarzenski and Mundorff, 1977).

The occurrence and extent of the different geologic units is rather well known at the surface, but uncertainties exist in many zones regarding the geologic units at some depth. Bestow (1960) and Joubert (1963) suggest, for example, that under Wajir town, basement rocks are directly overlain by (Tertiary and?) Quaternary rocks, whereas more recent reports (Swarzenski and Mundorff, 1977) and recent boreholes make plausible that the Mansa Guda Formation extends more southwestwards than previously thought and might occur under Wajir at a depth of 120 m.

Furthermore, equivalents of the Merti Beds, found in a wide belt from NW of Habaswein to Liboi at the Somali border (and beyond), are not mentioned by Joubert (1963), but Swarzenski and Mundorff (1977) suggest that they are present in the Wajir area as clays, sands and grits underlying a 20–25 m thick succession of Wajir Beds.

Finally, the sequence of Quaternary deposits varies considerably over short distances, which makes it difficult to recognise and correlate the different stratigraphic units in wells and boreholes. This fact has given rise to inconsistencies in this respect between various reports (Bestow, 1960; Joubert, 1963; Balasha-Jalon Consultants and Engineers Ltd., 1976; Swarzenski and Mundorff, 1977; Alexander Gibb & Partners, 1979).

2.2 Aquifer Geometry

The largest aquifer system in Kenya is located east of the Rift Valley where it stretches as a continuum all the way from the Marsabit through Garissa and terminates at Lamu within the Kenyan Coast. Geometrically, it stretches a distance of 740 km in NW–SE direction, around Marsabit town in which it attains a width of about 230 km; it constricts between Marsabit and Garissa to 75 km, slightly north of Garissa town where the aquifer attains 230 km, which to the east narrows and eventually enters Somali; and it is at around Lamu Town where the aquifer is broadest, that is, 250 km.

In addition to the large aquifer, there exists a relatively small aquifer to the NE, namely, Merti aquifer that stretches from Moyale town through Merti town and then extends laterally across the Somali border. Geometrically, this aquifer with subdued transmissivity values (1.5–20 m^2/day) runs from Moyale town south eastwards to Merti town. At Moyale, the aquifer attains a maximum width of about 100 km but narrows to less than 20 km between Moyale and Merti towns. At Merti, the aquifer attains a maximum width of 180 km up to the Somali border. The aquifer is associated with disappearance of the Ewaso Nyiro but its extent, particularly the lateral, has never been mapped before and appears for the first time in this study.

2.2.1 Aquifer Zones

The widely exploited shallow aquifer in the Wajir area consists of more or less permeable horizons in the Wajir Beds. It is commonly found at depths of around 6 m below ground level (bgl) and is assumed to be 2–15 m thick (Bestow, 1960; Joubert, 1963; Swarzenski and Mundorff, 1977). It is thought to consist of the arenaceous lower part of the 'impure limestone' and the sands or silts immediately underneath. The shallow aquifer is underlain by clay layers several tens of metres thick, with intercalated silty or sandy lenses.

Tentatively, a northern limit of the aquifer could be assumed at approximately 7–8 km north of Wajir town. The absence of shallow wells west of Wagalla suggests that also a lateral limit at some 20 km west of Wajir might exist. In the eastern and southern direction, the shallow aquifer probably extends over a larger (but still unknown) distance.

Within its lateral limits, the aquifer is assumed to be continuous. The preferential occurrence of livestock well fields in topographic depressions may be associated rather with history and convenience of well digging than with variations in the aquifer properties.

2.2.1.1 Well Yield

Pumping tests suggest that the sustained yield of shallow dug wells in the area may exceed some 5 m^3/h in roughly 40% and 2 m^3/h in 60% of the cases, provided that well depth and pump allow for 2 m of drawdown. About one-third of the tested wells have specific capacities higher than 1 l/s per metre of drawdown (after 24 h of pumping). Local variation is pronounced, which means that highly productive wells can be found close to almost unproductive ones.

2.2.1.2 Hydraulic Aquifer Properties

Pumping tests results show local 'apparent' transmissivities from less than 80 to 300 m^2/day or more. The effect of the relatively impervious saturated limestones on top of the sandy beds of the aquifer cannot easily be assessed quantitatively. It will, more or less, act as a confining layer during transient flow stages (during intermittent abstraction and recovery). On the other hand, it is assumed to be sufficiently pervious to allow direct recharge to take place.

2.2.1.3 Groundwater Levels, Groundwater Flow and Recharge

Groundwater levels are shallow: roughly between 1 and 11 m below surface in the wells. The water table follows closely the topographic surface and is sloping with the same dip of approximately 1.6 m/km from NNW to SSE.

The piezometric map (Figure 3) and the estimated average transmissivity indicate that a net flux of some 75,000 m^3 of groundwater annually moves SSE through each 1 km wide section perpendicular to the flow direction. The hypothesis is put forward that groundwater in this shallow aquifer does not follow one regional flow system, but circulates partly in small local flow systems, fed by local recharge and drained by wells and by deeply rooted trees and shrubs. Supporting evidence for this hypothesis is provided by the demonstrated occurrence of local recharge from rainfall; the assumed heterogeneity of the shallow limestone beds permeability; the presence of numerous 'cones of depression' around wells; and the, more or less, randomly scattered pattern of water quality variations, for example, EC and NO_3^- content.

Groundwater level hydrographs measured show that groundwater is directly recharged by local rainfall. On the other hand, tested yields of all deeper boreholes are disappointing at very large drawdowns. It is questionable, however, whether this was always due to poor aquifer characteristics or to poor well construction.

3 GROUNDWATER DISTRIBUTION AND AQUIFER CHARACTERISTICS OF SEDIMENTARY ROCKS OF COASTAL KENYA

There are three major aquifers in the coastal Kenya. These include Sabaki aquifer (at Baricho area), Tiwi aquifer and Msambweni aquifer both in the south coast. All these aquifers are found in similar geologic formations but different textural and, therefore, hydraulic characteristics.

3.1 Geologic and Structural Setting

The earliest description of the sedimentary rock sequence of the coastal Kenya was provided by Gregory (1921). However, the general geology of Kwale district has been described by Caswell (1953), Miller (1952) and Thompson (1956). Buckley (1981) gives a comprehensive summary including results of remapping undertaken by the Mines and Geology Department with the UK assistance.

According to these reports, the oldest rocks are the basement rocks of the Precambrian age found to the northwest of the area. Other rocks are the younger sedimentary rocks found towards the Shimba Hills and the Indian Ocean to the east. The regional strike of strata tends to parallel the coast on a SSW–NNE trend and the dip is to the east with the older Karoo sediments outcropping on the higher ground inland (Carruthers, 1985). As a result, the encountered rocks increase in age with distance from the seafront.

The Karoo rocks form part of an extensive sequence that includes both the arenaceous and argillaceous components. They appear to have limited groundwater potential and typically show poor yields of brackish water (Carruthers, 1985). The Triassic–Jurassic Mazeras Formation, which underlies the Msambweni area at depth and outcrop on the higher coastal range, was deposited under semiarid conditions in a landlocked basin. Within these sandstones, precipitated layers of mineral salts (carbonates, chlorides and sulphates) occur, giving rise to high salinities of groundwater. Due to the fine-grained texture of the rocks, yields were considered to be generally low (Caswell, 1953). The best groundwater prospects within the Karoo rocks lie in the upper units, especially the Mazeras Formation, where increased secondary permeability due to structural disturbance may be combined with the higher rainfall and recharge and a reduced risk of contamination from older waters, for example, in the Shimba Hills, which is a part of the coastal range (Carruthers, 1985).

Jurassic and Cretaceous Formations mainly consist of shales with subordinate limestones, which were formed in a marine/littoral environment and unconformably overlie the Mazeras Sandstone Formation. At outcrops, these shales form the 'Foot Plateau' between the Coastal Range and the Coastal Plain. They can be regarded as an impermeable barrier dipping seawards beneath the more recent sediments. Boreholes exclusively penetrating these clayey rocks are generally dry. Whereas the blackish, hard clays underlying the upper sediments of the Coastal Plain have often been interpreted as Jurassic shales, it has also been reported that their depth in the Tiwi area exceeds 200 m (Austromineral, 1980; Carruthers, 1985).

The shales are overlain by younger Cainozoic deposits, which comprise a wide range of unconsolidated and semiconsolidated sediments, derived from the extensive erosion and redeposition of Mesozoic Formations. The Tertiary Baratamu Formation tends to be clayey, although conglomerates, sands and limestones are found together with the marls and hard, shale-like clays.

Carruthers (1985) believes that the dark, compact 'claystone' generally underlying the reef complex is part of the Baratamu complex. In Pliocene times, the Marafa beds were formed, which consist mainly of fine-grained sediments north of Malindi but tend to have coarser lateral equivalents south of Mombasa. The possible similarity of these formations and their considerable lateral variation in facies make identification and correlation uncertain. Borehole yields tend to be low and suitable only for small-scale local supplies.

The Pliocene Magarini Sands were formed in an environment gradually changing from humid, coastal/deltaic to more arid, semicontinental. The formation consists of a wide range of sediments including medium- to coarse-grained sands (locally found in a clayey matrix), silty clays and bright red, fine windblown sands. The generally coarse-grained but poorly sorted material of the Lower Magarini suggests that deposition was associated with the erosion of a land surface rejuvenated by faulting. In contrast, the Upper Magarini is characterised by well-sorted dune sand, which is typically reddened by iron oxide. The latter facies is suspected to be exposed and mined over a large area on the boundary of the Coastal Plain and the Foot Plateau. In numerous quarries, red sands may be seen overlying a poorly sorted mixture of coarse sands in a compact, clayey matrix (Lower Magarini?). Judging from the outwash scattered over the surface, this sediment also contains extremely well-rounded quartz gravels.

The Magarini Sands generally are fairly pervious and their contact with the underlying Jurassic shales is often marked by seasonal springs. Although water from this formation is potable, it is frequently marked by objectionably high iron content.

The Pleistocene reef complex can be considered as two components, with a sandy back-reef facies (Kilindini Formation) lying inland of the reef limestone itself. This sand facies consists of sands, both course and fine, with clays and limestones, which were deposited in a lagoonal environment and may be an extension of the Lower Magarini (Carruthers, 1985). The poorly consolidated nature of this material gives it very good aquifer properties locally, as proved by boreholes into the Tiwi aquifer, although lateral variations associated with the location of the inflow channels and the reworking of the sediments can be expected. The Kilindini Formation is eroded and reworked sediments derived from the Magarini and Marafa Sands and occurs predominantly east of the Foot Plateau and appears to have been flushed into the back-reef depression by fluviatile action.

The highest potential for groundwater development in the coastal area is found in the Pleistocene Kilindini Formation and, to a lesser extent, in the coral reef facies and the Pliocene Magarini Sands. Although the Kilindini back-reef deposits are characterised by very favourable hydraulic properties (high conductivity, transmissivity and storativity), most boreholes experience severe problems due to the ingress of fine sands (200–600 mm) (Tables 2 and 3).

TABLE 2 Stratigraphic Table

Era	Period (Age)	Rocks and Sediments Deposited	Lithology	Environment	Climate	Events
Cenozoic	Recent (0.1 Ma)	Alluvial and coastal beach sands and coastal sand dunes	Superficial unconsolidated sands	Coastal and near-coastal	Present day	Gradual rise in sea level
	Pleistocene (2 Ma)	Coral reef and Lagoonal Kilindini Formation	Coral limestone Unconsolidated and cemented sands and clays	Marine, coastal and near-coastal	Pluvial and interpluvial	Fluctuating sea level
	Pliocene (7 Ma)	Magarini Sands:				Extensive erosion of Mesozoic deposits
		–Upper	Fine-grained red windblown sands	Continental	Arid	
		–Lower	Coarse quartz sands, gravels and silty clays	Fluviatile and deltaic	≠	Faulting
		Marafa Formation	Sandstone/sands/shales	,,	Wet	
	Miocene Oligocene Eocene (58 Ma)	Baratamu Formation	Marls/limestones/clays and sands	Shallow water Littoral	Warm	Deposition, extensive continental erosion

Mesozoic	Cretaceous (135 Ma)	Mtomku Formation:				Faulting?
		–Upper	Shales and limestones			
		–Middle	Shales, limestones and sandstones			
	Jurassic (195 Ma)	Mtomku Formation:				
		–Lower	Shales, siltstones and sandstones Limestones, shales and siltstones			Marine invasion of coastal plain and subsequent limestone deposition
		Kambe Limestone	Sandstones/arkoses	Marine	Warm	
		Mazeras Formation:	Sandstones			
		–Upper		Continental		
		–Middle				
	Triassic (225 Ma)	Mariakani Sandstone:				Continental subsidence to form basin of deposition
		–Upper	Sandstone	Deltaic and lacustrine		
		–Middle	Sandstone		Generally semiarid	
		–Lower	Sandstone			
		Maji ya Chumvi:		Deltaic		
		–Upper	Sandstone, shales and siltstone			

Continued

TABLE 2 Stratigraphic Table—Cont'd

Era	Period (Age)	Rocks and Sediments Deposited	Lithology	Environment	Climate	Events
Palaeozoic	Permian (286 Ma)	Maji ya Chumvi:		Lacustrine		
		–Lower	Black, shaly siltstones			Downwarping of continental margin
		Taru Grits:			Generally semiarid	
		–Upper	Arkoses/sandstones, siltstones, shales	Continental		
		–Lower	Arkoses/ conglomerates			

After Adams (1986).

TABLE 3 Typical Sequence of the Coastal Plain, Msambweni Area (Groundwater Survey Kenya, 1999)

Formation	Resistivity Range (Ω m)	Common Depth Range (mbgl)	Common Thickness (m)
Recent, windblown cover *sands*, dry, very loose	1000–12,500	0.5–2.5	0.5–2.5
Kilindini Formation, upper *sand* facies, dry	200–1000	0.5–10	1–10
Kilindini Formation, upper *sandy clay and clayey sand* facies	5–20	8–20	3–12
Kilindini Formation, lower *sand* facies, water bearing, bordering or interfingering with *coral limestone*	40–100 30–150	20–60	10–40
Kilindini Formation and lower *clay to sandy clay* facies	2–10	35–60	5–20
Mazeras *Sandstones*	500–3000	>50–60	Unknown

3.2 Groundwater Occurrence

The main aquifer of the Coastal Plain occurs in the vicinity of the Kwale town covering both the Msambweni and Tiwi aquifers. The aquifer zone marked by a red polygon on the transmissivity map (Figure 2) stretches over a distance of 80 km long the coastline and attains a width of about 80 km inland at its widest extent around Kwale town.

The aquifer zone is marked by saturated sandy layers within the Kilindini Formation. The sediments comprise a wide range of textures, which include coarse sands, clayey sands, sandy clays, heavy clays, coral debris and limestone bands. The Kilindini Formation is contemporaneous with the Pleistocene coral reef, and the boundary between the two units is a transitional one, with interfingering facies.

The Kilindini Formation has been largely deposited in the elongated depression west of the coral reef. The sediments are a combination of fluviatile, deltaic, aeolian, coastal and shallow marine deposits. At higher sea levels, rivers would reach the coast under a relatively steep gradient, dropping out a heterogeneous mixture of gravels, sands and clays in their deltas. Wave action reworked part of these deposits into well-sorted coarse beds, while elsewhere, clays were trapped in mangrove forests. During sea-level regressions, the river

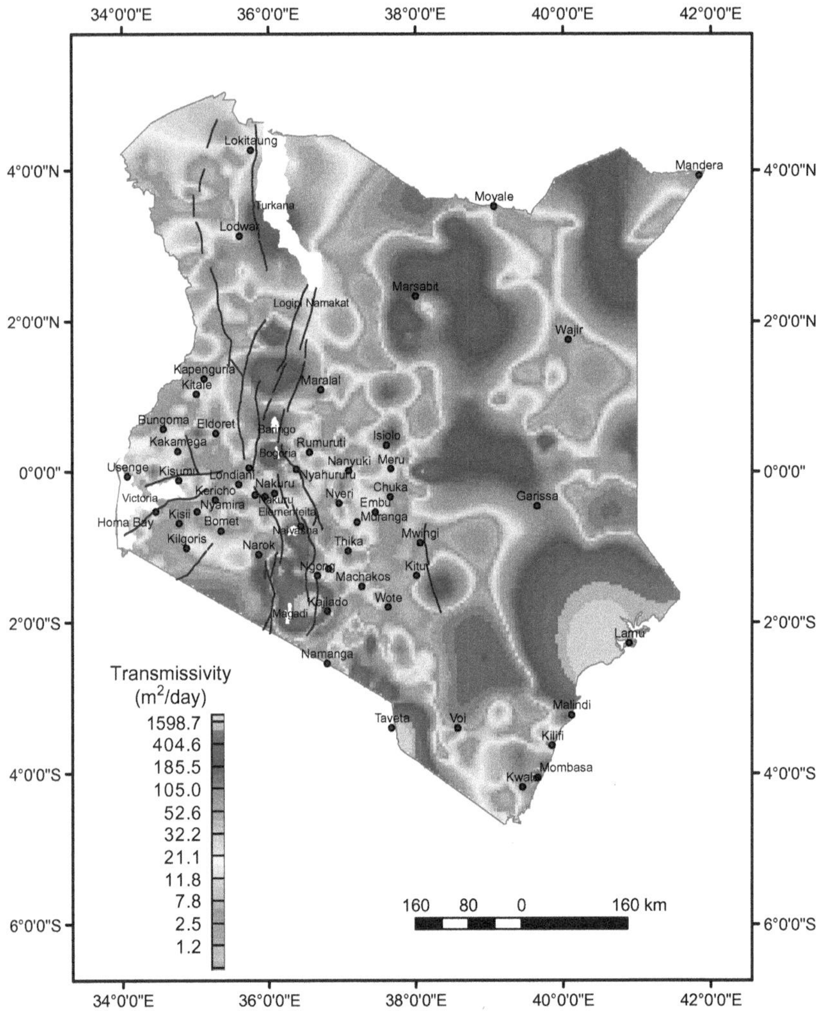

FIGURE 2 Transmissivity map of Kenya showing distribution of groundwater potential zones. (For colour version of this figure, the reader is referred to the online version of this chapter.)

systems were able to deposit thick layers of relatively clean sands. Coral growth was probably enhanced after rises in sea level or during humid periods, when little sediments were delivered from the hinterland.

The eastern boundary of the Kilindini Formation is formed by the coral reef, whose growth kept pace with the rate of sedimentation into the lagoon. To the west, the coastal sediments are spread like a series of blankets over the existing topography of Magarini Sands and Mazeras Sandstones. Because the Coastal Plain follows a rather monotonous slope towards the sea, the thickness of the back-reef sediments is to a large extent determined by the buried relief of the bedrock. Where the latter is deep (due to either structural features, lithology or ancient erosion patterns), the depression will have been filled entirely by a thick Kilindini sequence.

Compared to the Tiwi aquifer, approximately 25 km north of the Msambweni aquifer, the base of the back-reef channel at Msambweni is somewhat shallower. The average maximum depth to the underlying Mesozoic rock along the Msambweni cross sections is 56 m, while this is approximately 65–70 m at Tiwi (Groundwater Survey Kenya, 1996, 1998).

Based on resistivity data and field observations (Groundwater Survey Kenya, 1999), the bedrock in the Msambweni area is believed to be formed by Mazeras Sandstones. This is unlike the situation towards Mombasa, where a thick layer of younger, generally fine-grained rocks occurs between the sandstones and the Pleistocene series (Jurassic and Cretaceous shales of the Mtomku Formation and Baratamu 'claystones').

In the relatively shallow parts of the aquifer, the base of the lower sand unit (formed by a thick clayey layer) is often found at an elevation of −10 to 0 m above mean sea level (amsl). As a result, the available groundwater is stored at a relatively high level. At Tiwi, the water table in the area between the coral reef and the Foot Plateau ranges between 5 and 10 m amsl, while this is 15–25 m amsl in most parts of the Msambweni aquifer. Although the available storage in the aquifer sands is probably less than in Tiwi, the advantage is that most of it (an estimated 65%) is held above the mean sea level. This reduces the risk of saltwater intrusion into the freshwater zone. As a result, higher drawdown levels can be tolerated during abstraction.

The Msambweni aquifer can be subdivided into various zones with different hydrogeological characteristics. The highest potential for sustainable large-scale abstraction exists within the relatively deep central part, between Kidogoweni and Mkurumudzi Rivers.

It has been assumed that water retained within the clayey deposits and the coral reef is not available for abstraction. Consequently, the total storage is almost entirely made up of water contained in the lower sand member of the Kilindini Formation. Storage in the hydraulically connected Magarini Formation is not taken into account at this stage.

The characteristics of the three aquifer zones according to GSK (1999) are as follows:

Northern Aquifer Zone (Zigira–Magaoni)

Aquifer width	
Average maximum thickness of Kilindini Formation	49 m
Average maximum thickness of lower sand member (main aquifer)	18 m
Total estimated storage in lower sand member	20.1 Mm^3
Total storage retained above mean sea level	14.8 Mm^3 (74%)
Expected borehole yield	10–20 m^3/h

Central Aquifer Zone (Kidogoweni–Gongoni–Mkurumudzi)

Aquifer width	
Average maximum thickness of Kilindini Formation	72 m
Average maximum thickness of lower sand member (main aquifer)	59 m
Total estimated storage in lower sand member	92.0 Mm^3
Total storage retained above mean sea level	48.3 Mm^3 (53%)
Expected borehole yield	40–60 m^3/h

Southern Aquifer Zone (Koromojo–Milalani–Mabatani)

Aquifer width	
Average maximum thickness of Kilindini Formation	50 m
Average maximum thickness of lower sand member (main aquifer)	16 m
Total estimated storage in lower sand member	35.3 Mm^3
Total storage retained above mean sea level	32.1 Mm^3 (91%)
Expected borehole yield	Generally $<15\ m^3/h$

3.3 Aquifer Storage

Only water stored in the Kilindini Formation was considered to be safely abstractable (Groundwater Survey Kenya, 1999). Although there is expected to be a considerable volume of freshwater contained in the fossil reef, this zone forms an essential buffer between the sand aquifer and the lateral intrusion of saline water from the east. The amount of water drawn from the coral limestones should thus be kept to a minimum.

The total storage in the northern and central parts of the sandy facies of the Kilindini Formation, assuming an effective porosity of 30%, is in the order of 112 Mm^3. About 63 Mm^3 is retained above mean sea level. During extremely dry spells, this portion may be regarded as the buffer capacity of the aquifer, which can be temporarily dewatered without directly invoking drastic deterioration of water quality.

3.4 Safe Yield

Using a contribution area to the upstream catchment of 42 km^2, which includes the area of the hydraulically connected Magarini Sands, the bulk water balance for the Msambweni aquifer can be assessed as follows:

Mean annual rainfall	1300 mm
Effective precipitation	23%
Mean annual recharge	299 mm (0.3 m)
Total groundwater catchment area	42 km^2
Total recharge	12.6 Mm^3/year (34,500 m^3/day)

4 GROUNDWATER DISTRIBUTION AND AQUIFER CHARACTERISTICS OF SEDIMENTARY ROCKS OF LAMU ISLAND

The Lamu Island covers an area of about 50 km^2, out of which about 19 km^2 is covered by the double row of the longitudinal sand dunes along the southern border. At Kizingoni, dunes are 3 km wide and wedges out to 700 m at Shella area.

The topography varies from gently undulating flats to hilly, curvilinear and longitudinal sand dune at the southern beach. The dune morphology has a saddleback structure with the following altitudes: central parts, 23 m; eastern Shella, 51 m; and western Kizingoni, 63 m.

Geologically, the Lamu Island area is almost entirely covered by fine to medium grained sand and other rare and isolated rock types that include calcareous sandstone and pink-red coral limestone (Kuria et al., 2010a).

In Lamu Island, the aquifers are unconfined and found at relatively shallow depth (10–15 mbgl). The water table slopes gently towards the coast (sea), at a gradient of about 0.001–0.005. Drainage from the southern edge of the sand dune ultimately discharges into the sea.

A survey of groundwater occurrence using 2D electric imaging/topography revealed that the catchment zone covers an area of about 19,712,200 m^2. It is broad at Kizingoni area but wedges out towards Shella area with a width of only 600 m. The aquifer is linearly shaped though broader at Kizingoni area, where it includes small isolated aquifer and wedges out at Shella area. It is divided into four zones based on aquifer thickness and water quality (effect of saline intrusion). The main aquifer has a maximum depth of 25 m at Kizingoni area, whereas localised aquifer at Kipungani–Kizingoni area attains a thickness of about 50 m. To the further east, the aquifer becomes slightly deeper but thins out (5 m). The changes in aquifer thickness from 20 to 5 m are possibly related to over abstraction of the aquifer on the eastern side rather than availability of the aquiferous permeable materials. Where the aquifer is heavily exploited, that is, at Shella and Kizingoni areas (zones 1 and 4), not only the saline water intrusions are eminent (as compared to zone 2), but also the aquifer breaks, morphologically, into tongue-shaped narrow lenses.

The total storage of the entire the aquifer, assuming an effective porosity of 30%, is in the order of 18 Mm^3. This storage is retained above mean sea level. During extremely dry spells, a small volume of water within the aquifer may be regarded as the buffer capacity that can be safely abstracted without invoking seawater encroachment. Assuming a conservative aquifer width of 660 m and length of about 9093 m (sum of all the lengths of the aquifer between Shella and Kizingoni) and a thickness of about 25 m (unexploited parts of Kizingoni aquifer), it is apparent that the total effective storage would be in the order of 45 Mm^3. This indicates that only about 40% of the aquifer storage is currently available with over 60% having been depleted.

Using a contribution from the entire catchment area, the bulk water balance for the Shella–Kizingoni aquifer can be assessed as follows: mean annual rainfall, 1100 mm; total recharge, 2.85 Mm^3/year (7800 m^3/day); effective precipitation, 13.5%; mean annual recharge, 148 mm (0.15 m); and total groundwater catchment area, 19 km^2 (exempting dune slopes next to the beach area). If it is assumed that about 50% of the recharge can be safely abstracted, the sustainable yield of the aquifer amounts to 3600 m^3/day. This is equivalent to 94% of the maximum projected demand of 3494 m^3/day. However, it should be noted that the calculations were based on 50% abstraction and the entire aquifer. Therefore, the current well field covering only a small part of the aquifer (4 km compared to 19 km used in the calculation) is being exploited. This explains the reason as to why the problem of the saline water intrusion and upconing, particularly, for the southern wells is rampant.

5 GROUNDWATER DISTRIBUTION AND AQUIFER CHARACTERISTICS OF INTRUSIVE AND VOLCANICS OF RIFT VALLEY

5.1 Geologic and Structural Setting

The Rift Valley is a Cainozoic intracontinental rift system separating the Somalia subplate from the rest of Africa (Chorowicz, 1992; Delvaux et al., 1992; Smith and Mosley, 1993). The Rift Valley System has N–S geographic (submeridian) trend delineated by flank faults with individual faults oriented at diverse angles (Onywere, 1997). Steep scarps mark the flanks and long fault lines, while the intervening smaller faults are in places marked by low, narrow horst and graben structures (Baker et al., 1972). The rift system in the study area stretches from drainage basins of Turkana, Baringo–Bogoria, Nakuru–Elementaita–Naivasha and Magadi.

The system is bordered on the eastern shoulder by an abrupt topographic break, which is a surface trace of the rift bounding Sattima fault. On the western shoulder, the boundary of the sub-basin is a gradual rise across a series of roughly parallel fault scarps marking the Mau Escarpment. The Aberdare Detachment System sub-basin is linked to Baringo/Turkana sub-basin in the north, by N600W–N300W oriented zones of complex wrench and oblique subfaults forming the Marmanet fault and the Loboi–Kamasia cross structure, and is located around Lake Bogoria area and parallel to the Aswa lineament (Figure 1). It is linked to the Magadi/Natron sub-basin in the south, by Kedong–Ngiro cross structure, oriented approximately N300E and located just to the south of Suswa volcano. These linking zones are kinematically analogous to transform faults within the world rift system and have been referred to as accommodation zones by Chorowicz et al. (1987), Chorowicz (1992), Burgess et al. (1987) and Grimaud et al. (1994).

The project area lies on the northern slopes of the central rift dome within what has been described as Nakuru trough (Baker, 1970). The trough is bordered by the Naivasha trough in the south; the two are separated by the Eburru shield and to the north is the Bogoria trough. The area had been originally mapped at a scale of 1:25,000 (McCall, 1967).

McCall (1967) portrays the rift floor fault systems as a series of anastomosing and bifurcating faults with short segment cross structures. McCall

(1967) carried out geologic mapping of Nakuru area at a scale 1:25,000. He reported that this part of the Rift Valley is mainly composed of a succession of Late Tertiary and Quaternary volcanics. Found within the volcanic are intervening lacustrine beds and alluvium principally of reworked volcanic debris. Washbourn (1967) suggested that the larger part of the deposits is volcaniclastic material deposited after the Quarternary period. The lake had also been subjected to study by Baker et al. (1972). The Kenya Rift Valley was further subjected to intensive investigations. The earlier exploratory work involved geologic and geochemical sampling.

The history of the evolution of the Rift Valley in the Nakuru area is extremely complex with no rock exposures older than the Tertiary era. Firstly, the continental surface planed by the erosion cycle, which moulded the sub-Miocene erosion surface, was apparently warped down prior to the first eruptions, along a zone roughly corresponding with the present Rift Valley trend.

The first volcanic eruptions were the Samburu series of thin basalt and picrite lava flows and subordinate pumice showers. These were in part erupted from swarms of dykes and in part from central sources. This phase of eruption was followed by earth movements, which led to unconformity between the Samburu volcanics and the overlying plateau phonolites. The plateau phonolites form an extensive series of flows with no pyroclastic horizons intervening. A continuous succession of more than 600 m of these phonolites is exposed on Ngelesha. However, they thin out rapidly towards the median line of the rift and away from the rift shoulder to the east.

These eruptive series were followed by major faulting in the north of the area. The vertical displacement on these earlier faults, now represented by dissected scarps, may have been as much as 1370 m in the Lake Bogoria sector.

After a long erosional interval, there followed a second series of volcanic eruptions. These were initiated by basalt eruptions, for example, Kwaibus basalts, which outcrop in the area and consist of cinder cones marking old vents and numerous successive thin flows of lava. These basalts were followed by phonolite eruptions from centres in Menengai and Kiplombe (to the west of the area) and from fissure sources away from the volcanic centres. The eruptives included pumice tuff, lava and ignimbrites. These eruptions were followed by major faulting that roughed out the Rift Valley as we know it today to the south (outside the area).

Following these fault movements, an erosional period formed an unconformity between outliers of late pumice tuff, welded tuff 'ignimbrite', sediments and the Kinangop and Bahati tuffs. A third major faulting episode succeeded, forming Menengai caldera. A minor volcanicity in Menengai followed with the eruption of trachyte cinder cones and lava flows. This was followed by the deposition of Solai tuffs associated with pumiceous sediments and tuffs and fluviatile sediments of Mugurin. Minor faulting followed creating new fractures and renewing the older ones in Solai area.

Our discussion of the groundwater distribution within the Rift Valley will concentrate on Nakuru area (Olobanita aquifer). This area is underlain by lava flows and pyroclastics. Towards the end of this chapter, a summary of the geology of Nairobi area will also be provided. The geology of Nairobi area is very similar to that of the Rift Valley, and therefore, an understanding of the lithostratigraphy of Nairobi area gives impression on the aquifer zones in this area mainly located between different lava flows.

5.2 Groundwater Occurrence

The prevailing geology favours groundwater occurrence in

- lacustrine sediments,
- weathered and/or fractured zones in the volcanic rocks and
- sediments interbedded between volcanic rocks.

The regional groundwater system in the area is locally recharged by infiltration of rainwater through the permeable volcanic soils, whose percolation into the deep-lying aquifers is facilitated by the open fault and fissure zones acting as groundwater conduits. However, ample recharge takes place along the high-lying areas, for example, Bahati forest and Kiplombe Hill located to the east and west of the of Nakuru area, respectively. Maximum mean annual rainfall in these recharge areas is approximately 1200 mm.

5.3 Aquifer Characteristics

5.3.1 Turkana Aquifer

The Rift System, as marked by major boundary faults, is characterised by well-defined aquifer systems whose location corresponds to the location of the Rift Lakes, that is, the Turkana aquifer roughly has a rectangular shape and occupies both the western and eastern part of the Rift Lakes. It measures 220 km in N–S direction and 135 km in E–W direction. A narrow arm of the aquifer stretches over a distance of 110 km in southwestwards and enters Uganda. This Turkana aquifer is isolated from the small northern aquifer located northwest of Lokitaung town. The transmissivity values of this aquifer range from 50 to 404 m^2/day (Figure 2).

5.3.2 Baringo–Bogoria Aquifer

The Baringo–Bogoria aquifer stretches from the area south of Lake Logipi and terminates weakly at the area south of Lake Bogoria. The aquifer has an anastomosing structure stretching over a distance of 220 km in N–S direction but highly variable width in E–W direction: the upper part is oriented in NW–SE direction, which deflects to NE–SW direction, a point where the aquifer attains the broadest segment (138 km); around Lake Baringo, the aquifer is oriented almost E–W and at Lake Baringo, it is oriented almost N–S.

Characteristically, the Baringo–Bogoria aquifer has very high groundwater potential zones with transmissivity values of about 1600 m^2/day located as follows: a few kilometres south of Lake Logipi, the area west of Maralal town and west of Lake Bogoria. The rest of the aquifer has transmissivity values above 100 m^2/day.

5.3.3 Nakuru Aquifer

The Nakuru aquifer is weakly connected to the Baringo–Bogoria aquifer through a southeastern arm. The Nakuru aquifer has a crescent-shaped structure, which is concave towards Lake Nakuru and stretches over a distance of about 110 km around the Lake Nakuru. Apart from the crescent-shaped structure, on the western part, the aquifer is elongated in N–S direction past the western boundaries of the Lake Elementaita close a distance of 42 km and attains a width of about 25 km. The entire aquifer is marked by moderate transmissivity values ranging from 50 to 100 m^2/day.

A consequence of localised study, three aquifers have been identified in Nakuru drainage basin; these include Olobanita, Bahati and Baharini aquifers, each of which has a well field.

5.3.3.1 Examples of Aquifer Piezometry of the Larger Lake Nakuru Aquifers

Menengai Crater—North: Groundwater elevations adjacent to the Menengai Crater are high, indicating significant groundwater flow towards Olobanita area. Groundwater elevations vary between 1901 and 1623 m, indicating significant flow in a northwest direction. Confining pressures are generally low; hence, the aquifer encountered in this area may result from permeable horizons within the Menengai volcanic deposits, rather than the Bahati tuffs (Figure 3).

Bahati Escarpment: In the area north of the Menengai Crater, groundwater elevations show a fairly even gradient from east to west down the escarpment. In Solai area, the groundwater flow occurs towards a small Lake Solai and further west of the region. The groundwater elevations range between 2143 and 1509 m. Confining pressures are low, indicating the possibility of unconfined or semiconfined conditions occurring, if the local groundwater levels are lowered by borehole abstraction.

From the results, it is inferred that groundwater from the Menengai Crater flows in the northwest direction, augmenting the flow from the Bahati forest. The combined flow moves towards Olobanita Swamp. In the central part of the zone, Menengai Crater and Solai groundwater flow occurs in an east–west direction towards Kisanana. The groundwater elevations range between 2143 and 1496 m at Solai and Legisianan, respectively.

Rongai Area: Groundwater level conditions in this area are not clearly understood because boreholes drilled in this area are either dry or produce

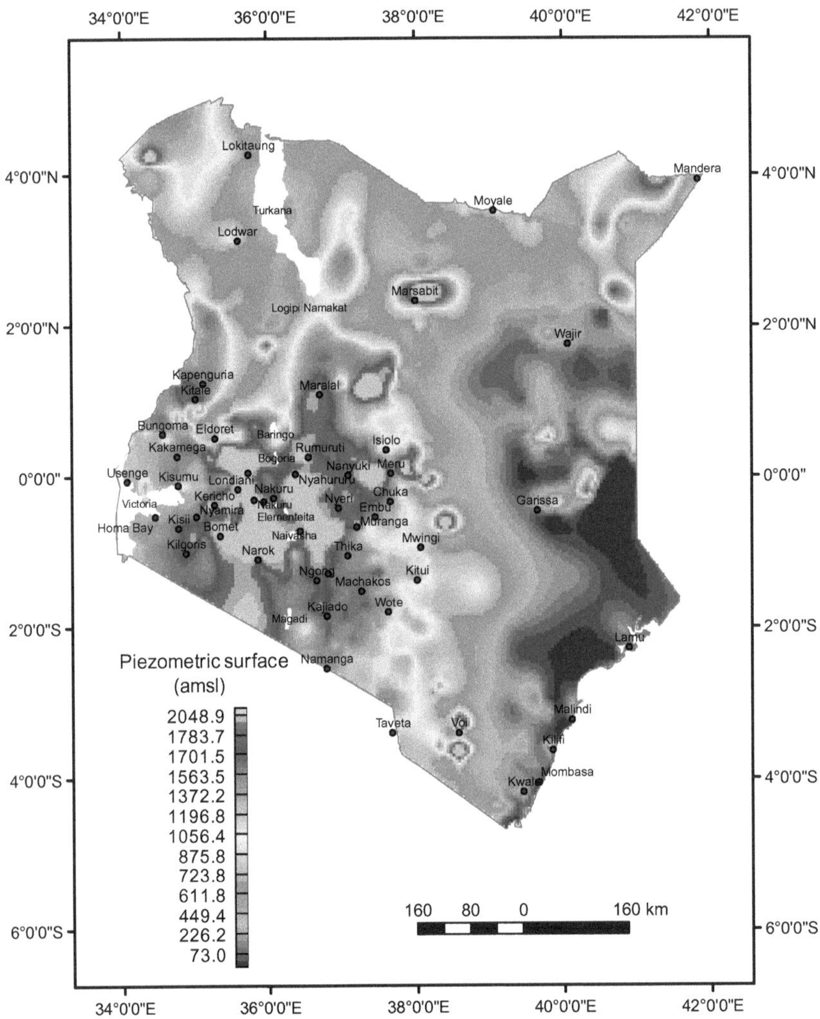

FIGURE 3 Piezometric surface of Kenya showing the groundwater flow mechanism. (For colour version of this figure, the reader is referred to the online version of this chapter.)

steam. The groundwater elevations range between 1919 and 1739 m amsl: this indicates a flow in NNE direction towards Mogotio and Lomolo areas. The confining pressures are low, indicating unconfined conditions.

Lomolo–Mogotio Area: Groundwater elevation in this area is low ranging between 1425 and 1417 m; this indicates groundwater inflow from other areas. This receives inflow from Rongai to the southwest, Olobanita to the southeast, Solai to the west and Kiplombe to the east. In the northern part towards Chemogoch area, the flow occurs in a northwest direction. From the groundwater results, it can be postulated that the groundwater of

Lomolo–Mogotio area escapes through the fault systems (identified from aerial photograph) forming a part of the flow towards Lake Baringo. The confining pressures are high, indicating confined conditions resulting in high yields for the boreholes in the area.

5.3.3.2 Aquifer Geometry and Aquifer Parameters

The boreholes drilled in the area have revealed that the aquifers within either the volcanic or the sediments are confined. Even though the area has been affected by intense faulting, it supports a regional aquifer system. Three aquifers exist in the area identified on the basis of borehole-specific capacities, groundwater flow map and isochore map:

- Lomolo–Mogotio and Olobanita aquifer
- North Solai aquifer
- Rongai aquifer

5.3.3.2.1 Lomolo–Olobanita Aquifer The regional aquifer system in Lomolo/Olobanita area is composed of weathered volcanic rocks, fractured volcanic rocks, fractured and weathered volcanic rocks and lacustrine sediments. The weathered tuffs underlying the Waseges flows (phonolites and trachytes) of the Rumuruti group and overlying the Samburu basalts form the best aquifers in Lomolo/Olobanita area. This system is recharged locally by rainwater and also by rivers flowing along fault zones, for example, Molo and Rongai rivers and other streams to the west of the area and Olobanita river to the east. From the western side, this zone is further recharged by high rainfall occurring in the area of Kiplombe Hill.

The characteristics of this aquifer are as follows:

- Groundwater in this zone is confined.
- Average depth of boreholes is 180 mbgl.
- Average depth to groundwater is 150 mbgl.
- Average water rest level is 118 mbgl.
- Average tested yield is 15.7 m^3/h.
- Average specific capacity is 378 m^2/day, which indicates that the zone has high groundwater potential.
- Lomolo/Mogotio aquifer is very extensive covering about 90 km^2 and as such is favourable for development of a well field.

5.3.3.2.2 North Solai Aquifer The regional aquifer system of this zone is composed of weathered volcanic rocks, fractured volcanic rocks, fractured and weathered volcanic rocks, sediments and old land surfaces with sediments forming the best aquifers. However, these sediments have mainly been deposited in grabens; hence, the aquifers are localised. This system is recharged locally along the fault and fissure zones by both rainwater and rivers. Several streams flow off the Solai escarpment and disappear underground in the area around Milton's sidings. The zone is further recharged regionally from the south in Bahati forest.

The characteristics of this aquifer are as follows:

- Groundwater in this zone is confined.
- Average depth of boreholes is 105 mbgl.
- Average depth to groundwater is 88 mbgl.
- Average water rest level is 52 mbgl.
- Average tested yield is 10 m^3/h.
- Average specific capacity is 103.3 m^2/day, indicating that the zone is of medium groundwater potential.

5.3.3.2.3 Rongai Aquifer The regional aquifer system is composed of weathered volcanic rocks, fractured volcanic rocks, fractured and weathered volcanic rocks, and old land surfaces. This system is recharged locally and from the south in the areas of Njoro.

The characteristics of this aquifer are as follows:

- Groundwater in this zone is confined.
- Average depth of boreholes is 225 mbgl.
- Average depth to groundwater is 159 mbgl.
- Average water rest level is 166 mbgl.
- Average tested yield is 5.0 m^3/h.
- Average specific capacity is 4.4 m^2/day, indicating that this area is of low groundwater potential.

5.3.4 Magadi Aquifer

The Magadi aquifer stretches from the area south of Lake Naivasha covering Suswa area and terminates south of Lake Magadi. This is the area covered by a southward flow of water from Lake Naivasha, that is, Figure 3.

The aquifer shows an irregular trapezium with the small part located around Lake Naivasha. The highest groundwater potential zone is located SE of Lake Naivasha with transmissivity values of above 400 m^2/day. This roughly isolated aquifer zone measures 35 km in E–W direction and 25 km in NW–SE direction.

The larger Magadi aquifer is deflected from the Lake Naivasha aquifer by Aswa Shear Zone (Kuria et al., 2010b). Whereas the Lake Magadi aquifer is narrow in the northern part, it attains a width of about 85 km covering the area east of Lake Magadi. The transmissivity values for this aquifer zone range from 400 to 1000 m^2/day.

The recharge area to the Lake Magadi aquifer includes Ngong Hills to the NE, Narok to the NW and Nguruman Escarpment to the west. The piezometric surface indicates a drop from 2000 m for all the recharge areas to 1190 m at the floor of the rift, which indicates a high hydraulic gradient.

6 GROUNDWATER DISTRIBUTION AND AQUIFER CHARACTERISTICS OF METAMORPHIC ROCKS

In the areas covered by the basement rocks, there is no single identified aquifer with significant amount of groundwater. According to the piezometric map (Figure 3), the aquifer within the basement rocks is localised; examples of these aquifers occur around this area: Isiolo, Meru, Embu, Wote and Kitui.

6.1 Groundwater Occurrence

Under the geologic conditions of the basement terrains, the groundwater occurs in

- weathered zones above the crystalline basement rocks,
- fractured zones within the crystalline bedrocks and
- alluvial deposits along the main drainage channels.

Especially in the Precambrian shields, metamorphic rocks are compact and have virtually no intergranular (or primary) porosity. On the other hand, while mostly solid, nonporous and absolutely impervious at the scale of a hand sample, these hard rocks have a type of porosity that can be termed as fracture (or secondary) porosity. This implies that they can hold water in a network of fissures, cracks, joints, fractures or faults.

The secondary porosity is caused by systems of 'macropores' and 'micropores'. Macropores are fissures and fractures of structural origin. Micropores are interstices created by weathering. These two types of secondary porosity are usually closely related, as cracks and fissures facilitate the percolation of water and hence more intensive weathering. Secondary porosity particularly controls the mineralization of the groundwater. The aquifers in the metamorphic rocks are characterised generally by a very low primary and (depending on the degree of fracturing) variable secondary porosity of the metamorphic rocks.

The thickness and mineral characteristics of the weathered layer play an important role in the determination of the amount of groundwater it can hold. Topography, drainage pattern, rainfall and evaporation are some of the major factors that determine the amount and occurrence of groundwater.

REFERENCES

Adams, B., 1986. Tiwi Aquifer Study, Final Report. Prepared on behalf of British Geological Survey for the Ministry of Water Development, Kenya and British Technical Cooperation.

Alexander Gibb & Partners, 1979. Wajir Airfield Water Supply. Technical Report.

Austromineral, 1980. Groundwater Exploration in Taita Hills and Mombasa South Coast Areas. Unpublished Report for the Government of Kenya, Ministry of Water Development and Ministry of Environment and Natural Resources, Mines and Geology Department.

Baker, B.A., 1970. Tectonics of the Kenya Rift Valley. PhD Thesis. Department of Geology, University of Nairobi, Nairobi, Kenya.

Baker, B.H., Mohr, P.A., Williams, L.A.J., 1972. Geology of the Eastern Rift System of Africa. Geol. Soc. Am. Spec. Paper 136, 67 pp.

Balasha-Jalon Consultants and Engineers Ltd., 1976. Preliminary Design Report. Wajir Water Supply Project.

Bestow, T.T., 1960. Northern Province Groundwater Survey. Explanatory Note on Sheet 13—Lower Ewaso Ngiro.

Buckley, D.K., 1981. Report on a Visit to Assess Groundwater Potential of the Kenya Coast South of Malindi (including proposals for the South Coast Groundwater Resources Project). Institute of Geological Sciences (IGS), Wallingford, UK.

Carruthers, R.M., 1985. Report on Geophysical Studies Relating to the Coastal Aquifer of the Mombasa District, Kenya. Rep. Reg. Geophys. Research Group. British Geol. Survey No. RGRG 85/4. Keyworth, Nottingham.

Caswell, P.V., 1953. Geology of the Kilifi-Mazeras area. Geological Survey of Kenya, Report No. 34, 54 pp.

Chorowicz, J., Vidal, G., 1987. Intérpretation géologique d'une image SPOT XS du Kenya: les réseaux de failles du Gregory Rift. Nouvellesde SPOT 9, pp. 16–17.

Chorowicz, J., 1992. The role of ancient Structures in the genesis and evolution of the East African Rift. Bull. Soc. Geol. Fr. 163 (3), 217–227.

Delvaux, D., Levi, K., Kajara, R., Sarota, J., 1992. Cenozoic paleostress and kinematic evolution of the Rukwa. North Malawi rift valley (East African rift system) Bull. Centres Rech. Explor.- Prod. Elf Aquitaine 16 (2), 383–406.

Gregory, J.W., 1921. The Rift Valleys and Geology of East Africa. John Murray, London, United Kingdom, p. 424.

Grimaud, P., Richert, J.P., Rolet, J., Teircelin, J.J., Xavier, J.P., Morley, C.K., Coussement, C., Karanja, S.W., Renaut, R.W., Guern, G., le Turdu, C., Michael-Noel, G., 1994. Fault geometry and extension mechanism in the Central Kenya Rift Valley, East Africa. A 3D remote sensing approach. Bull. Centres Rech. Explo. Prod. Elf Aquatine 18 (1), 59–92.

Groundwater Survey Kenya, 1996. Second Mombasa and Coastal Water Supply Engineering and Rehabilitation Project - Tiwi Aquifer Development: Additional Hydrogeological Investigations. Report for National Water Conservation & Pipeline Corporation.

Groundwater Survey Kenya, 1998. Second Mombasa and Coastal Water Supply Engineering and Rehabilitation Project - Tiwi Aquifer Development: Borehole Drilling and Associated Works; Borehole Completion Report. Volume I-III. Report for National Water Conservation & Pipeline Corporation.

Groundwater Survey Kenya, 1999. Hydrogeological investigation of Tiomin Mining Site, Kwale District. Technical Report No. 608.

Joubert, P., 1963. Geology of Wajir–Wajir Bor Area. Mine and Geology Department, Kenya. Report No. 57, 34 pp.

Kuria, Z.N., Woldai, T., Barongo, J.O., van der Meer, F.D., 2010a. Active fault segments as potential earthquake sources: inferences from integrated geophysical mapping of the Magadi fault system, southern Kenya Rift. J. Afr. Earth Sci. 57, 345–359.

Kuria, Z.N., Woldai, T., Opiyo Akech, N., 2010b. Imaging saltwater intrusion into coastal aquifers with electrical resistivity tomography at Lamu Island, South Coast Kenya. Afr. J. Sci. Technol. 11 (2), 57–72.

McCall, G.J.H., 1967. Geology of the Nakuru-Thomson's Falls-Lake Hannington area. Mines and Geology Department, Kenya, 78, p. 122.

Miller, J.M., 1952. Geology of the Mariakani-M ackinnon road area. Geogical Survey of Kenya, Report number, 20.

Onywere, S.M., 1997. Structural analysis of the drainage basin of Kenyan Rift Valley Lakes within the Aberdare Detachment using satellite data, GIS and Field Observations. Unpublished PhD Thesis, University of Nairobi.

Smith, M., Mosley, P., 1993. Crustal heterogeneity and basement influence on the development of the Kenya Rift, East Africa. Tectonics 12 (2), 591–606.

Swarzenski, W.V., Mundorff, M.J., 1977. Geohydrology of Northeastern Province, Kenya. USGS Water Supply Paper 1757-N.

Thompson, A.O., 1956. The Geology of the Malindi Area. Mine and Geology Department, Kenya. Report no. 36, 63 pp.

Thompson, A.O., Dodson, R.G., 1960. Geology of the Bur Mayo-Tarbaj area. Ministry of Commerce and Industry, Geological Survey of Kenya. Report no. 47.

Washbourn, C.K., 1967. Lake levels and quaternary climates in the Eastern Rift Valley of Kenya. Nature, vol. 216, pp. 672–673.

Chapter 9

Coastal Waters

Christine A. Omuombo*, Daniel O. Olago*,† and Eric O. Odada*
*Department of Geology, University of Nairobi, P.O. Box 30197–00100, Nairobi, Kenya
†Institute for Climate Change and Adaptation (ICCA), University of Nairobi, P.O. Box 30197–00100, Nairobi, Kenya

1 COMPOSITION AND TEMPERATURE

Off the Kenyan coast is the Indian Ocean, which is dominated by wave action of the equatorial currents that transport air masses westward along the equator. To the southwest of the Kenyan coast, the Indian Ocean subtropical gyre, a combination of the seasonal monsoon winds to the north and south of the area influences the sea surface temperatures (SSTs) displaying a north–south migration of temperatures. The movement of the currents is accelerated during the South Equatorial Current and slowed down during the North Equatorial Current under the influence of monsoon winds. The deflection of the South Equatorial Current northward when it reaches the mainland south of Tanzania–north of Mozambique generates the East African Coastal Current (EACC) that flows northward throughout the year (Obura, 2001; Schott and McCreary, 2001; Abuodha, 2003; Tomczak and Godfrey, 2003). The seasonal monsoon winds play an important role in the coastal processes such as the wave transport of sediments. These warm winds are high in water vapour and thus influence humidity conditions along the coast (Obura, 2001; Schott and McCreary, 2001; Abuodha, 2003; Tomczak and Godfrey, 2003).

Cloud cover and the circulation of cold deeper waters brought to the surface by the monsoon winds influence the mean annual SSTs of 28.4 °C, while the sea surface salinity values range between 32 and 33 PSS (Practical Salinity Scale) (Obura, 2001) with localized variations associated with rainfall or heavy discharges of the large deltas of Tana and Sabaki to the ocean and limited influence from the smaller rivers. Maximum temperatures have been noted to correspond to the transition periods between the monsoons when the winds are light and variable and the insolation is high. The first maximum occurs in October–November when the reef water temperatures rise to 30 °C, while the second maximum occurs in March–April when reef water temperatures rise to 32 °C, with localized midday peaks of 36 °C in shallow still

Developments in Earth Surface Processes, Vol. 16. http://dx.doi.org/10.1016/B978-0-444-59559-1.00009-8

waters; these are associated with high solar insolation (Schott and McCreary, 2001; Tomczak and Godfrey, 2003).

The Indian Ocean is thermally stratified and its density varies with temperature, salinity and pressure. The warm less dense and less saline waters compose the surface waters. The Indian Monsoon has upwelling effects such that during the predominance of the southwest monsoon winds, the upwelling is slightly active along the East African coast with large effects felt in the Arabian Sea. Existing records of nutrient flux show that nitrogen and phosphorus concentrations in Gazi and Malindi are higher in the rivers that drain the agricultural lands than in the ocean waters with higher impacts during the rainy seasons (Ohowa et al., 1997; Schott and McCreary, 2001; Kitheka et al., 2003; Tomczak and Godfrey, 2003).

2 POLLUTION

There is a lack of long-term environmental monitoring records for Kenya's coastal waters. Increases in turbidity have been noted at the deltas of Rivers Tana and Sabaki due to the presence of terrigenous material resulting from erosion in the inland agricultural areas of the upper catchment and the coastal forests in the lower catchment, and these have effects on the coral reefs, the aquatic trophic chain, sea grasses and the mangroves as they reduce the amount of sunlight and hinder the growth of several aquatic organisms (McClanahan and Obura, 1997; Abuodha, 2004). It has been noted, for example, that input from the Sabaki River has caused a reduction in coral cover and diversity over a period of 8 years in the Malindi reefs (Abuodha, 2004).

High turbidity in the rivers occurs during the long rains in April–May and the short rains in October–November. These rivers supply nutrients as they drain their catchments to the ocean. The EACC transports these sediments northward, thus off-loading on the northern banks of the Kenyan coast; hence, they have minimum influence on the water quality in the south (Obura, 2001; Schott and McCreary, 2001; Abuodha, 2003; Tomczak and Godfrey, 2003; Ballot et al., 2006).

Although the discharges of Galana and Tana Rivers are not monitored, McClanahan and Obura (1997) noted that the incorporation of sediments into porites in the past three decades increased in several sites along the coast with an estimated sedimentation rate of approximately of 3–4 $mg/cm^2/day$ while the visibility is <9 m. Further investigations into the coastal waters by Everaats and Nieuwenhuize (1995) showed high copper and cadmium concentrations in the Gazi area (mean copper concentration of 5–30 μm/g and cadmium concentration 0.01–0.34 μm/g, dry weight) in shallow (<200m) and deep (>2000m) waters, respectively, thus showing the enhancement of these elements in the coastal waters. These conclusions were supported by copper, cadmium and zinc in crustaceans varying from 45 to 90 μg/g, 1.0 to 8.5 μg/g and 49 to 102 μg/g of dry weight.

High concentrations of Co, Cr, Cu, Ni, Pb, V and Zn have been measured in the surface samples of sediments from Makupa Creek, Mombasa City, and have been attributed to urban development in the industrial city (Williams et al., 1997; Muohi et al., 2003).

The human activities along the coast contribute to pollution of the coastal waters. The increase in population, lack of knowledge and awareness of the waste management system and impacts of these wastes on the environment and poverty among the local communities have led discharge of domestic wastes from urban and informal settlements to mangrove forests or to the open sea because of the current stress on the existing dysfunctional sanitation facilities. This is evident in Tudor and Mtwapa Creeks where during storm events, the wastes are flushed in large quantities from the inland settlements into the creeks (WRI, 2007; Government of Kenya, 2009, 2010).

The biggest seaport along the East African coast is located in Kenya, making it an economic shipping hub for the region. Oil spill events have generated devastating effects on the ecosystems; in 1988, for example, about 10 ha of mangrove forests was lost, while in 2005, an oil spill from Kipevu oil terminal affected 234 ha of mangroves at Port Reitz (Abuodha and Kairo, 2001; Kairo et al., 2005). The recovery of this ecosystem has been slow and a full recovery has not yet been noted. The full impacts of oil pollution are still unknown, although it is common to sight discharges from ships and tankers that are washed offshore (Government of Kenya, 2009).

Potential pollution from the export processing zone, industries and mining has not been recorded, although these are activities that contribute to pollution of the coastal waters as the industries discharge untreated solid and liquid effluents into the sea (Government of Kenya, 2009, 2010; Hashim et al., 2004).

3 TIDAL MOVEMENTS

The tides tend to flow in the direction of the prevailing winds (southeast and northeast monsoon winds) rather than in that of the offshore EACC. The tidal regime is semidiurnal, and due to the presence of submerged and reef habitats, it diffuses the energy providing 6 h of calm conditions with a short residence time of 3–4 h causing strong tidal reversal of the currents in the mangroves (Pugh, 1978; Ruwa, 1983; Kitheka, 1997; Kitheka et al., 2003; Ballot et al., 2006). The significant wave height is 1 m, although the tidal variations are between 0.8 and 3.4 m due to strong monsoon influence. The larger wave heights ($>$1.5 m) occur during the SE monsoon, while lower wave heights occur during NE monsoon ($<$1.5 m). It is during extreme climatic events such as cyclones that the wave heights exceed 3 m (Kitheka, 1997). No significant changes in the tidal conditions have been noted through the years (Pugh, 1978; Kitheka, 1997; Kitheka et al., 2003).

4 A NOTE ON COASTAL MANAGEMENT

The coastal zone supports numerous land-based activities that affect the coastal waters. The management of the Kenyan coast is sectoral with the formulation of legislation to regulate fisheries, water, tourism, forests and biodiversity conservation through various Acts of Parliament (land planning, tourism industry, tourism industry licensing, tourism development and cooperation, wildlife conservation and water). These acts focus on the management and conservation of the various resources and play a role in the management of the coastal resources.

In realizing the importance of development along the coastal zone, the Coast Development Authority was established under an Act of Parliament with the mandate to plan and co-ordinate the implementation of development of projects throughout the Coast Province (Government of Kenya, 2009, 2010). In 2006, Beach Management Units (BMUs) were implemented through the Ministry of Fisheries in a co-management approach. The BMUs empower the local communities as decision-makers in the management of the resources and thus work with the government to enforce legislation on fishing and conservation (McClanahan, 1997; Cinner et al., 2012).

The Environmental Management and Co-ordination Act (EMCA) of 1999 provides a legal and institutional framework for integrated coastal zone management (ICZM) that can bring together all the sectoral management frameworks. EMCA provides important steps towards sustainable coastal zone management, and it is through the existing framework that a shoreline management strategy that outlines the important legal and institutional basis for ICZM has been outlined. The shoreline management strategy is the basis for the formulation of the National Coastal Zone Management Plan. The success in the implementation of the marine protected areas as a tool for protection and management of ecosystems in the coastal zone has influenced the shoreline management strategy that forms the basis of ICZM to use a site-specific approach. In this approach, issues affecting predefined individual sediment cells are identified and prioritized as areas of concern with regard to management (Government of Kenya, 2009, 2010; Muthiga, 2009).

REFERENCES

Abuodha, P.A.W., 2003. Geomorphology of the Kenyan Coast: Not as a Result of Sea-Level Change Alone. Buletim Geologico, *no*. 43 Mozambique, 2–7. Paper presented during the 'Impact of Sea Level Rise: Past Present–Future' Inhaca Island, Mozambique 4–8 November 2002. Organized and sponsored by the Commission of the Holocene of the International Union for Quaternary Research (INQUA). [Online] http://www.oceandocs.org/bitstream/1834/407/1/1Abuodha.pdf, cited 15th April 2013.

Abuodha, J.O.Z., 2004. Grain size distribution and composition of modern dune and beach sediments, Malindi Bay coast, Kenya. J. Africa Earth Sci. 36, 41–45.

Abuodha, P.A.W., Kairo, J.G., 2001. Human-induced stresses on mangrove swamps along the Kenyan coast. Hydrobiologia 458, 255–265.

Ballot, J., Hoyng, C., Kateman, I., Smits, M., de Winter, R., 2006. Report on Coastal erosion project, Masters Thesis, Delft university of Technology.

Cinner, J.E., Daw, T.M., McClanahan, T.R., Muthiga, N., Abunge, C., Hamed, S., Mwaka, B., Rabearisoa, A., Wamukota, A., Fisher, E., Jiddawi, N., 2012. Transitions toward co-management: the process of marine resource management devolution in three east African countries. Glob. Environ. Chang. 22, 651–658.

Everaats, J.M., Nieuwenhuize, J., 1995. Heavy metals in surface sediment and epibenthic macroinvertebrates from the coastal zone and continental slope of Kenya. Mar. Pollut. Bull. 31 (4–12), 281–289.

Government of Kenya, 2009. State of the Coast Report: Towards Integrated Management of Coastal and Marine Resources in Kenya. National Environment Management Authority (NEMA), Nairobi, p. 88.

Government of Kenya, 2010. Shoreline Management Strategy for Kenya. National Environment Management Authority (NEMA), Nairobi, p. 87.

Hashim, N.O., Rathore, I.V.S., Kinyua, A.M., Mustapha, A.O., 2004. Natural and artificial radioactivity levels in sediments along the Kenyan coast. Radiat. Phys. Chem. 71, 805–806.

Kairo, J.G., Bosire, J.O., Omar, M., 2005. Assessment of the Effects of Oil Spill on the Mangrove Forest of Port Reitz, Mombasa, 12 pp.

Kitheka, J.U., 1997. Coastal tidally driven circulation and the role of water exchange in the linkage between tropical coastal ecosystems. Estuar. Coast. Shelf Sci. 45, 177–187.

Kitheka, J.U., Ongwenyi, G.S., Mavuti, K.M., 2003. Fluxes and exchange of suspended sediment in tidal inlets draining a degraded mangrove forest in Kenya. Estuar. Coast. Shelf Sci. 56, 655–667.

McClanahan, T.R., 1997. Primary succession of coral-reef algae: differing patterns on fished versus unfished reefs. J. Exp. Mar. Biol. Ecol. 218, 77–102.

McClanahan, T.R., Obura, D., 1997. Sedimentation effects on shallow coral communities in Kenya. J. Exp. Mar. Biol. Ecol. 209, 103–122.

Muohi, A.W., Onyari, J.M., Omondi, J.G., Mavutia, K.M., 2003. Heavy metals in sediments from Makupa and Port–Reitz Creek systems Kenyan Coast. Environ. Int. 28, 639–647.

Muthiga, N.A., 2009. Evaluating the effectiveness of management of the Malindi–Watamu marine protected area complex in Kenya. Ocean Coast. Manag. 52, 417–423.

Obura, D.O., 2001. Kenya. Mar. Pollut. Bull. 42 (12), 1264–1278.

Ohowa, B.O., Mwashote, B.M., Shimbira, W.S., 1997. Dissolved inorganic nutrient fluxes from two seasonal rivers into Gazi Bay, Kenya. Estuar. Coast. Self Sci. 45, 189–195.

Pugh, D., 1978. Sea levels at Aldabra Atoll, Mombasa and Mahé, western equatorial Indian Ocean, related to tides, meteorology and ocean circulation. Deep sea Res. 26A, 237–258.

Ruwa, R.K., 1983. Field demonstration of tidal component in the rhythmic vertical migrations of Grapsus maculatus (Brachyura: Grapsidae) on a rocky shore in Kenya. Kenya J. Sci. Technol. Ser. B 4, 49–52.

Schott, F.A., McCreary, J.P., 2001. The monsoon circulation of the Indian Ocean. Prog. Oceanogr. 51, 1–123.

Tomczak, M., Godfrey, J.S., 2003. Regional Oceanography: an Introduction, second ed. Daya Publishing House, Delhi. [online], cited 15th April 2013. http://www.es.flinders.edu.au/~mattom/regoc/pdfversion.html.

Williams, T.M., Rees, J.G., Ferguson, A., Herd, R.A., Kairu, K.K., Yobe, A.C., 1997. Metals, petroleum hydro-carbons and organochlorines in inshore sediments and waters of Mombasa, Kenya. Mar. Pollut. Bull. 34 (7), 570–577.

WRI, 2007. Nature Benefits in Kenya: an Atlas of Ecosystem and Human Well-Being. World Resource Institute, Department of Resource Surveys and Remote Sensing, Ministry of Environment and Natural Resource, Kenya Central Bureau of Statistics, Ministry of Planning and National Development, Kenya and International Livestock Research Institute.

Chapter 10

Renewable Energy Resources in Kenya

Christopher Oludhe
Department of Meteorology, University of Nairobi, Nairobi, Kenya

1 INTRODUCTION

Energy is an important input in a country's development process since it provides the stimulus, drive and momentum for socioeconomic development. It forms one of the foundations on which the three key pillars of the Kenya Vision 2030 (Economic, Social and Political Governance) are anchored on. The country therefore needs to generate more energy and increase efficiency in energy consumption in order to realize Vision 2030 and the overall socioeconomic growth in the country. Kenya is endowed with significant amounts of renewable energy resources such as wind, solar, geothermal, small hydro and biomass. However, few renewable energy resources in the country have been fully assessed, mapped and appraised for their technical and economic viability. If harnessed, these resources can play a significant role in the country's energy supply mix. Kenya has made a significant effort to assess wind, solar and small-hydro potential in the country. However, comprehensive assessment, mapping and appraisal of all the renewable energy resources in the country have not been fully done in order to determine their technical and economic viability. The sections below present the renewable energy situation in Kenya.

2 SOLAR ENERGY

Kenya's geographic location within the equator region gives it a unique opportunity for a vibrant solar energy market. The country receives good solar insolation all year round coupled with moderate to high temperatures estimated at 4–6 $kWh/m^2/day$. Figure 1 below presents the global horizontal radiation in Kenya. It can be seen that significant solar energy potential is found in the eastern and northeastern parts of the country including the coastal areas of Kenya.

Solar energy can be used for lighting, heating, drying and generating electricity. Solar water heating (SWH) systems are mainly used in homes, hotels, hospitals and learning institutions. The solar energy in Kenya is widely used

Developments in Earth Surface Processes, Vol. 16. http://dx.doi.org/10.1016/B978-0-444-59559-1.00010-4

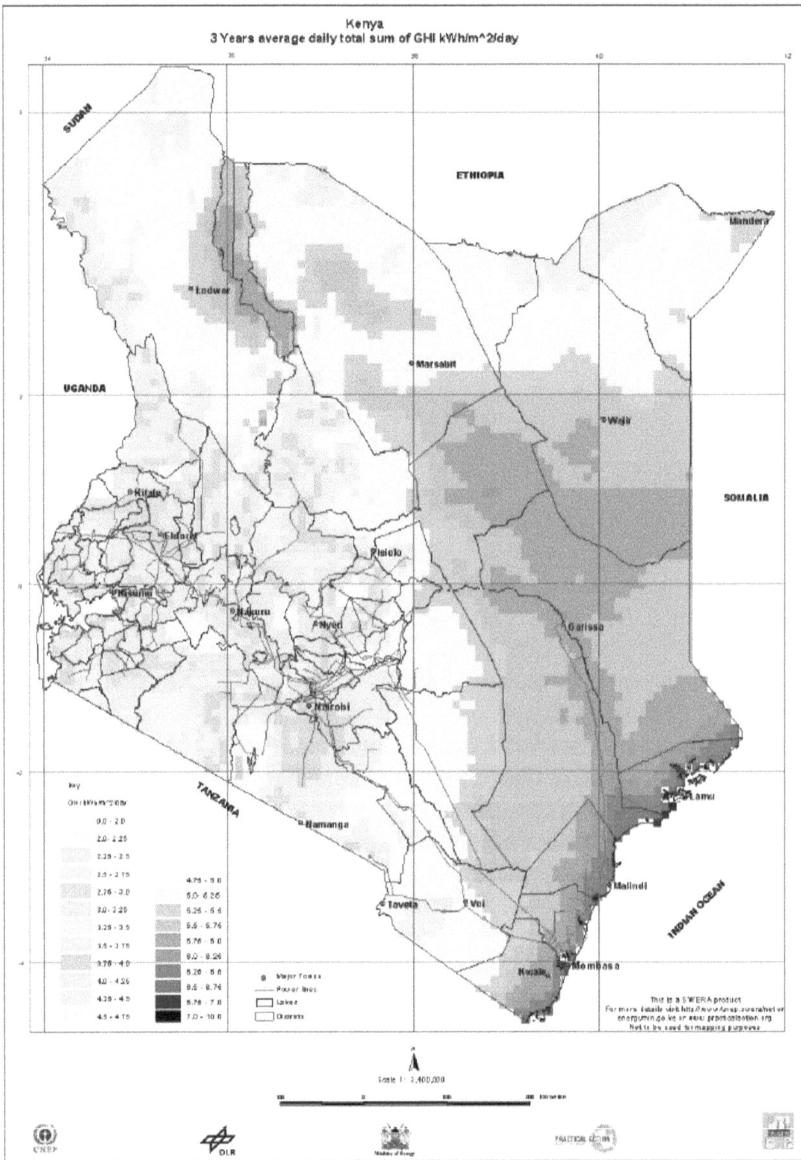

FIGURE 1 Global horizontal radiation. *Source: SWERA Atlas, MoE (2008)*. (For colour version of this figure, the reader is referred to the online version of this chapter.)

for drying crops, fish, hides and skins; for water heating and for electricity generation using photovoltaic (PV) systems. The demand for SWH is projected to grow at a rate of 20% per annum. This demand will mainly be from domestic, institutional and small commercial consumers spurred by the operationalization of the energy (SWH) regulations, 2009 (MoE, 2012).

Kenya has a large-scale market-driven penetration of small PV systems with capacity of 12–50 Wp (Watt Peak) consisting of low-cost amorphous silicon modules and both mono- and polycrystalline silicon modules. The Ministry of Energy (MoE) has been promoting solar electricity generators or PV systems across the country through its energy demonstration centers.

3 WIND ENERGY

Kenya has a proven wind energy potential of as high as 346 W/m^2 and speeds of over 6.0 m/s in most parts of Marsabit, Kajiado, Laikipia, Meru, Nyandarua, Kilifi, Lamu, Isiolo, Turkana, Samburu, Uasin Gishu, Narok and Kiambu areas, among others. Kenya first developed an indicative Wind Atlas in 2008 that was then improved to produce a high-resolution solar and wind energy resource assessment (SWERA). This atlas shows areas with good wind regimes for power generation in various parts of the country. To augment the information contained in this Wind Atlas, the Ministry of Energy (MoE), with the assistance of development partners, installed 55 wind masts and data loggers in 2011 to assess the wind regimes in the high-potential areas identified in the Wind Atlas. Figure 2 below depicts the wind power density across Kenya.

Wind energy has been used in Kenya since the beginning of the nineteenth century having been introduced by the white-settler farmers, primarily for domestic water pumping and irrigation purposes, but its use declined with the advent of oil-fired combustion engines. However, with the rising cost of oil, the exploitation of wind energy is becoming more attractive particularly in areas remote from the power grid and oil supply outlets. Areas where windmills were first installed include Laikipia, Ngong, Nanyuki, Eldoret, Kilifi, Ukunda, Thika, Turkana, Mbooni, Garissa and Amboseli. Kenya has three operational wind-generating farms, two in Ngong owned by Kenya Electricity Generating Company (KenGen) with a total capacity of 5.45 MW and one in Marsabit owned by Kenya Power with a total production of 0.55 MW. There are proposals for additional wind farms with a production capacity of 625 MW as shown in Table 1 (source: Kenya National Power Development Plan (1986–2006)).

Name of Wind Farm	Capacity (MW)
Aeolus Kinangop	60
Aeolus Ngong	100
Aperture Green	60
Daewoo Ngong	30
KenGen	15
Lake Turkana	300
Osiwo Ngong	60
Total	625

4 HYDROPOWER

Kenya's drainage system consists of five major basins: Lake Victoria, Great Rift Valley, Athi/Sabaki River, Tana River and Ewaso Ng'iro North River

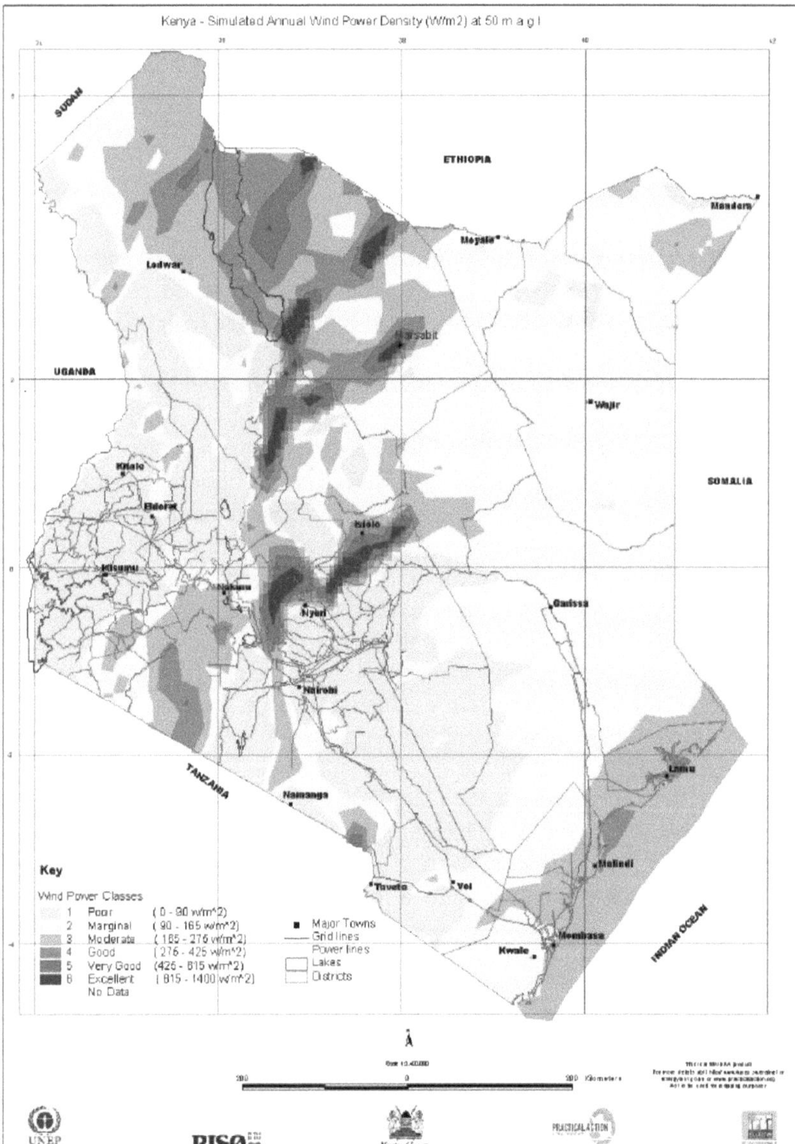

FIGURE 2 Wind power density at 50 m AGL. *Source: SWERA Atlas, MoE (2008).* (For colour version of this figure, the reader is referred to the online version of this chapter.)

(see Figure 3). These basins contain the bulk of the country's hydro resources for power generation. Kenya has considerable hydropower potential consisting of large hydros (sites with capacity of more than 10 MW) and small hydros. Of the large hydros, 750 MW has been exploited and accounts for

TABLE 1 Existing Geothermal Power Plants in Kenya

Station	Entity	Location	No. of Units	Rating per Unit	Installed Capacity (MW)	Year Installed
Olkaria I	KenGen	Olkaria	3	15	45	1981
Olkaria II	KenGen	Olkaria	2	35	70	2003
Olkaria II	KenGen	Olkaria	1	35	35	2010
Olkaria III	Orpower 4	Olkaria	3	4–5	13	2000
Olkaria III	Orpower 4	Olkaria	1	35	35	2009
Total					198	

Source: MoE (2012)

about 50% of installed electric power generation capacity as of 2011 while about 1450 MW remains unexploited. This hydropower potential is located in five geographic regions, representing Kenya's major drainage basins: Lake Victoria (295 MW), Great Rift Valley (345 MW), Athi River (84 MW), Tana River (570 MW) and Ewaso Ng'iro North River (146 MW).

The potential for small-, mini- and micro-hydro systems (with capacities of less than 10 MW each) is estimated at 3000 MW nationwide, 45% of which is located in the Lake Victoria drainage basin. Only about 30 MW had been developed. However, the installed grid-connected small-scale hydroelectric projects contribute about 15.3 MW, though there are several other small-hydro schemes under private and community generation especially in the tea estates across the country, which are not grid-connected. Kenya's electricity-generating capacity as of June 2011 was 1593 MW. This is expected to grow to 15,000 MW by 2030. The current installed capacity is made up of hydro, 48%; thermal (fossil fuel), 38%; geothermal, 12%; co-generation (bagasse), 2% and wind, 0.3%. Hydro generation dominates the electricity supply mix but is highly vulnerable to weather conditions and climate change.

5 GEOTHERMAL POWER GENERATION

Geothermal energy in Kenya is generated using natural steam tapped from volcano-active zones in the Great Rift Valley. Some 127 MW is fed into the national grid from the three plants at Olkaria. Studies have shown that Kenya has the potential to generate 7000 MW of electricity using geothermal its resource, but only 130 MW has been harnessed, and arrangements are under way to harness an additional 210 MW by the year 2012. A feasibility study

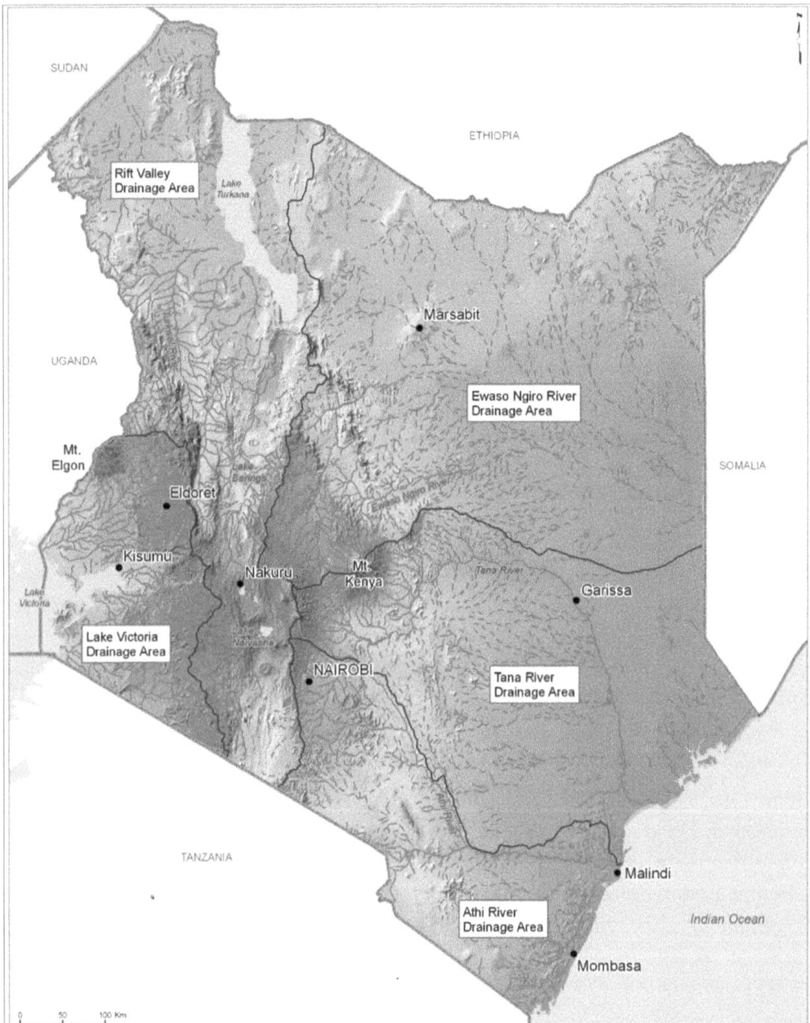

FIGURE 3 Drainage basins in Kenya. (For colour version of this figure, the reader is referred to the online version of this chapter.)

carried out in 2009 concluded that the Greater Olkaria Geothermal Area reservoirs could sustain exploitation of 430 MW worth of steam for 25 years. Of this capacity, 115 MW is in operation by KenGen at Olkaria I and II, and 35 MW was commissioned in May 2010. The remaining 280 MW is available for development.

The Government of Kenya plans to increase exploration and exploitation of geothermal resources to cushion the country from expensive diesel-generated power and over reliance on hydropower. More than 14 high-temperature potential sites occur along the Kenyan Great Rift Valley with an estimated potential of more than 15,000 MW (see Figure 4). Other locations include Homa Hills in Nyanza, Mwananyamala at the Coast and Nyambene Ridges. These prospects are at different stages of development.

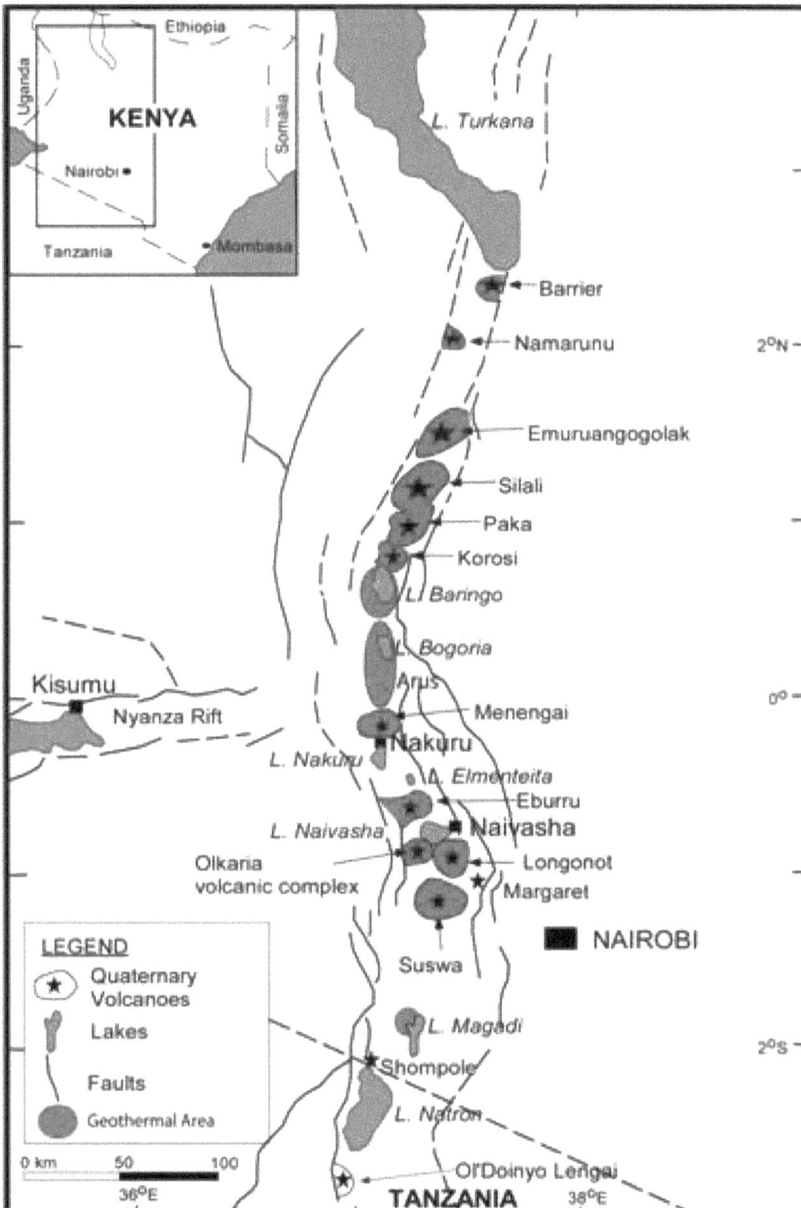

FIGURE 4 Geothermal sites in Kenya. (For colour version of this figure, the reader is referred to the online version of this chapter.)

6 TIDAL

No information for Kenya as yet.

7 BIOGAS AND BIOFUELS

Biogas is a clean form of energy that has great potential to serve as an alternative to wood fuel. Biogas technology has been actively promoted in the country since early 1980s. It has the benefit of improving community health since it is clean and also agricultural productivity by providing slurry manure. Kenya's MoE supports individual farmers and communities in the construction of biogas digesters through technical advice, training of artisans, awareness campaigns and demonstrations. Surveys are currently being undertaken to determine the uptake of biogas technology in the country. A feasibility study on promotion of biogas digester technologies in Kenya has showed that it is possible to construct 6500 biogas digesters over the coming 5 years.

REFERENCES

MoE, 2008. Ministry of Energy, National Energy Policy.

MoE, 2012. Ministry of Energy, National Energy Policy. Ministry of Energy (MoE), Kenya. http://www.energy.go.ke/index.php/universal-access/download/finish/3-ministerial-documents/6-national-energy-policy-third-draft-may-11-2012.

Chapter 11

Major Soil and Data Types in Kenya

Christian T. Omuto
Department of Environmental and Biosystems Engineering, University of Nairobi, P.O. Box 30197-00100, Nairobi, Kenya

1 INTRODUCTION

Soil is a natural body consisting of layers (soil horizons) of weathered mineral materials, organic material, air, and water. It is the result of combined interaction of climate, relief, organisms, parent materials (original minerals), and time. It supports food production; acts as repository for seeds, germplasms, and genes for flora and fauna; supports millions of living organisms in soil; absorbs, purifies, transports, and releases water in the hydrologic cycle; and also acts as construction material and foundation to support building infrastructures. In general, soil is a very important resource for life on earth.

Soil map is a tool scientists and policymakers use to identify different types of soil in the landscape so that their potential for different uses can be maximised. It is developed by combining different input data, predictive mapping methods, tools (computer, infrared spectrometry, remote sensing, etc.), and expert knowledge. The input data consist of products of traditional soil survey (samples of soil properties and/or classified soil profiles), parent material, vegetation, climate, and topography (Jenny, 1964). Traditional soil survey provides the georeferenced and classified soil profiles or soil properties that are then linked with the other input data using the predictive mapping method. In this predictive method, the classified profiles/soil properties are the dependent variables, while the other input variables are the independent variables (Figure 1).

Many approaches exist in the literature for the predictive mapping methods such as kriging, regression, and classification (Lagacherie et al., 2007). The two most important factors, which determine the quality and scale of the resultant soil map, are the mapping method used and the number and spatial distribution (or resolution) of the dependent and independent input data. The resolution of the final soil map is usually directly related to the resolution of input data and the geographic area to be mapped. There are many references in the literature showing the mathematical linkage between number of points for the dependent

Developments in Earth Surface Processes, Vol. 16. http://dx.doi.org/10.1016/B978-0-444-59559-1.00011-6

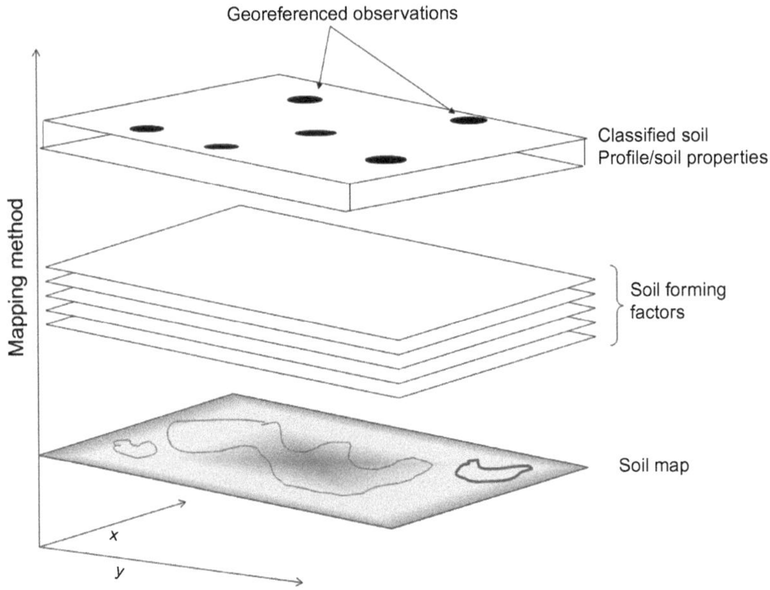

FIGURE 1 Concept of soil mapping tool. (For colour version of this figure, the reader is referred to the online version of this chapter.)

variables, mapping area, and resolution of the input independent variables (http://www.itc.nl/personal/rossiter; Hengl et al., 2007). For the input independent data, high-resolution remote sensing images exist for some of the data such as vegetation and topography. This may not always be the case with the other input data. Although there are some countries with equally high-resolution soil surveys, climate, and parent material, the majority of countries in the world do not have high-resolution input data (Omuto et al., 2012). Consequently, it is not uncommon to find low-resolution soil maps in such countries.

In Kenya, the input data for soil mapping can be found from Kenya Soil Survey, ISRIC (www.isric.nl), and other government departments, NGOs, research institutes, and private organisations. Some of these datasets have low resolution such as the 1:1M geologic map, sparsely distributed climate data, and few georeferenced and classified soil profile pits. There is a need to assemble all available information about soil mapping input data to enable development of a better soil map of the country than what is currently existing. The objective of this chapter is to describe publicly available soil data and new tools that can be used to produce higher-resolution soil map of Kenya than what currently exists at 1:1M scale (Sombroek et al., 1982).

2 MATERIALS AND METHODS

2.1 Data

Publicly available classified soil profile data used for mapping soil in Kenya can be obtained from www.isric.orgm, postgraduate study reports, and soil survey reports from different departments within the Government of Kenya and numerous NGOs working in Kenya.

Apart from classified soil profile data, other input data such as rainfall data, digital elevation model (DEM), land use, Normalised Difference Vegetation Index (NDVI) image, landform, and geologic maps are also important in soil mapping. They are mainly auxiliary information for spatial prediction of the soil attributes. A 90 m DEM map can be downloaded from http://srtm.usgs.gov. It is a map of surface elevation in metres above sea level at a spatial resolution of 90 m. Land use map can be obtained from www.africover.org. It contains 73 land use/cover types mapped at the scale of 1:200,000. NDVI data are the image of average vegetation index from 16-day composite MODIS images. They are available for dates beginning from January 2000 and have a spatial resolution of 250 m. The images can be downloaded from http://pekko.geog.umd.edu/usda/apps. The landform map is obtained from the DEM using the approach given by Iwahashi and Pike (2007). Using that approach, a 90 m DEM produces a landform map with 18 landform types at 1000 m spatial resolution. The geologic map can be obtained from www.rcmrd.org. It contains 42 classes of lithology mapped at the scale of 1:1,000,000. The rainfall data can be obtained from the Kenya Meteorological Department (www.kmd.go.ke).

2.3 Prediction of Soil Types

In order to couple the input data together in a soil mapping process, it is necessary to make them uniform in terms of spatial resolution and extent. This entails converting all vector data types into spatial raster data format. Then, all the raster data types have to be resampled to a common spatial resolution (i.e. pixel size). Some of the vector data types among the input data included rainfall, land use, and geologic maps. The rainfall data are always converted to rainfall image by use of a simple kriging method (Nielsen and Wendroth, 2003). The land use and geologic data can be converted to images using vector-to-raster algorithms such as those in ERDAS Imagine® (ERDAS, 2003; Teng et al., 2008). After the conversion, the input raster datasets are resampled a common resolution to ensure consistency with all input data.

Spatial prediction of soil types uses the classes in the classified soil profiles as the dependent variables, with mean annual rainfall, NDVI, land use, landform, and geology as the predictors. These predictors are chosen to conform to the soil-forming factors in Jenny's equation (Jenny, 1964; McBratney et al., 2003). In this study, they were modelled with a generalised linear mixed-effects model (GLMM) for the deterministic part of the regression kriging (Faraway, 2006). This model is given in Equation (1):

$$\begin{aligned} \text{Soiltype}_{ij} &= g(\eta)\eta \\ &= (\beta_o + b_o) + (\beta_1 + b_{1i})^*\text{landform}_{ij} + (\beta_2 + b_{2i})^*\text{ndvi}_{ij} \\ &\quad + (\beta_3 + b_3)^*\text{rain}_{ij} + (\beta_4 + b_4)^*\text{landuse} + e_{ij} \end{aligned} \tag{1}$$

where (η) is the link function between the soil types and the predictors. Equation (1) was fitted for each lithology class in the geologic map using the Poisson link function (Faraway, 2006). Its residuals were treated in the same way as was done for the residuals in the clay content example.

Results from Equation (1) and the small-scale autocorrelation trends were used to predict the soil types in the unsampled locations in the study area. The prediction accuracy was also tested just like in the prediction of soil properties: three-quarters of the data was used to build the models and the remaining one-quarter was used to validate the models. After accuracy testing, the entire dataset was used to produce a map of major soil types in Kenya.

3 RESULTS AND DISCUSSIONS

3.1 Publicly Available Soil Data Types in Kenya

The soil profile used in this study consisted of 574 georeferenced observations of classified soil profiles (Figure 2). Three hundred fifty of these observations were obtained from www.isric.org on 4 October 2012, 19 from Tana and Athi River Development Authority (TARDA) on 19 September 2010 (Tana and Athi River Development Authority, 1987), 200 from postgraduate study reports at the University of Nairobi, and the remaining 5 samples from the Ministry of Livestock Development (MOLD) on 11 October 2010 (Touber, 1991). The soil profile data were reclassified according to WRB (2006) classification system to make them uniform.

The ISRIC dataset also contained soil spectral reflectance data. Soil spectral reflectance is the reflectance of light shone on soil samples taken from the field. It is obtained by scanning soil samples with FieldSpec FTR spectroradiometer (Shepherd and Walsh, 2002). The scanning produces spectral reflectance signatures with wavelengths between 0.35 and 2.5 μm (Figure 3).

Soil spectral reflectance is influenced by the chemical and physical properties of soil (Baumgardner et al., 1985). Due to this relationship, many researchers have used it to predict a number of soil properties and to assess threats such as degradation to soil and soil classification (Ben-Dor and Banin, 1995; Islam et al., 2003; Awiti et al., 2008).

3.2 Major Soil Types in Kenya

Mixed-effects model for spatial prediction of soil types produced 31 major soil types for Kenya (Figure 4). The final output map showed that most of the arid areas (northeastern and northwestern Kenya) were classified as

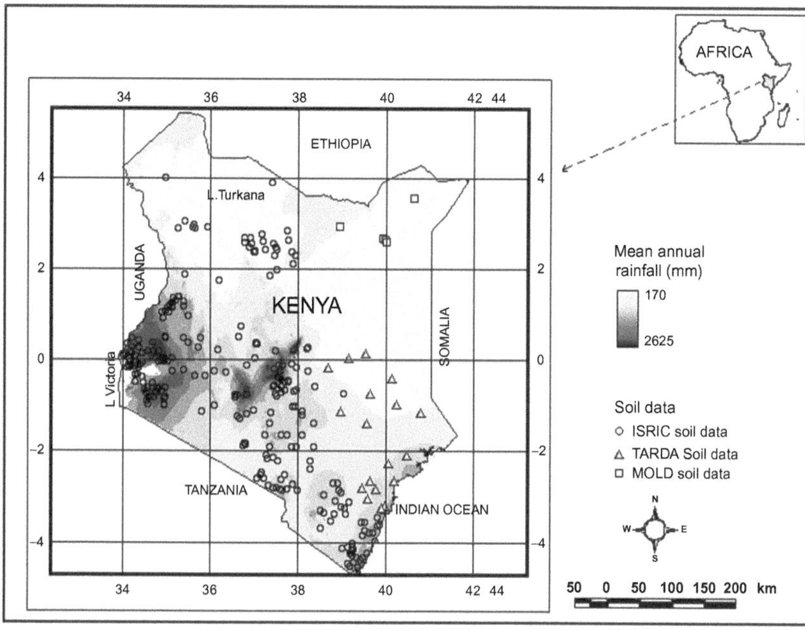

FIGURE 2 Spatial distribution of georeferenced soil data.

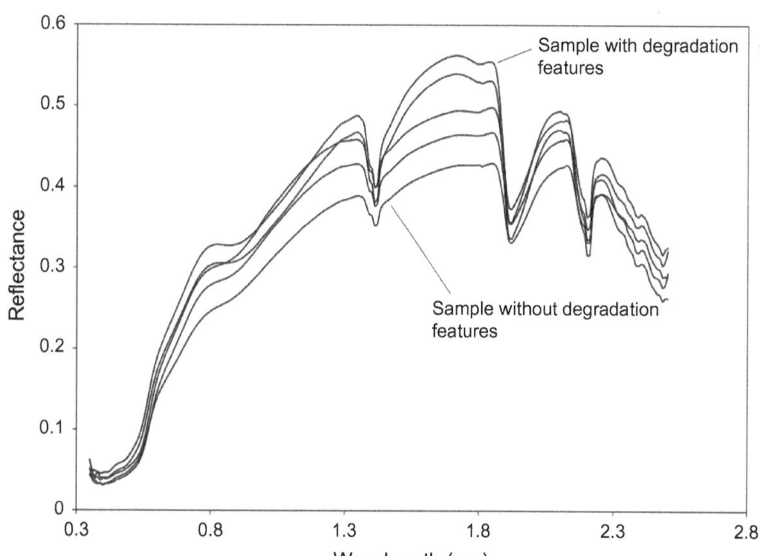

FIGURE 3 Sample spectral reflectance information of Kenyan soil.

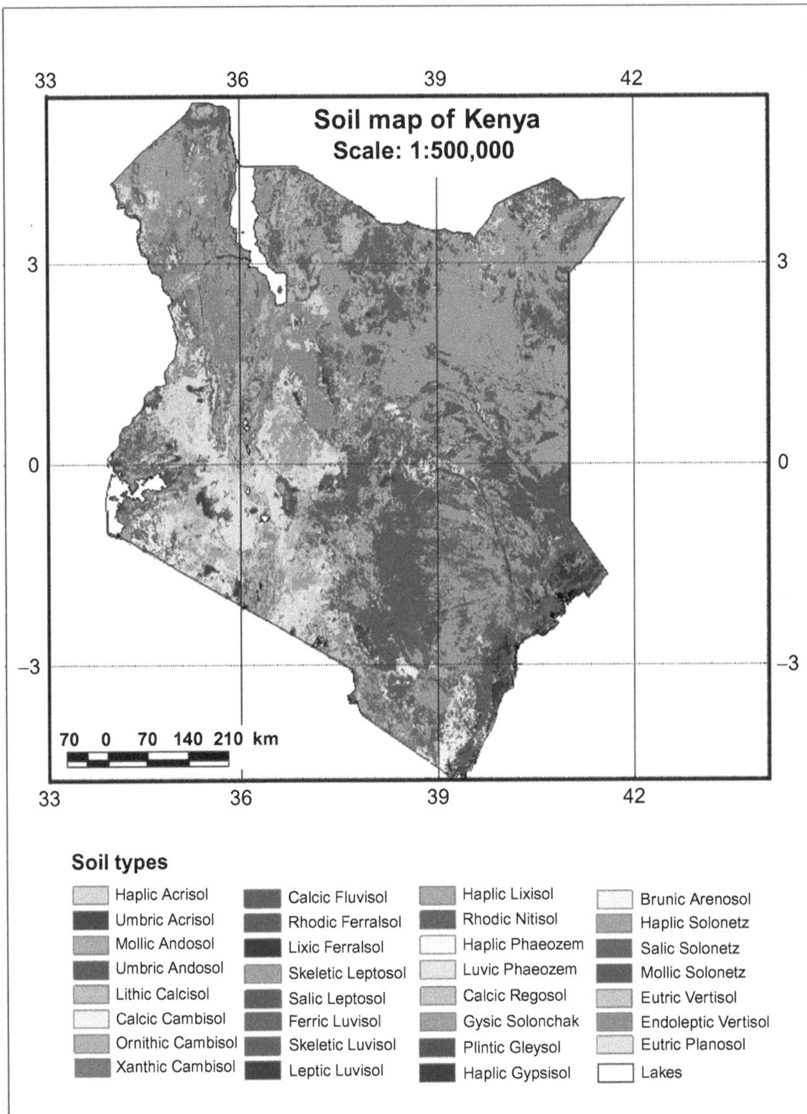

FIGURE 4 Soil map of Kenya. (For colour version of this figure, the reader is referred to the online version of this chapter.)

Solonetz. Nitisol largely dominated the humid areas while majority of the highland areas were classified as Cambisol. In the lower parts of Rift Valley, the prediction had Andosols as the major soil type while the coastal region had Arenosols, Ferralsols, and Solonetz. Its comparison with holdout classified soil profiles had a kappa index of 68%.

The mixed-effects modelling process did not predict well certain soil types. For example, an Alisols soil profile in the coastal area (4°24′S and 39°18′E) was predicted as Ferralsols, a Cambisol profile in Kapenguria (1°12′N and 35°6′E) was predicted as Phaeozem, and a Solonchak profile in Lopusiki in northwestern Kenya (4°1′N and 35°E) was classified as Solonetz. Attempts were made to establish the causes for these misclassifications. According to WRB (2006), Alisols mainly occur on hilly or undulating topography, and therefore, the classification of a profile in the flat areas of the coastal region as Alisols did not seem right. Due to this conflict, it was difficult to identify the misclassification with either the mixed-effects model or the classified soil profile. In the case of the misclassification near Kapenguria, two neighbouring soil profiles (one at 1°6′N and 35°6′E and another at 1°10N and 35°8′E) were classified as Phaeozems. Owing to their close proximity and due to the spatial resolution of the input predictors, the mixed-effects model tended to incline its classification towards Phaeozems. In general, it can be said that the misclassification by mixed-effects model was largely due to errors in soil profile classification and the accuracy and scale of the input predictors. The misclassification also shows that the northeastern parts of the country have been poorly studied in terms of soil characterization. This is particularly important for a scenario where national soil information system is envisaged.

In Figure 4, Solonetz seems to be the majority, occupying about 15.5% of the country. The others are Cambisols (9.3%), Regosols (8.9%), Planosols (8.6%), Luvisols (8.3%), Ferralsols (7.5%), and Arenosols (5.6%). Most of these dominant soil groups can be found in the arid and semiarid areas of the country. They may have strong crop production limitations but are good for the development of pastoral resources (e.g. pasture, water harvesting, and salt licks) for extensive livestock/animal production (WRB, 2006). The soil groups with high crop production potential and intensive livestock production are few and occupy a small proportion of the country. Table 1 gives the characteristics, potential, and limitations of the major groups of soil types in Kenya.

3 CONCLUSIONS

This study has shown where public soil information can be obtained for Kenya. The soil information from these sources is publicly available. The spatial distribution of the locations of these soil information sources showed that the northeastern parts of the country have been poorly represented in scientific research. This is an important point for consideration if a national soil information system is to be established.

Using the available soil data, this study developed a new soil map of Kenya at a higher scale than the currently available area-class map. This soil map depicts the country as consisting of 22 main soil groups according to the FAO-UNESCO classification. These groups are dominated by soil types that have strong crop production limitations under rain-fed agriculture but are good

TABLE 1 Characteristics and Rain-Fed Agricultural Potential of Major Soil Types in Kenya

Soil Type	Characteristics	Agricultural Potential
Acrisol	It is a clay-rich soil. It has low levels of plant nutrients, high levels of aluminium, and high susceptibility to erosion	Silviculture and low-intensity pasture. If climate is permissible, then tea, coffee, and sugarcane can be supported
Andosol	Rich soil but may have problem of phosphorous fixation	Tea, wheat, orchids, vegetables, and potato
Calcisol	Desert soil with significant accumulation of secondary calcium carbonates	Low volume grazing
Cambisol	Young soil and good for agriculture	A wide variety of crops
Fluvisol	Occurs in periodically flooded areas such as alluvial plains, river fans, valleys, and tidal marshes	Paddy rice and many dry land crops
Ferralsol	Good soil physical properties but low in nutrients and strong P fixation	A wide variety of crops and soybeans
Leptosol	Shallow soil over continuous rocks	Wet season grazing and some tree crops
Luvisol	Soil with a subsurface horizon of high activity clay accumulation and high base saturation. Susceptible to erosion	A wide variety of crops
Lixisol	Soil with a subsurface horizon of low activity clay accumulation and high base saturation. Susceptible to erosion	Low volume grazing
Nitisol	Deep well-drained soil with stable structure and good nutrient content	A wide variety of food crops, coffee, and pineapple
Phaeozem	Porous fertile soil. Susceptible to erosion	A wide variety of crops
Regosol	It is a very weakly developed mineral soil in unconsolidated materials with only a limited surface horizon having been formed. It has low moisture retention	Small grains, grazing, fruit trees, groundnuts, and cassava

TABLE 1 Characteristics and Rain-Fed Agricultural Potential of Major Soil Types in Kenya—Cont'd

Soil Type	Characteristics	Agricultural Potential
Solonchak	It is a strongly saline soil type with high concentration of soluble salts. It occurs where saline groundwater comes near to the surface or where the evapotranspiration is considerably high. It is very low in inherent fertility and devoid of moisture and vegetation throughout the year	Grazing
Gleysol	It is a wetland soil that develops a characteristic gleyic colour pattern if saturated with groundwater for long	Paddy rice
Gypsisol	It is soil with accumulation of gypsum. Commonly found in dry lands	Grazing
Arenosol	It consists mainly of unconsolidated sand deposits. It has very low water-holding capacity and extremely low in all essential nutrients	Root and tuber crops, extensive grazing, grains, pulses, and melons
Solonetz	It has a subsurface horizon (subsoil) that is higher in clay content than the upper horizon. It also has more than 15% exchangeable sodium	Grazing
Vertisol	It is soil with high content of expansive clay known as montmorillonite that forms deep cracks when dry	Cotton, wheat, sorghum, and rice
Planosol	Soil in seasonally waterlogged flat areas	Extensive grazing, wheat, and silviculture

for the development of pastoral resources. This implies that rain-fed crop production in the country cannot adequately sustain the consumptive demand of over 40 million people unless improved farming methods are applied.

ACKNOWLEDGEMENTS

The data used in this study were obtained from ISRIC (www.isric.org), the University of Nairobi, Tana and Athi River Development Authority (TARDA), and the Government of Kenya. A special acknowledgement is given to different authors and people who participated in generating the data. This study was financed by the International Foundation for Science (www.ifs.se) through the project number C/3953-2.

REFERENCES

Awiti, A.O., Walsh, M.G., Shepherd, K.D., Kinyamario, J., 2008. Soil condition classification using infrared spectroscopy: a proposition for assessment of soil condition along a tropical forest-cropland chronosequence. Geoderma 143, 73–84.

Baumgardner, M.F., Silva, L.F., Biehl, L.L., Stoner, E.R., 1985. Reflectance properties of soils. Adv. Agron. 38, 1–44.

Ben-Dor, E., Banin, A., 1995. Near-infrared analysis as a rapid method to simultaneously evaluate several soil properties. Soil Sci. Soc. Am. J. 59 (2), 364–372.

ERDAS, 2003. ERDAS Imagine 8.7. Leica Geosystems GIS and Mapping LLC. Georgia.

Faraway, J.J., 2006. Extending the Linear Model with R: Generalized Linear Models, Mixed Effects and Nonparametric Regression Models. Taylor and Francis, Boca Raton, FL.

Hengl, T., Heuvelink, G.B.M., Rossiter, D.G., 2007. About regression kriging from equations to case studies. Comput. Geosci. 33, 1301–1315.

Islam, K., Singh, B., McBratney, A., 2003. Simultaneous estimation of several soil properties by ultra-violet, visible, and near-infrared reflectance spectroscopy. Aust. J. Soil Res. 41 (6), 1101–1114.

Iwahashi, J., Pike, R.J., 2007. Automated classifications of topography from DEMs by an unsupervised nested-means algorithm and a three-part geometric signature. Geomorphology 86, 409–440.

Jenny, H., 1964. Factors of Soil Formation: A System of Qualitative Pedology. Dover Publications Inc., New York.

Lagacherie, P., McBratney, A.B., Voltz, M., 2007. Digital Soil Mapping. An Introductory Perspective. Elsevier, Amsterdam.

McBratney, A.B., Santus, M.L.M., Minasny, B., 2003. On digital soil mapping. Geoderma 117, 3–52.

Nielsen, D., Wendroth, O., 2003. Spatial and Temporal Statistics: Sampling Field Soils and Their Vegetation. Catena-Verlag, Reiskirchen.

Omuto, C.T., Nachtergaele, F.O., Vargas-Rojas, R., 2012. State of the Art Report on Global and Regional Soil Information: Where Are We? Where to Go? FAO, Rome.

Shepherd, K.D., Walsh, M.G., 2002. Development of reflectance spectral libraries for characterization of soil properties. Soil Sci. Soc. Am. J. 66 (3), 988–998.

Sombroek, W.G., Braun, H.M.H., van der Paow, B.J.A., 1982. Exploratory soil map and agro-climatic zone map of Kenya. Scale 1:1000000. Kenya Soil Survey, Ministry of Agriculture. Nairobi.

Tana and Athi River Development Authority, 1987. Development Plan: Report Vol II. TARDA, Nairobi.

Teng, J., Wang, F., Liu, Y., 2008. An efficient algorithm for raster-to-vector data conversion. Ann. GIS 14, 54–62.

Touber, L., 1991. Wet Season Restrictions in Accessibility of Livestock in Mandera and Wajir Districts. Ministry of Livestock and Development, Government of Kenya, Nairobi, Kenya.

WRB, 2006. World Reference Base for Soil Resources 2006: A Framework for International Classification Correlation and Communication. FAO, Rome.

Chapter 12

Vegetation

Andreas B. Brink*, Catherine Bodart*, Graeme Buchanan*, Marco Clerici*, François Donnay*, George Eshiamwata*, Jean-Marie Gregoire*, Bernard K. Kirui*,†, Andrea Lupi*, Ilaria Palumbo* and Jean-François Pekel*

*Joint Research Centre—European Commission, Via E. Fermi, 2749 Ispra, Italy

†Department of Natural Resources, Egerton University. P.O. Box 536, 20115, Egerton, Kenya

In Kenya, as well as in many other African countries, the population and economy are heavily dependent on primary products or natural resources such as freshwater, forests, croplands, pastures and fisheries (Homer-Dixon and Blitt, 1998). Population increase associated with economic development and land management practices and last but not least climate change can give rise to pressure and shortages in the provision of these ecosystem goods and services and finally lead to an increased risk of social and economic destabilisation (UNEP, 2009).

During the last decades, not only major human but also natural disturbances such as drought, civil disturbances, large population increases and the impacts of 'globalisation' had important implications for land-use requirements, with subsequent impacts on natural vegetation cover, biodiversity, socioeconomic stability and food security. The modification of the vegetation cover, with a predominant clearing of natural vegetation, may have a long-term impact on sustainable food production, freshwater and forest resources, the climate and ultimately human welfare (Foley et al., 2005).

1 ECOREGIONS, VEGETATION ZONES AND LAND COVER

The following description of Kenya's ecoregions and vegetation zones and analysis of the various land cover types are based on two key data and information sources: the White ecoregion map for Africa (White, 1983) and the Africover land cover map for Kenya (FAO, 2000). Both maps have been used for this study, because the first one provides a broad overview of the key vegetation zones (or phytochoria) and its relation to climate and rainfall, while the second one characterises the detailed land cover and partly land use within the different ecoregions for a specific year as derived from Earth observation. In addition to these two 'static' information layers, long-term vegetation dynamics profiles have been derived from SPOT-VEGETATION satellite

Developments in Earth Surface Processes, Vol. 16. http://dx.doi.org/10.1016/B978-0-444-59559-1.00012-8

data over the 1998–2011 period and Africa Rainfall Climatology data in order to characterise the vegetation phenology and to highlight the strong rainfall–vegetation relation.

Definition of data and terms:

White vegetation map of Africa	A vegetation classification and map for continental Africa, consisting of 17 major vegetation types that we will refer to as ecoregions
Africover land cover map for Kenya	Land cover map derived from medium-resolution Landsat satellite images from the year 1999
Land cover	Land cover is the observed (bio)physical cover on the Earth's surface
Land use	Land use is characterised by the arrangements, activities and inputs people undertake in a certain land cover type to produce, change or maintain it. Definition of land use in this way establishes a direct link between land cover and the actions of people in their environment
NDVI	The Normalised Difference Vegetation Index (NDVI) is an estimate of the photosynthetically absorbed radiation over the land surfaces. NDVI thus allows monitoring of vegetation photosynthesis through time and enables easy temporal and spatial comparisons. The NDVI is computed as (NIR − Red)/(NIR + Red), where NIR and Red are the amount of near-infrared and red light, respectively, reflected by a surface and measured by satellite sensors
Vegetation phenology	The seasonal activity/dynamics of vegetation
SPOT-VEGETATION	The VEGETATION sensor is a multispectral instrument flown aboard the SPOT 4 and SPOT 5 satellite platforms. It provides accurate measurements of basic characteristics of vegetation canopies on an operational basis The sensor has 4 spectral bands: blue (0.43–0.47 μm), red (0.61–0.68 μm), near-infrared (NIR, 0.78–0.89 μm) and shortwave infrared (SWIR, 1.58–1.74 μm) Spatial resolution, 1 km; field of view, 2200 km swath width; and spatial coverage, about 90% of the equatorial areas are imaged each day; the remaining 10 % being imaged the next day. For latitudes higher than 35° (north and south), all regions are acquired at least once a day

1.1 Ecosystems of Kenya

Kenya encompasses a wide range of climatic and ecological regions resulting in a wide variety of phytochoria and hence vegetation dynamics. These range from the dry and sparsely vegetated northern part of the country to the low-land and more humid rain forest remaining areas along the east coast to the fertile areas in the central part and around the Lake Victoria region and finally the high biodiverse mountain forest areas of Mount Kenya and Mount Elgon, the Aberdares, the Cherangani Hills and the Mau Forest zone.

White (1983) stratifies the country into four main ecological regions (see Figure 1). From its total land area of about 58 Mha (UNEP, 2009), the highest proportion is covered by the Somalia–Masai ecoregion, followed by the

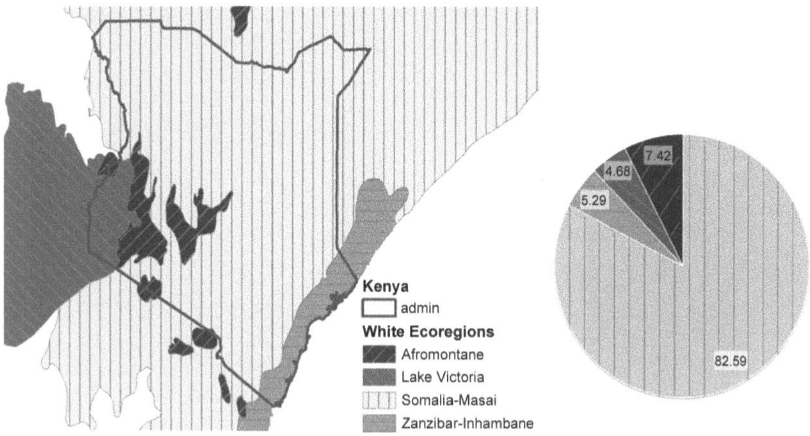

FIGURE 1 Kenya's ecoregions as defined by White (1983): Somalia–Masai, Zanzibar–Inhambane, Lake Victoria and Afromontane. The pie chart represents the percentage that each ecoregion covers within the country.

Afromontane, the Zanzibar–Inhambane and the Lake Victoria ecoregions, respectively.

1.2 Somalia–Masai Ecoregion

The Somalia–Masai ecoregion, which is characterised by arid to semiarid type of climate, is the most predominant in the country covering about 83% or 48 Mha of the total country area. In its entirety, the Somalia–Masai region covers the countries of Somalia, eastern and southern Ethiopia (except the mountain areas), southeast Sudan, a small part of northeast Uganda and the dry lowlands of north and central Tanzania (White, 1983). The vegetation structure is predominantly open to closed deciduous shrubs and woody vegetation including tree and shrub savanna, which become semievergreen and evergreen bushland and thicket on the lower slopes of the mountain areas. About 72% of the Somalia–Masai ecoregion in Kenya is covered by this land cover type (FAO, 2000), not only mainly in the northern part but also in the eastern part bordering the coastal area and to some extent in the southern end of the country. Not only more than 90% of the total area of wood and shrubland in the country is situated in this ecoregion, but also this land cover type is covering about 65% of the total country area, as highlighted in Figures 2 and 3.

Two distinct rainfall seasons influence the woodlands and shrublands in Kenya, but depending on the geographic location, these are more or less well defined. The top right graph of Figure 3 shows an example, which is representative for Kenya's southeastern coasts. Here, the two rainy seasons and hence growing seasons are evident, but they are very close in time and rainfall occurs almost throughout the year. The long-term average vegetation

FIGURE 2 Two examples of wood and shrub type of savanna in Kenya. (For colour version of this figure, the reader is referred to the online version of this chapter.)

phenology, as derived from the SPOT-VEGETATION sensor, follows this general rainfall trend, highlighting also a relative small seasonal effect (relative little NDVI variation) and modest interannual variability (this is represented by the upper and lower dotted light green line in the graphs, which correspond to the long-term minimum and maximum NDVI). Moving inland, the two growing seasons become more distinct as shown in the two bottom graphs of Figure 3. A longer dry season and a shorter dry season are evident from the rainfall patterns and reduced rainfall amount. This is reflected also in

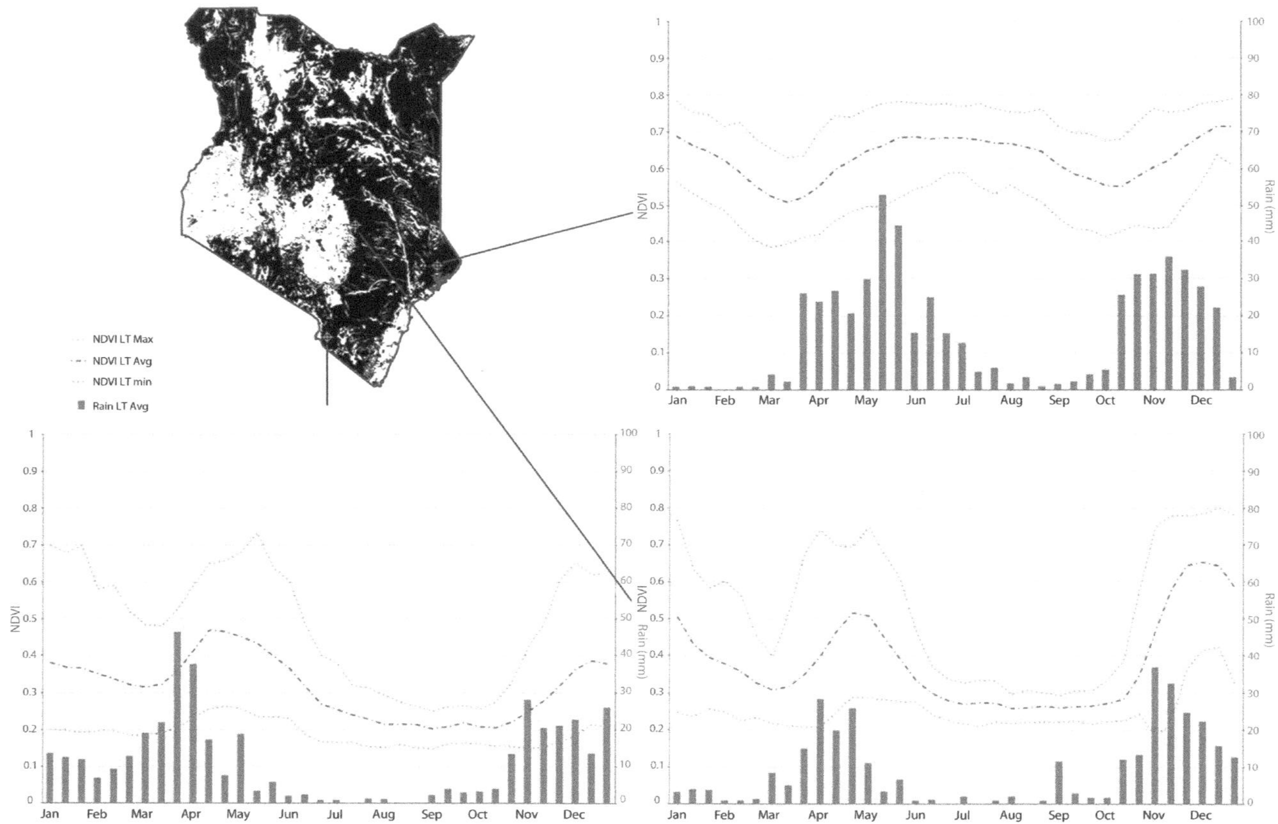

FIGURE 3 Wood and shrub savanna distribution in Kenya (in black on top left map) and three examples of characteristic vegetation growing seasons. (For colour version of this figure, the reader is referred to the online version of this chapter.)

general lower NDVI values and stronger variation throughout the year. In addition, the pronounced difference between the long-term minimum and maximum NDVI highlights the fact that interannual variability is strong, and hence, climatic oscillations are of great concern for these areas. The understorey of the wood and shrub savanna is characterised by open to closed grasslands that cover in total almost 10% of the ecoregion, predominantly in the northern part of the country. Very little natural grasslands grow outside the Somalia–Masai ecoregion; in fact more than 98% of the country's natural grasslands are distributed in this ecoregion. The Lake Turkana area receives the lowest amount of rainfall in Kenya and is therefore the most sparsely vegetated zone of the country. Most of the nearly 3% bare areas of the ecoregion are located here as well. Similar to the grasslands, the total amount of bare areas is concentrated here, and some minor bare lands can be found, for example, on top of high mountains such as Mount Kenya. The northern part of the Somalia–Masai ecoregion in Kenya is also the most affected by recurrent droughts (Gommes and Petrassi, 1996), such as the 2011 severe drought where erratic and below average rainfall for two consecutive growing seasons negatively impacted both natural vegetation and agricultural production. The consequences were low pastoral performance with a high number of livestock death and food shortages that led to critical food insecurity associated with population displacement and a dramatic humanitarian crisis especially in Somalia (AMESD, 2011). While most of the arid and semiarid areas of the Somalia–Masai region in the country are home to pastoral livelihoods, the central and western parts of the ecoregion are dominated by rain-fed croplands (FAO, 2000) (Figure 4).

FIGURE 4 Ploughed fields ready for cultivation in the central part of the Somalia–Masai region. (For colour version of this figure, the reader is referred to the online version of this chapter.)

In particular, the areas northeast of Nairobi and in the Great Rift Valley around the Lakes Naivasha and Nakuru are the most fertile of the ecoregion, also because they represent the more reliable rainfall zones. Maize is the most widely grown cereal crop, followed by beans and cassava and cereal crops such as wheat, millet and sorghum. Tomatoes, onions, potatoes, carrots and cabbage are important minor crops (WRI, 2007). Towards the mountain areas, food crops are replaced by cash crops and high-value crops such as coffee plantations and fruits and vegetables. In total, the agriculture cropland represents about 10% of the Somalia–Masai ecoregion but nearly 50% of the total agriculture area in the country.

Rainfall is not a limiting factor in the southwestern part of the country (see top right graph in Figure 5) where relative abundant and reliable distribution is guaranteed over the yearly cycle. The two typical rainy seasons are not well pronounced and seasonality is very limited. The less significant seasonality and interannual variability is also illustrated by the NDVI behaviour. The long-term average NDVI values are relatively stable and high throughout the year, with a minor drop in March. Indeed, the growing cycle including harvest is not apparent, which could be due to permanent crops such as coffee or tea plantations or a mixture of rain-fed herbaceous crops with superimposed cycles. This region is one of the most intensively cultivated areas in Kenya and due to favourable climatic conditions is very stable in agriculture production. This is not the case for other areas in the country, where agriculture production suffers from unreliable rainfall amounts and patterns as emphasised by the strong seasonality and interannual variability in the two lower graphs of Figure 5. In particular, the agriculture zone east of Nairobi is of concern due to this pattern (lower right graph in Figure 5), especially when considering that agriculture expansion is encroaching even more towards drier and less productive territories eastwards (see following chapter 2 on land cover change).

1.3 Zanzibar–Inhambane Ecoregion

The coastal belt of Kenya is distinguished by the Zanzibar–Inhambane ecoregion. The region stretches over the entire 536 km of the country's coastline with a variable width between max. ca. 100 km and min. ca. 50 km. 3 Mha is characterised by this ecoregion that corresponds to 5% of the total country area. Rainfall is mostly between 800 and 1200 mm per year with well-defined but short and mild dry seasons, since relative humidity is high and no month is absolute dry (White, 1983). The region as a whole stretches from the coast of southern Somalia through a relative narrow coastal belt along Kenya and Tanzania and terminates almost at the southern end of the Mozambique coast. The typical climax vegetation of the Zanzibar–Inhambane ecoregion is forest, but this has been largely replaced by cultivated areas and secondary grasslands and woodlands and shrublands (White, 1983; FAO, 2000). Indeed, from about Malindi to the southern end of the ecoregion in Kenya, the dominant

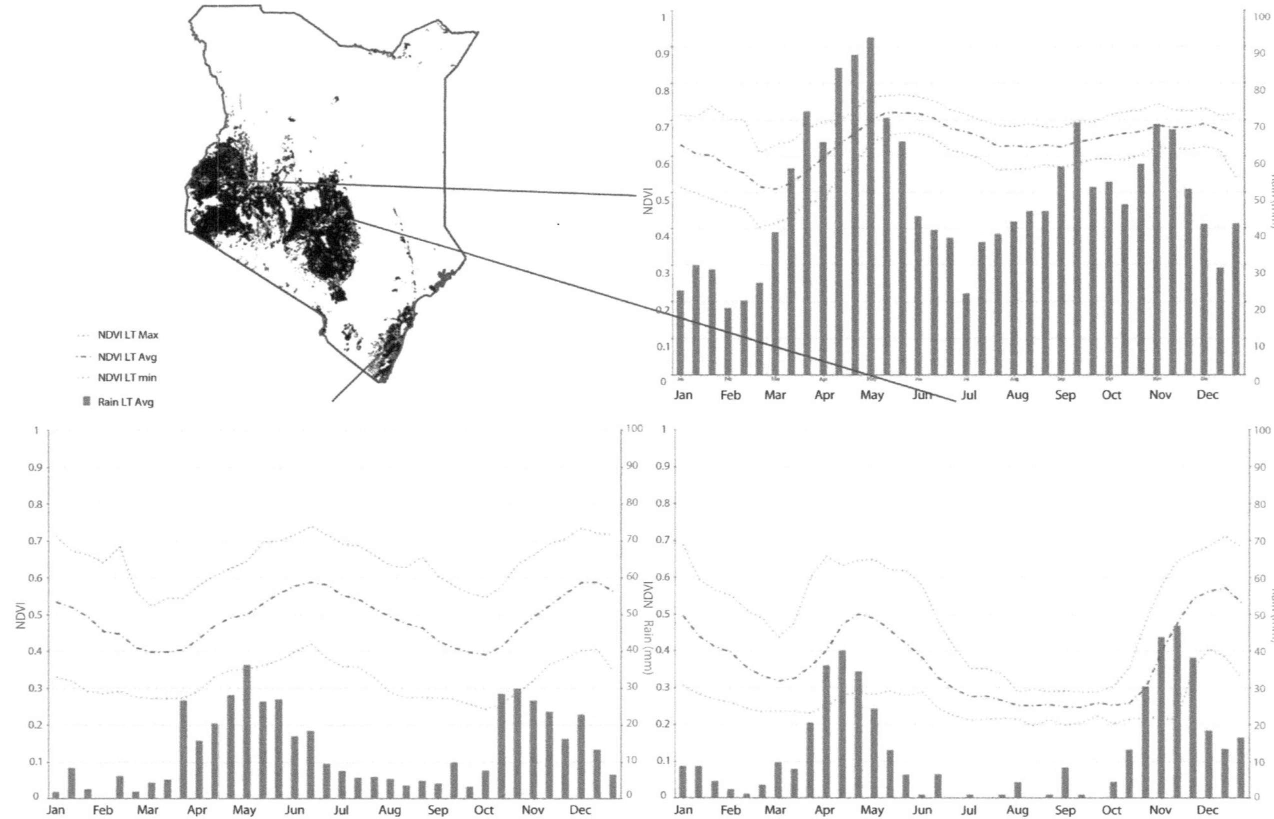

FIGURE 5 Agriculture distribution in Kenya (in black on top left map) and three examples of characteristic vegetation growing seasons. (For colour version of this figure, the reader is referred to the online version of this chapter.)

FIGURE 6 Mangrove forest along Kenya's coast. *Source: Kirui.* (For colour version of this figure, the reader is referred to the online version of this chapter.)

land cover type is agriculture (see Figure 5) land including rain-fed herbaceous crops and tree and shrub crops and forest plantations (FAO, 2000), covering nearly 30% of the ecoregion and almost 10% of the total agriculture area in Kenya. The few remaining closed tree cover can be found in the isolated and disconnected protected areas of Arabuko–Sokoke and Shimba Hills and to some extent in the central and northern part of the region, although tree cover is becoming more open and fragmented in this area. The closed forests that dominated the landscape in the past now represent less than 8% of the ecoregion (excluding mangrove forest) (FAO, 2000); nevertheless, they still contribute 18% to the total closed forest areas in the country. Woodlands and shrublands are predominant in the central and northern part and to some extent in the far southwestern end of the region, covering the largest proportion of the Zanzibar–Inhambane ecoregion in Kenya (nearly 60%). As common along many tropical and subtropical coastlines, mangrove forests are widespread also along Kenya's bays, creeks, estuaries and river deltas (Kirui et al., 2012). A recent study from Kirui et al. (2012) estimates the total mangrove forest area for the year 2000 at 46,930 ha. This corresponds to 1.5% of the ecoregion (Figure 6).

1.4 Lake Victoria Ecoregion

At the southwestern fringe of the country lies the Lake Victoria regional mosaic ecoregion. White (1983) describes this region as complex in its physiography with associated steep climatic gradients with annual rainfall ranging from 700 to 2000 mm. Locally, the high and well-distributed rainfall throughout the year

FIGURE 7 Intensive wheat production as it can be found on the highlands and the Lake Victoria region. (For colour version of this figure, the reader is referred to the online version of this chapter.)

is sufficient to support rain forest—well visible in Figure 9, bottom left graph—elsewhere semievergreen forest and woodland and shrubland represent the climax. The entire ecoregion includes most of Uganda, the whole of eastern Rwanda and Burundi and small parts of Kenya, Tanzania and the Democratic Republic of Congo (DRC). The Kenyan part of the Lake Victoria region covers an area of just above 2.7 Mha, which represents almost 5% of the country. The Lake Victoria itself comprises nearly 15% of the ecoregion in the country, whereas the great majority of the region is covered by rain-fed and to some extent irrigated cropland, adding up to 78% of the area in the ecoregion (FAO, 2000). Indeed, the Lake Victoria region in Kenya has the highest proportion of cropland in the country, representing more than 20% of the total land under cultivation in Kenya and also the highest cropland intensity along with some of the highland areas in central and western Kenya (Figure 7).

Intensively produced crops include wheat, tea, sugarcane, irrigated rice and high-yielding maize (WRI, 2007). The few remaining but important forest areas in the region are concentrated mainly in the Kakamega Forest Reserve and to some extent around the Mount Elgon area, covering less than 2% of the ecoregion. The 4% of woodlands and shrublands, which characterise the ecoregion, are concentrated in and around the Ruma National Park, in the far southeastern part of the region and to some extent in the extreme north of the Lake Victoria region of Kenya (FAO, 2000) (Figure 8).

1.5 Afromontane Ecoregion

The fourth ecoregion in Kenya is represented by the Afromontane region. This zone is an archipelago-like centre of endemism (White, 1983), which

FIGURE 8 Mountain forest in the Aberdare Range. (For colour version of this figure, the reader is referred to the online version of this chapter.)

is common in tropical Africa on elevations above 2000 m. Depending on climate and latitude, this limit can decrease to elevations around 1200 m and in a few cases even to just a few hundred metres above sea level as in the Cape Region in South Africa. Physiography and climate of this island type of formations are extremely varied, but in general, mean annual rainfall is above 1000 mm in the forest zone, diminishing with increasing altitude outside the forest belt (White, 1983). In Kenya, the main Afromontane 'islands' are located in the central and western part of the country. Two additional small mountain areas are located at the southern border next to the Masai Mara National Park and the Mount Elgon region bordering Uganda. Finally, a small portion of the Kilimanjaro mountain zone is also part of Kenya's landscape. The Afromontane ecoregion covers the second largest proportion of the country, summing up to a total area of 4.3 Mha, which represents slightly more than 7% of the country. Kenya's highland and mountain zone is home to the country's key forest resources. The four largest forest blocks in the country are all situated in Kenya's Afromontane region and include the Mau Forest complex, the Mount Kenya forest area, the Aberdare Range and the Cherangani Hills. Altogether, they cover an area of about 740,000 ha, which represents 57% to the country's total forest area, although covering only 17% of the ecoregion (FAO, 2000).

Figure 9 shows the closed evergreen and partly semievergreen forest distribution over the country, highlighting also the fact that the rainfall pattern is very different in terms of both amount and temporal repartition, as shown in the three exemplary graphs of Figure 9. This impacts strongly the temporal behaviour of the forest. Abundant and nearly regular rainfall during the year is available in the five main 'water towers' of Kenya—Mount Kenya, Aberdare Range, Mau Complex, Cherangani Hills and Mt Elgon—these play a critical role as water catchments and provide most of the nation's water. The bottom left graph of Figure 9 is an example of this type of mountain forest. NDVI values are very high and nearly constant, showing a very small seasonality and interannual variability thanks to abundant rainfall all along the year. As seen before, moving away from the favoured rainfall areas in the western part of the country and on the highlands, the typical two rainy season and two dry season patterns become apparent. This is clearly observable, in particular, in the forest areas in the north of the country (top right graph of Figure 9), where NDVI values are lower compared to the mountain 'rain' forest and seasonality and especially interannual variability is well visible. In total, Kenya's closed forests cover only 2.2% of the land area, yet they provide crucial ecosystem services such as regulating the water catchment, providing a variety of wood and nontimber products, reducing soil erosion and functioning as carbon sinks and reservoirs (UNEP, 2009). The latest FAO Forest Resource Assessment report from 2010 (FAO, 2010) ascribes 6% of the country area to forest. This difference is due to the fact that in the earlier figure, only closed forests were taken into account, while FAO includes all land with tree crown cover of more than 10%. In general, the forest belt of the Afromontane region represents the lowermost vegetation zone, diminishing in stature reaching higher altitudes. Below the forest belt, a natural transition zone towards the Somalia–Masai phytochoria and the Lake Victoria vegetation zone in the west of the country is expected, but this has been replaced almost completely by intense agriculture. In fact, cropland is the dominant land cover type of the Afromontane region in Kenya covering nearly 60% of the ecoregion (FAO, 2000) and contributing with 25% to the total agriculture area of the country. Tea plantations are commonly found right to the edge of the forest zone, followed by intensively farmed land at lower altitudes (Figure 10).

An exception is given by the small Afromontane zone at the southern edge of the country, where natural vegetation such as closed tree cover followed by open tree cover and woodland and shrubland is the predominant land cover type.

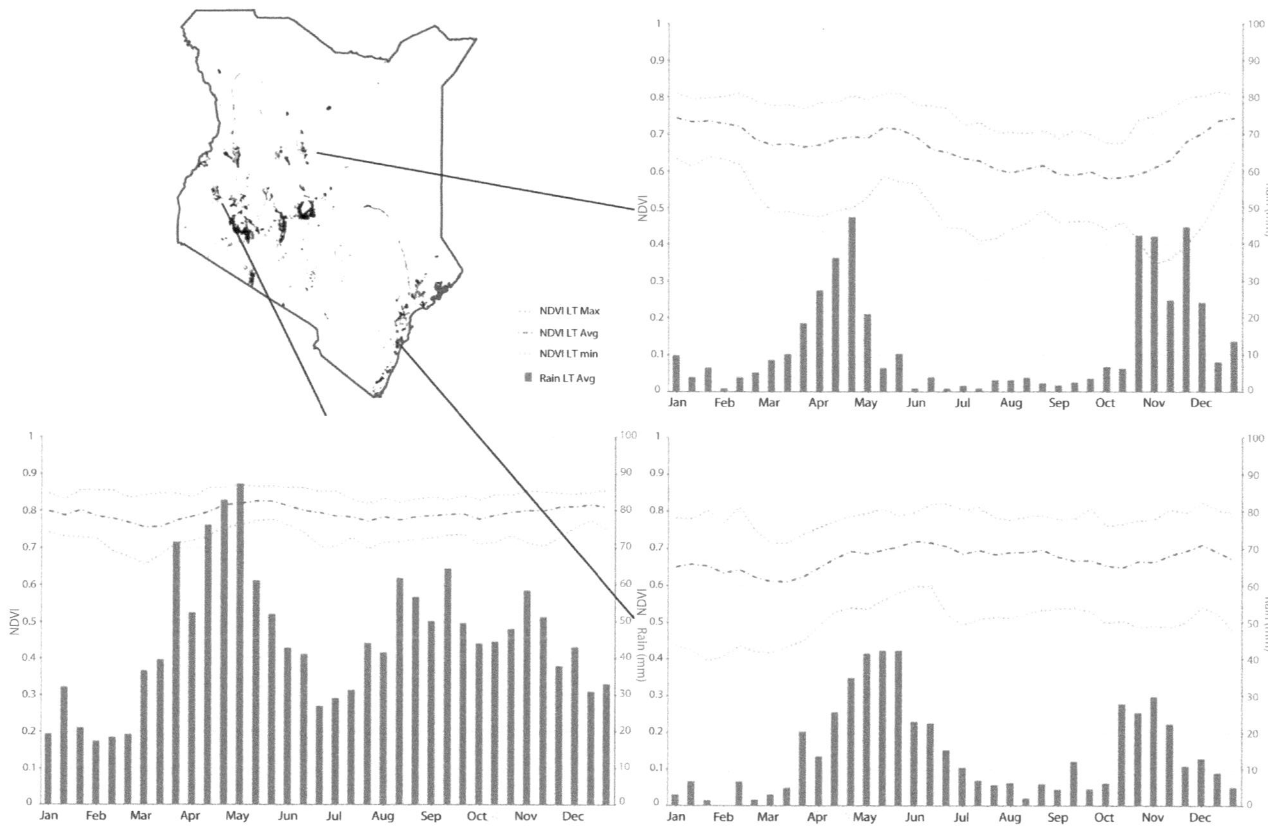

FIGURE 9 Closed trees distribution in Kenya (in black on top left map) and three examples of characteristic vegetation growing seasons. (For colour version of this figure, the reader is referred to the online version of this chapter.)

FIGURE 10 Typical example of agriculture landscape in the Afromontane region. Cash and high-value crops include tea, pineapple, banana and forest plantation are grown on the fertile highlands. (For colour version of this figure, the reader is referred to the online version of this chapter.)

2 LAND COVER CHANGE IN KENYA

The previous section characterised Kenya's ecoregions and its associated land cover distribution and highlighted the seasonality of vegetation dynamics over the yearly cycle. As opposed to the intra-annual vegetation variability, in this section, we assess permanent land cover changes in the country over a period of 20 years. Land cover and land cover change are one of the most evident and detectable indicators of land surface characteristics and associated human-induced and natural processes (GTOS, 2009). Hence, monitoring land cover changes over time provides important information on issues such as compliance with and success of land management schemes, predicting threats and pressure to biodiversity from shifting human populations and changes in agricultural practices, quantifying the fragmentation and loss of remnant habitats and calculating economic metrics such as carbon storage.

In this study, a systematic sampling of high-resolution Landsat and DMC imagery is used to assess and quantify land cover dynamics in Kenya. The method is part of the global TREES-3 project developed by the Joint Research Centre of the European Commission in collaboration and support to the FAO FRA 2010 Remote Sensing Survey (FAO, 2010) to monitor tree cover and its changes over the tropics (Achard et al., 2009). A systematic sample of 189 satellite imagery subsets (Landsat and DMC) was taken at each half degree of the geographic latitude and longitude rectilinear grid, covering about 13% of the total country area (Figure 11). The samples, 20 km × 20 km in

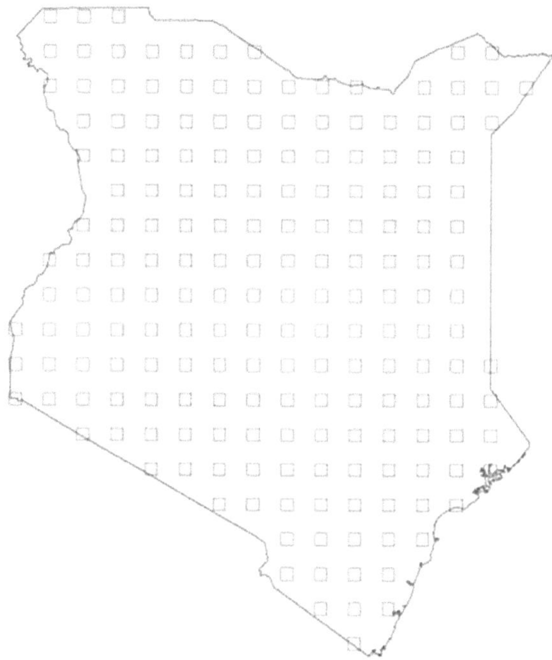

FIGURE 11 The systematic 20 × 20 km sampling framework over the study area.

TABLE 1 Land Cover Classes and Its Definition

Class Name	Definition
Dense tree cover	≥70% tree cover portion in segment
Open tree cover	30–70% tree cover portion
Other wooded land	≥70% shrubs and forest regrowth
Other land cover	Including croplands, herbaceous and bare lands
Inland water	Permanent water

size, were analysed for three reference years: 1990, 2000 and 2010. At each sample site and for all years, dense tree cover, open tree cover, other wooded land and other vegetation cover were identified from the analysis of satellite imagery (see class definition in Table 1).

The methodology is based on multidate image segmentation and automatic classification steps followed by visual control by national forestry experts. A detailed description of the whole methodology is described in the papers of Beuchle et al. (2011), Bodart et al. (2011) and Raši et al. (2011).

3 RESULTS

According to our study, Kenya has lost 739,558 ha of natural vegetation over the 1990–2000–2010 period including tree cover and other wooded land. This equals to nearly 7% of forest and other wooded land converted into other vegetation cover (agriculture, grassland including pasture, barren and artificial surfaces). More than 80% of this loss is attributed to the change of other wooded land towards other land cover, and almost 20% of the figure is related to the loss of tree cover to other wooded land and other vegetation cover. However, the highest proportional changes are detected in the tree cover class, since from the original 2,563,390 ha tree cover identified in 1990, only 2,484,819 ha of this land cover type remains in the year 2010. This equals to a total loss of over 120,000 ha or 5% of Kenya's forest cover over the analysed 20-year period. The left part of Figure 12 in the succeeding text shows that this reduction is nearly constant over the entire time period, increasing just slightly from a total deforestation rate of 2.3% in the 1990–2000 period to 2.6% over the 2000–2010 epoch. Assuming a linear trend over time, the net annual deforestation rate amounts to 0.23% for the 1990–2000 time frame and 0.26% for the following 2000–2010 interval. Evaluating the total annual tree cover changes in Table 2 shows that the gross annual tree cover loss is counterbalanced by a relatively high gross annual gain in tree cover. Indeed, nearly 80% of the gross forest cover lost in the 1990–2000 period and about 70% of the forest cover lost in the next 2000–2010 time frame have recovered or regrown, respectively.

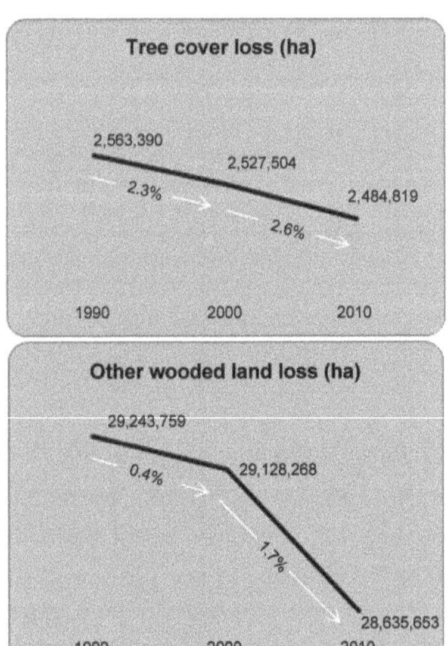

FIGURE 12 Tree cover loss and loss in other wooded lands in hectares for the 1990–2000–2010 time interval and percentage of total change.

TABLE 2 Net and Gross Annual Percentage Changes Between 1990–2000 and 2000–2010

	1900–2000	2000–2010
Tree cover —*net* annual loss %	0.23	0.26
Tree cover —*gross* annual loss %	1.03	0.88
Tree cover—*gross* annual gain %	0.81	0.62
OWL—*net* annual loss %	0.04	0.17
OWL—*gross* annual loss %	0.20	0.14
OWL—*gross* annual gain %	0.16	0.005

As mentioned previously, from an area point of view, the highest total changes in land cover in the country were detected in the woodland and shrubland category. Nearly 620,000 ha of other wooded lands were converted into other lands over the 20-year time span. As illustrated in the right section of Figure 12, in total, 2.1% of other wooded lands were lost over the entire period, but in contrast to a nearly stable deforestation trend, the conversion and degradation of other wooded land towards other land have increased considerably from 0.4% in the 1990–2000 epoch to 1.7% in the subsequent 2000–2010 period. Distributing the percentage change rates linearly over the two time frames results into a net annual loss of other wooded land of 0.04% and 0.17% for the 1900–2000 and 2000–2010 intervals, respectively. The gross percentage annual change figures as shown in Table 2 highlight again a strong pattern of dynamism in both directions (loss and gain) for the 1990–2000 period where more than 80% of the converted other wooded land coverage into other land cover has regrown again to other wooded land. However, the 2000–2010 time interval indicates a dramatic reduction in the regrowth of other wooded land, since the gross annual percentage gain is close to zero, and therefore, only about 4% of other land cover has been reconverted into other wooded land.

Looking at the geographic distribution of the land cover changes, we note a strong bias towards the southern part of the country (Figure 13). Strongest loss of natural vegetation can be identified in particular not only at the fringes but also inside of the Afromontane ecoregion—here, the loss is associated mostly with deforestation—and the central and southern portion of the Somalia–Masai ecoregion where conversion of natural vegetation to other land cover is predominantly related to the loss of woodland and shrubland. The Zanzibar–Inhambane ecoregion exhibits a mixture of both tree cover and other wooded land loss. Little or no loss of natural vegetation is detected

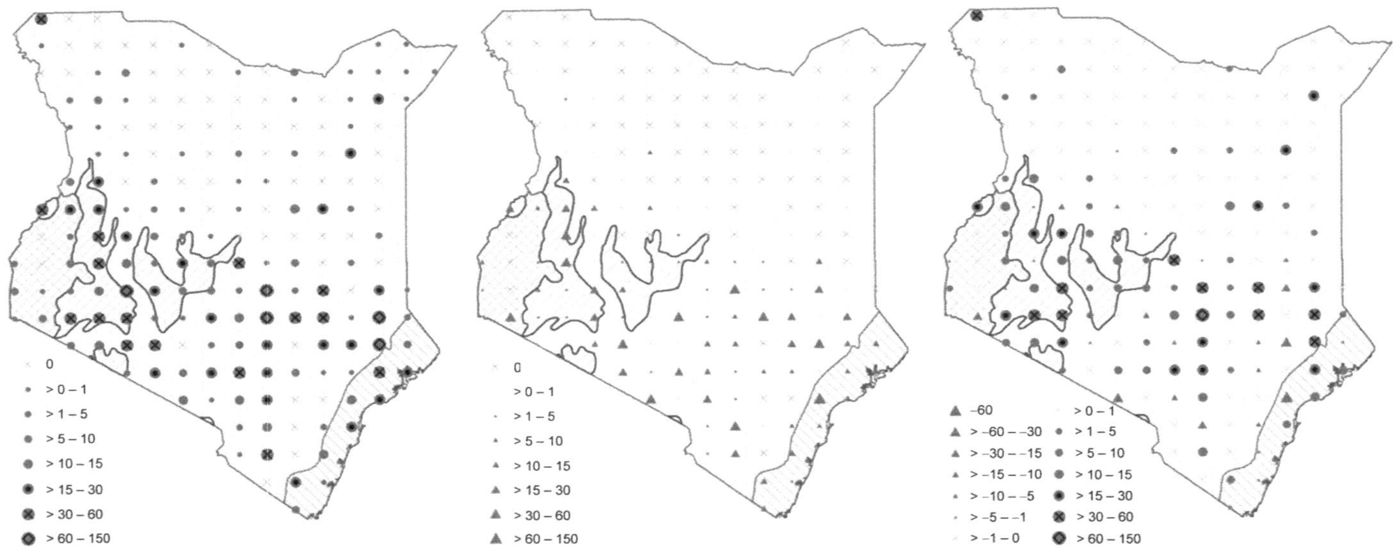

FIGURE 13 Net cumulative changes in natural vegetation (tree cover and other wooded land) from 1990 to 2010, left image showing total gross loss, central images showing total gross gain and right image representing total net changes. Eight categories from smaller to bigger circles (loss) and triangles (gain): [0], [>0–1], [>1–5], [>5–10], [>10–15], [>15–30], [>30–60] and [>60–150]. Figures are in square kilometre. For ecoregion description, see Figure 1.

in the Lake Victoria ecoregion and in the central northern and northern part of the country (left image of Figure 13). Natural vegetation regrowth is most apparent in the woodland and shrubland domain of not only the central southern section of the Somalia–Masai ecoregion but also to some extent in the Zanzibar–Inhambane and in the western part of the Afromontane ecoregion (central image on Figure 13).

4 DISCUSSION

Making use of a regular sample of high-resolution Landsat and DMC data, we were able to report unique figures on major land cover changes occurring in Kenya between 1900–2000 and 2000–2010. We estimate the loss of natural vegetation to be just about 740,000 ha over the observed 20-year period—almost 620,000 ha of loss in other wooded land and over 120,000 ha of tree cover loss—this is more than four times the size of the Masai Mara National Park.

The underlying human-driving forces of the previously described land cover changes are an interlinkage of three key variables (according to Geist and Lambin, 2002): agriculture expansion, wood extraction and infrastructure development. These are a direct response to the demand generated by a fast-growing population in the country. Kenya has an annual population growth rate of 2.7% compared to a global average of 1.2% according to FAO statistics (FAOSTAT) and a rural population of nearly 80% although migration and increase in urban areas are strong, driving even more the need for food and energy mostly in the form of charcoal for cooking. This increasing pressure on natural resources is well described in Figure 14, where the population distribution, density and connectivity are strongly correlated to the vegetation dynamics and in particular the loss of forest and other wooded land. Figure 14 demonstrates that the largest losses of natural vegetation are located at the border of the most populated areas in the central and western part of the country, highlighting the growing population expansion and hence the need for wood as construction material and energy source and more land for agriculture production. In fact, the population distribution is highly correlated to the expansion of agriculture areas (see Figure 5).

Despite the steady deforestation trend, the importance and value Kenya's forests play in supporting the economy, environment and human well-being has been highly recognised (UNEP 2012). In particular, the montane forest of Kenya—the so-called water towers—provides important ecosystem services far beyond the harvest of timber and fuelwood. According to a recent report published by UNEP (UNEP 2012), the cumulative negative effect of deforestation on the economy through reduction in regulating services was an estimated KSh 3652 million/year, more than 2.8 times the cash revenue of deforestation. However, while much attention has been given in recent years to the country's water towers, far less interest has been given to the vast tree and wood savanna type of landscape suffering from increased pressure

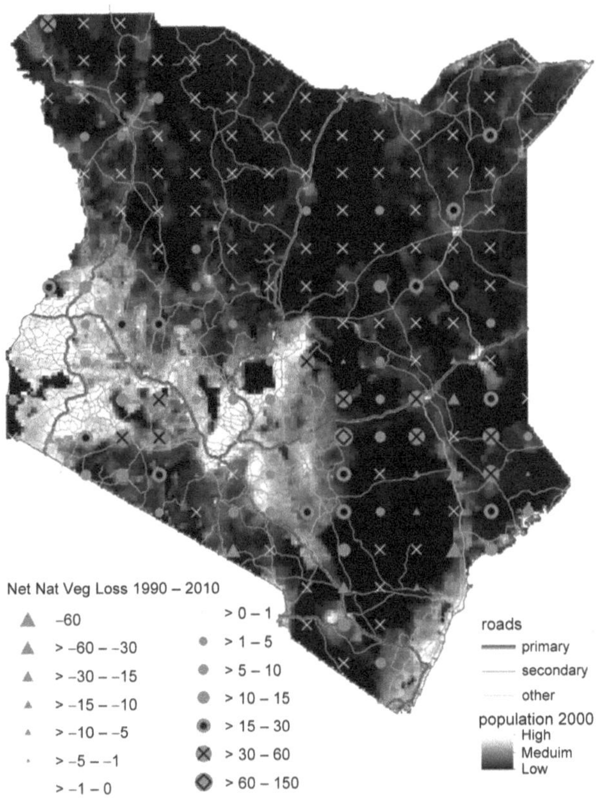

FIGURE 14 Net natural vegetation loss 1990–2010 overlaid over the road network and the population data for the year 2000.

caused by agriculture expansion and even more by harvesting and degradation due to wood fuel and charcoal production. Indeed, some studies show that the contribution of land cover change caused by wood fuel and charcoal production can be considerably higher compared to the conversion of natural land to agriculture (Malimbwi et al., 2001). This means also that forest and woodland degradation for firewood and charcoal production is not only a by-product of bigger forces such as logging for timber and agriculture expansion but also a specific need to satisfy energy demand. In Kenya, about 70% of the population relies on wood fuel as their main source of energy and the charcoal industry in the country employs about 200,000 in production alone, explaining the significant importance this topic has on the use and degradation of natural resources (Mugo and Ong, 2006). Nevertheless, despite these facts, the wood fuel and charcoal 'industry' has been kept out of the formal economies of the country and attracts little attention by policymakers (Mugo and Ong, 2006). However, firewood and charcoal will remain the main cooking fuel for most people in the region's towns and villages for the foreseeable future because it is affordable. Hence, with the ongoing rapid urbanisation, the demand for charcoal will ever increase, putting increasing pressure on the already degraded forests and woodlands.

5 PROTECTED AREAS IN KENYA

Kenya is famous for its parks and nature reserves and the species that occur within them. These areas are often protected areas and, as such, have been identified according to international definitions. The definition of a protected area was produced by IUCN and states that a protected area is 'a clearly defined geographical space, recognised, dedicated and managed, through legal or other effective means, to achieve the long-term conservation of nature with associated ecosystem services and cultural values' (Dudley, 2008). The sites that conform to this definition together are the protected area network, which is composed of sites that benefit from some degree of legislative protection and forms the cornerstone of international site-based conservation (Chape et al., 2005). Protected areas differ in the reasons for designation, degree of protection afforded and management at sites, reflecting differences in approaches to and need for conservation management.

Protected areas that are designated by national governments can be divided on the basis of the reason for designation, access to people and management and placed in categories devised by IUCN. These definitions are used across the globe to produce an international catalogue of protected areas, the World Database of Protected Areas (WDPA) (IUCN and UNEP-WCMC, 2013). This holds information on the status and locations of protected areas and is the international standard, allowing comparisons to be made within and between countries. In applying the categories system, the first step is to determine whether a site meets the definition of a protected area given in the preceding text and then, if it does, to decide on the most suitable category. In some cases, areas of land may not meet the criteria for these categories. In others, it may not yet have been assessed to fit any of the categories. Some protected areas are identified using international and regional criteria, including UNESCO MAB Biosphere Reserves, Wetlands of International Importance (Ramsar) or World Heritage sites.

According to the WDPA, and as such may be an underestimate, there are some 279 national terrestrial protected areas in Kenya (Figure 15), covering 69,100 km^2 (11.8% of land). There are also 11 internationally designated areas in Kenya, identified variously as UNESCO MAB Biosphere Reserves (four sites), Wetlands of International Importance (five Ramsar sites) or World Heritage sites (two) that together cover 13,345 km^2. Of the sites designated by national governments, the first forest reserves were designated back in 1932, but it was not until 1946 that Nairobi was designated as the first national park, which is the first site that fits with any of the IUCN protected

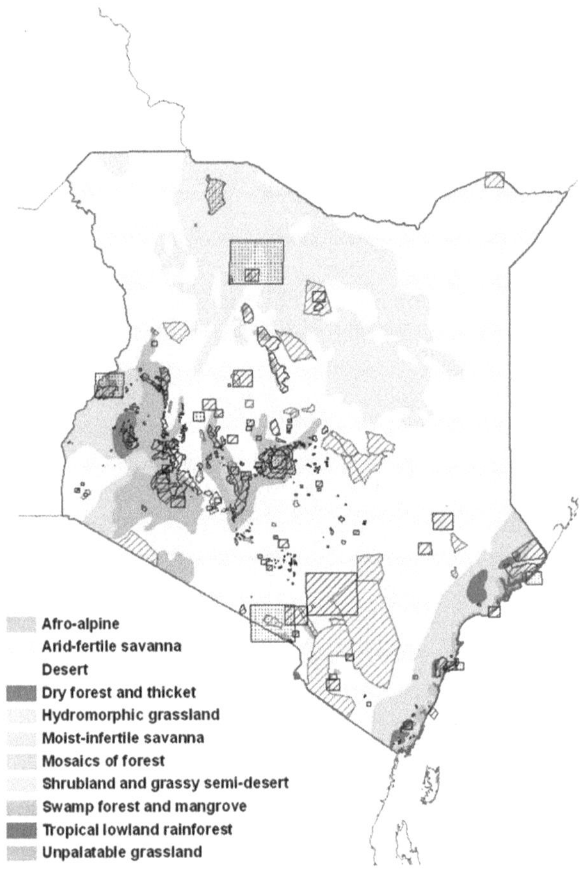

FIGURE 15 Distribution of protected areas in Kenya, overlaid on White's Africa vegetation map (White, 1983). International protected areas are stippled, national uncategorised protected areas are black hatching and type II are blue, IV purple and VI red hatch. (For interpretation of the references to colour in this figure legend, the reader is referred to the online version of this chapter.)

area categories. The area of land under protection increased steadily through forest and private reserves until 1946, after which there was a rapid increase in the area protected. There was another rapid increase in the area protected in the 1970s, and most of the land under protection is within types I–VI (42,100 km^2 or 7.2% of land, Figure 15). Twelve marine protected areas cover 10.5% of sea area. Kenya is a signatory of the Convention on Biological Diversity, the broad objectives of which include the conservation of biological diversity, the sustainable use of its components and the fair and equitable sharing of the benefits arising out of the utilisation of genetic resources. One component of the convention required signatories to protect at least 10% of their land area by 2010. Kenya exceeded this requirement, but under the targets for 2020, it is expected to increase the coverage of protected areas to 17% of the land area. Expansion of the network to areas of objectively defined biodiversity, such as Important Bird Areas (IBAs), could help address the poor coverage of many globally threatened species in Africa by the PA network (Beresford et al., 2011).

World-renowned parks such as the Masai Mara, Amboseli or the Mount Kenya National Parks (just to name a few) not only have an important role in the conservation of nature but also provide also a major contribution to the country's GDP. Tourism generated $700 million in Kenya in 2005, with about 70% of visitors to Kenya citing nature-based activities as the reason for their trip (World Resources Institute, 2007). In 2005, Kenya's parks and reserves were visited by 2.1 million people, and more than 80 of Kenya's top 120 tourist destinations are national parks and wildlife reserves (GoK, 1995). The visitors are not spread evenly though, with the most popular parks receiving 100,000 and 350,000 visits per year, while other sites that are similarly rich in wildlife receive fewer than 10,000 visits annually (KWS 2005 in World Resources Institute, 2007). The pressure on protected areas that these visitors generate can be significant, with vegetation being degraded in some game parks, wildlife behaviour altered and resources have been overused (Ikiara and Okech, 2002). Appropriate management of resources, including protected areas, is needed if tourism is to be maintained. Directing the proceeds of tourism to local people and communities is also seen by some as critical as is developing a strategy that sees visitors spread more evenly across the country (World Resources Institute, 2007).

6 LAND COVER ON PROTECTED AREAS

Kenya's protected areas are distributed within the four ecoregions described at the beginning of the chapter (Figure 1), and hence, also its vegetation structure mirrors roughly the general climatic and vegetation pattern of the respective ecoregion. Land cover within protected areas was quantified by overlaying the WDPA (IUCN and UNEP-WCMC, 2013) boundaries onto the 30 m resolution Africover map of Kenya (FAO, 2000). Considering just the nationally designated protected areas (not least to avoid the overlap in boundaries between these and internationally designated protected areas), shrub is the most widespread land cover type, covering 45% of the nationally designated protected area (Figure 16). Grassland/herbaceous is next, then woodland and forest combined would be about as prevalent as shrub, but the majority of both these land cover types is found in protected areas that have not been assigned an IUCN category.

6.1 Monitoring

Although protected areas are, by definition if not in practice, legally protected, monitoring is key to effective biodiversity conservation as it allows problems to be identified, solutions to be developed and their effectiveness to be assessed. Not only can monitoring identify problems and help develop

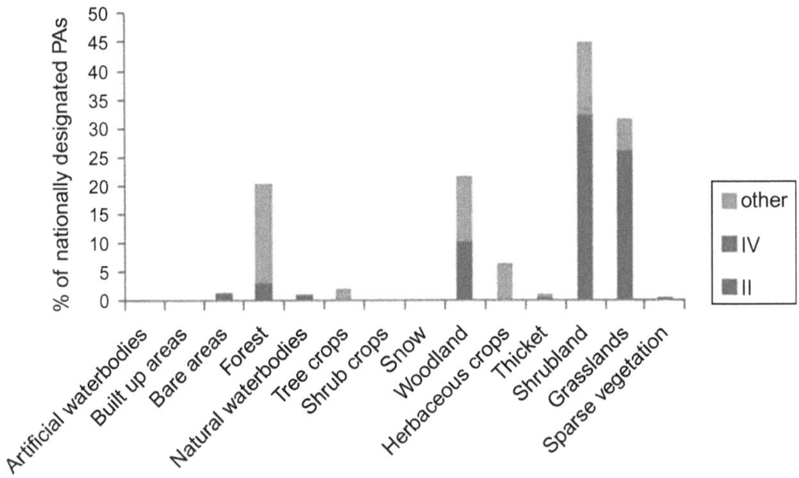

FIGURE 16 Percentage composition of land cover within protected areas in Kenya, divided into those with IUCN categories (II and IV, not I or III) and uncategorised PAs (others), based on Africover (FAO, 2000). (For colour version of this figure, the reader is referred to the online version of this chapter.)

solutions, but it is also a requirement of the CBD. The information required covers the *state* of the park (e.g. information on the populations of species within parks and the vegetation composition of parks) and the *pressure* on the park (e.g. from logging or hunting) such that the information can be combined to produce a *response* to any problems (Balmford et al., 2005). Monitoring needs to be repeated frequently if degradation of sites is to be detected in time in order to permit remedial action.

Producing a system that produces objective assessments of all that are comparable not only within Kenya but also more widely across Africa will maintain objectivity in assessments allowing national and regional prioritisations. The Digital Observatory for Protected Areas (http://dopa.jrc.ec.europa.eu/) provides a mechanism for producing comparable information on the state of, and pressures to, protected areas in Kenya and more widely across Africa, assessing as it does each protected area on a number of criteria (Hartley et al., 2007). Combining these indices can help prioritise conservation actions (response).

The assessment of state in DOPA is based on vegetation (land cover and bioclimatic characteristics) and species of vertebrates that may be present. The irreplaceability of vegetation is mapped based on nine variables including percentage composition of broad land cover, vegetation phenology, topography and water availability, primarily assessed from remote sensing data. It identifies, on a continuous scale, the similarity of all areas within the same ecoregion (White, 1983) to the conditions found in the protected area. This can be used as a measure of the uniqueness of the park. In the north and east of the country, the PAs were more similar to the arid habitats of Ethiopia and other parts of the Horn of Africa (e.g. Sibioli, Figure 17a), and in the west

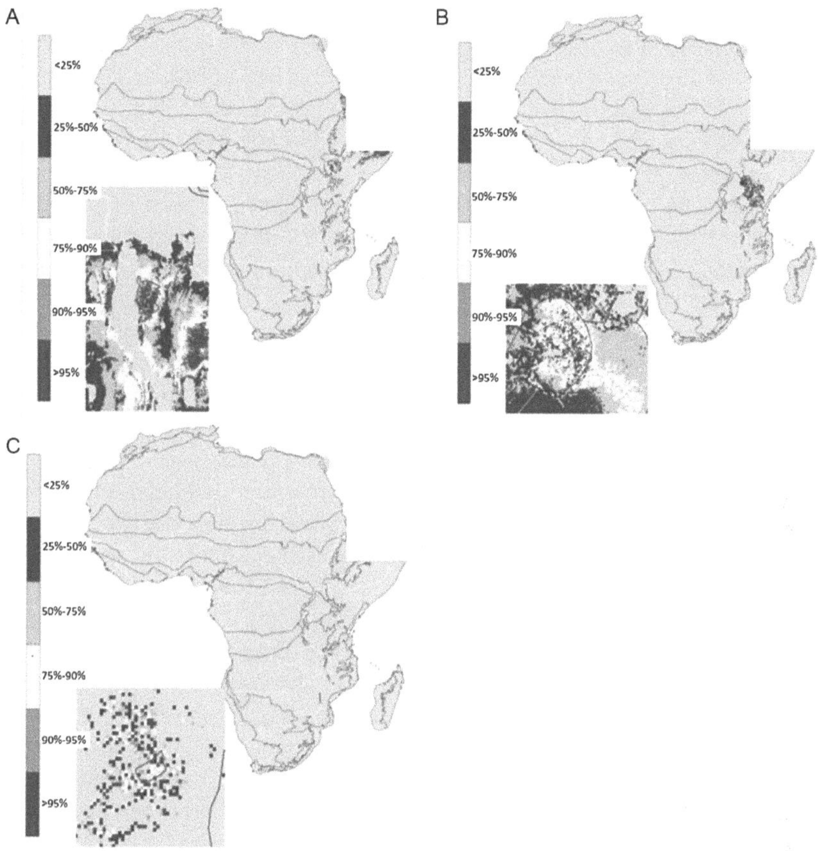

FIGURE 17 Irreplaceability of PAs based on nine characteristics, with greener colours indicating greater similarity for (a) Sibioli, (b) Mount Elgon and (c) Kakamega. (For colour version of this figure, the reader is referred to the online version of this chapter.)

(e.g. Mount Elgon, Figure 17b), they shared characteristics with western Kenya and Eastern Uganda. However, Kakamega in the west had a high irreplaceability score and little similar habitat, suggesting conditions there are not found in many other parts of Kenya (Figure 17c). In addition to habitat uniqueness, state is also assessed by species irreplaceability. The distribution of species of vertebrates across sites cannot be readily assessed from satellite data and would ideally be done from field data (Rodrigues, 2011). Unfortunately, field data are not available from all areas, and so species extent of occurrence maps was used to overlap with the PA boundaries for mammals (Boitani et al., 1999) and amphibians (IUCN, 2006) and known occurrences in globally threatened birds. The limitations of species occurrence maps, such as commission errors, are recognised (Rodrigues et al., 2004b; Beresford et al., 2011). However, they are the best available data, and over time, refinements can be made to these maps as species distributions are better understood, or data from other sources such as online databases become available. For birds, the distribution of globally threatened species in IBAs that overlapped PAs was noted, based on information from BirdLife (www.birdlife.org). Irreplaceability was calculated as the inverse of the number of other PAs that the species range was mapped as overlapping. Pressures on PAs are also important in determining conservation priorities, and so assessments of these were made using continent-wide data on human population densities, road networks and populated places. This allowed continent- and country-wide comparisons to be produced. These are crude again but are perhaps the best available data for regional assessments. Combining these indicators enabled coarse assessments of the state of and pressures to protected areas in Africa. They suggested that Nairobi National Park, Lake Nakuru, Aberdares, and Kakamega are among PAs in Kenya under most pressure.

By its very nature, broad-scale assessments will generally be crude and not always identify subtle changes in the state of or pressures to PAs. This can best be achieved through dedicated site-by-site monitoring. Methods are available for monitoring protected areas (http://www.wdpa.org/ME/), but as yet, Kenya-wide assessments are not available. Other site-based monitoring approaches are also available, and in Kenya, the IBA network has been monitored for a number of years. Using teams of local stakeholders (Site Support Groups), trained in the collection of basic field data, trends in the state of and pressures to Kenya's IBAs between 1999 and 2005 have been assessed (Mwangi et al., 2010), and monitoring is still ongoing. Of the 36 IBAs considered, 20 were entirely or largely within national protected areas, including 6 national parks, 1 game reserve, 5 forest reserves, 1 national monument and 7 mixed, including ungazetted forest. The condition of protected IBAs was found to have improved during the study period, while the condition of the unprotected IBAs had deteriorated, although the scores for the eventual condition were not actually significantly different. The response to pressures was better on protected IBAs, as might be expected, with a doubling of the number of comprehensive and up-to-date management plans in existence and a 25% increase in the level of conservation management being undertaken on these protected IBAs over the length of the study. This compares to just one comprehensive management plan, and no action, on unprotected IBAs. The pressures on the protected IBAs were slightly but not significantly lower than the pressures on the unprotected IBAs. Together, these results indicate both the value of monitoring and also that designation as protected areas apparently improves the overall condition of sites of recognised biodiversity importance or value.

Such dedicated field surveys (although ideal) are not always possible due to lack of resources or remoteness of sites or even political instability. In these circumstances, a simple remote sensing-based approach could potentially provide information on the condition of land cover on sites (Buchanan et al., 2009). One of the simplest approaches is tracking changes in land cover, but this only indicates changes in vegetation and does not track threats such as hunting or slow vegetation degradation.

A dedicated analysis of land cover using visual assessment (Bastin et al., 2013; Beresford et al., 2013) was trialled on 14 protected IBAs in Kenya, tracking land cover between around 1990 and 2008. These represented a diverse range of land cover types, from mainly forest (Mau Forest) through a mix of forest and shrub (Aberdares) to a mix of sparse vegetation and herbaceous and shrub cover (Amboseli) (Figure 17). By the time of the repeat survey, the percentage cover of natural land cover decreased on average by 0.012% of the park per annum (Table 3), although in Mau Forest,

TABLE 3 Changes (per Annum) in the Composition of Land Cover on Protected Areas in Kenya

	Forest	Herb/ Shrub	Sparse/ Bare	Fresh Water	Agric/ Urban
Aberdare Mountains	–0.05	–0.02	0.00	0.00	0.07
Amboseli National Park	–0.04	0.08	–0.17	0.13	0.00
Cherangani Hills	–0.16	0.16	0.00	0.00	0.00
Chyulu Hills Forests	–0.15	0.15	0.00	0.00	0.00
Kakamega Forest	0.34	0.19	0.00	0.00	–0.53
Kikuyu Escarpment	0.12	0.29	0.00	0.00	–0.41
Lake Naivasha	0.00	0.12	0.00	–0.18	0.06
Mau Forest Complex	–0.41	0.01	0.00	0.00	0.40
Meru National Park	–0.06	0.06	0.00	0.03	–0.03
Mt Elgon	–0.08	0.02	0.00	0.00	0.06
Mt Kenya	–0.10	0.02	0.00	0.00	0.08
Nairobi National Park	0.00	–0.19	0.00	0.00	0.19
Samburu and Buffalo	0.00	0.00	0.00	0.00	0.00
South Nandi Forest	–0.28	0.00	0.00	0.00	0.28

Negative values indicate a decrease in the percentage cover of each land cover and positive values indicate an increase.

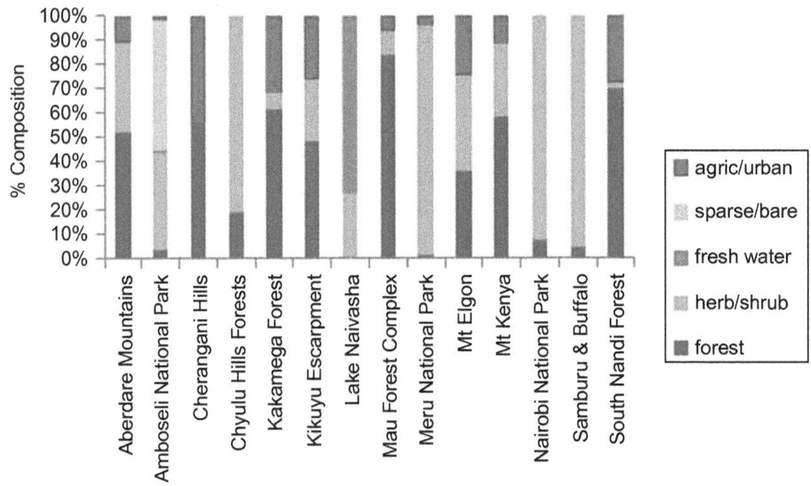

FIGURE 18 Land cover on Kenyan protected areas (assessed by sample-based visual interpretation of satellite images) in around 2008. (For colour version of this figure, the reader is referred to the online version of this chapter.)

the extent of agriculture increased by 0.4% of the park per annum (although this area was only designated as a park in 2001, halfway through the study period). In addition to an overall increase in agriculture, the percentage composition of forest decreased by on average 0.06% as it was converted to both agriculture and herbaceous and shrub cover. This was particularly noticeable in Meru National Park. The loss of natural vegetation, whether deforestation or increase in agriculture, was lower than the 0.7% loss per annum rate estimated over a slightly earlier time period from across Africa (Brink and Eva, 2009). There is evidence that protection or management is having a beneficial effect in some parks, with the extent of agriculture in Kakamega decreasing from 31% to just 20% between 1987 and 2006 (equating to −0.5% per annum). However, despite these notable successes, the general picture suggests that the protection of parks is not complete and that regulations need to be enforced more rigorously if the natural vegetation within Kenya's protected areas is to be conserved (Figure 18).

REFERENCES

Achard, F., Beuchle, R., Bodart, C., Brink, A.B., Carboni, S., Eva, H.D., Mayaux, P., Raši, R., Simonetti, D., Stibig, H.-J., Achard, F., Beuchle, R., Bodart, C., Brink, A.B., Carboni, S., Eva, H.D., Mayaux, P., Raši, R., Simonetti, D., Stibig, H.-J., 2009. Monitoring Forest Cover at global scale: the JRC approach. In: Proceedings of the 33rd ISRSE Conference, p. 4.

AMESD Newsletter No. 7, May to September 2011. For download at: http://au.int/amesd/ (accessed on 22. 11. 2012).

Balmford, A., Bennun, L., ten Brink, B., Cooper, D., Côté, I.M., Crane, P., Dobson, A., Dudley, N., Dutton, I., Green, R.E., Gregory, R.D., Harrison, J., Kennedy, E.T., Kremen, C., Leader-Williams, N., Lovejoy, T.E., Mace, G., May, R., Mayaux, P., Morling, P., Phillips, J., Redford, K., Ricketts, T.H., Rodríguez, J.P., Sanjayan, M., Schei, P.J., Jaarsveld van, A.S., Walther, B.A., 2005. The convention on biological diversity's 2010 target. Science 307 (5707), 212–213, 14 January 2005.

Bastin, L., Buchanan, G., Beresford, A., Pekel, J.-F., Dubois, G., 2013. Open-source mapping and services for Web-based land-cover validation. Ecol. Inform. 14, March 2013, 9–16, ISSN 1574–9541.

Beresford, A., Buchanan, G., Donald, P., Butchart, S.H.M., Fishpool, L.D.C., Rondinini, C., 2011. Minding the protection gap: estimates of species' range size and holes in the protected area network. Anim. Conserv. 14, 114–116.

Beresford, A.E., Eshiamwata, G.W., Donald, P.F., Balmford, A., Bertzky, B., et al., 2013. Protection reduces loss of natural land-cover at sites of conservation importance across Africa. PLoS ONE 8 (5), e65370. http://dx.doi.org/10.1371/journal.pone.0065370.

Beuchle, R., Eva, H.D., Stibig, H.-J., Bodart, C., Brink, A., Mayaux, P., Johansson, D., Achard, F., Belward, A., 2011. A satellite data set for tropical forest area change assessment. Int. J. Remote Sens. 32, 7009–7031.

Bodart, C., Eva, H., Beuchle, R., Raši, R., Simonetti, D., Stibig, H.-J., Brink, A., Lindquist, E., Achard, F., 2011. Pre-processing of a sample of multi-scene and multi-date Landsat imagery used to monitor forest cover changes over the tropics. ISPRS J. Photogramm. Remote Sens. 66, 555–563.

Boitani, L., Corsi, F., de Biase, A., D'Inzillo Carranza, I., Ravagli, M., Reggiani, G., Sinibaldi, I., Tra-panese, P., 1999. African Mammals Databank: A Databank for the Conservation and Management of the African Mammals. Instituto di Ecologia Applicata and European Commission.

Brink, A.B., Eva, H.D., 2009. Monitoring 25 years of land cover change dynamics in Africa: a sample based remote sensing approach. Appl. Geogr. 29, 501–512.

Buchanan, G., Donald, P.F., Fishpool, L.D.C., Arinaitwe, J.A., Balman, M., Mayaux, P., 2009. An assessment of land cover and threats in Important Bird Areas in Africa. Bird Conserv. Int. 19, 49–61.

Chape, S., Harrison, J., Spalding, M., Lysenko, I., 2005. Measuring the extent and effectiveness of protected areas as an indicator for meeting global biodiversity targets. Phil. Trans. Biol. Sci. 360 (1454), (February 2005), 443–455, ISSN 1471-2970.

Dudley, N. (Ed.), 2008. Guidelines for Applying Protected Area Management Categories. IUCN, Gland, Switzerland, x + 86pp.

FAO, 2000. Africover Multipurpose Land Cover Databases for Kenya. FAO, Rome. www.africover.org.

FAO, 2010. Global Forest Resources Assessment 2010. Forestry Paper 163. Rome.

FAO, FAOSTAT-PopSTAT (http://faostat.fao.org/site/550/default.aspx#ancor).

Foley, J.A., Defries, R., Asner, G.P., Barford, C., Bonan, G., Carpenter, S.R., et al., 2005. Global consequences of land use changes. Science 309, 570–574.

Geist, H.J., Lambin, E., 2002. Proximate causes and underlying driving forces of tropical deforestation. BioScience 52, 143–150.

Gommes, R., Petrassi, F., 1996. Rainfall Variability and Drought in Sub-Saharan Africa Since 1960. FAO Agrometeorology Series 9, Food and Agriculture Organization of the United Nations.

Government of Kenya (GoK), 1995. The Study on the National Tourism Master Plan in the Republic of Kenya, vol. 1. GoK, Nairobi.

GTOS, 2009. Assessment of the Status of the Development of the Standards for the Terrestrial Essential Climate Variables. Global Terrestrial Observing System, Rome.

Hartley, A., Nelson, A., Mayaux, P., Gregoire, J.-M., 2007. The Assessment of African Protected Areas. JRC Scientific and Technical Reports. Office for Official Publications of the European Communities, Luxembourg, EUR 22780 EN.

Homer-Dixon, T.F., Blitt, J. (Eds.), 1998. Ecoviolence: Links among Environment, Population, and Security. Rowman and Littlefield Publishers Inc., New York.

Ikiara, M., Okech, C., 2002. Impact of Tourism on Environment in Kenya: Status and Policy. Kenya Institute for Public Policy Research and Analysis (KIPPRA) Discussion Paper No. 19. KIPPRA, Nairobi. Online at http://www.kippra.org/resources/publications.asp.

IUCN and UNEP-WCMC, 2013. The World Database on Protected Areas (WDPA). [On-line]. UNEP-WCMC, Cambridge, UK. Available at: www.protectedplanet.net (accessed 01.09.13.).

IUCN, Conservation International, and NatureServe, 2006. Global Amphibian Assessment.

Kirui, K.B., Kairo, J.G., Bosire, J., Viergever, K.M., Rudra, S., Huxham, M., Briers, R., 2012. Mapping of mangrove forest land cover change along the Kenya Coastline using Landsat imagery. Ocean Coast. Manage. In press, corrected proof.

Malimbwi, R.E., Misana, S., Monela, G., Jambiya, G., Nduwamungu, J., 2001. Charcoal Potential in Southern Africa. Sokoine University of Agriculture, Morogoro, Tanzania.

Mugo, F., Ong, C., 2006. Lessons of eastern Africa's unsustainable charcoal trade. ICRAF Working Paper No. 20. World Agroforestry Centre, Nairobi, Kenya.

Mwangi, M.A.K., Butchart, S.H.M., Munyekenye, F.B., Bennun, L.A., Evans, M.I., Fishpool, L.D.C., Kanyanya, E., Madindou, I., Machekele, J., Matiku, P., Mulwa, R., Ngari, A., Siele, J., Stattersfield, A.J., 2010. Tracking trends in key sites for biodiversity: a case study using Important Bird Areas in Kenya. Bird Conserv. Int. 20, 215–230.

Rodrigues, A.S.L., 2011. Improving coarse species distribution data for conservation planning in biodiversity-rich, data-poor, regions: no easy shortcuts. Anim. Conservat. 14 (2), 108–110, April 2011.

Rodrigues, A.S.L., Andelman, S.J., Bakarr, M.I., Boitani, L., Brooks, T.M., Cowling, R.M., Fishpool, L.D.C., da Fonseca, G.A.B., Gaston, K.J., Hoffmann, M., Long, J.S., Marquet, P.A., Pil-grim, J.D., Pressey, R.L., Schipper, J., Sechrest, W., Stuart, S.N., Under - Hill, L.G., Waller, R.W., Watts, M.E.J., Yan, X., 2004b. Effectiveness of the global protected area network in representing species diversity. Nature 428, 640.

Raši, R., Bodart, C., Stibig, H.-J., Eva, H., Beuchle, R., Carboni, S., Simonetti, D., Achard, F., 2011. An automated approach for segmenting and classifying a large sample of multi-date Landsat imagery for pan-tropical forest monitoring. Remote Sens. Environ. 115, 3659–3669.

UNEP, 2009. Kenya: Atlas of Our Changing Environment. Division of Early Warning and Assessment (DEWA) United Nations Environment Programme (UNEP), Nairobi, Kenya.

UNEP, 2012. The Role and Contribution of Montane Forests and Related Ecosystem Services to the Kenyan Economy. United Nations Environment Programme (UNEP), Nairobi, Kenya. February 2012.

White, F., 1983. Vegetation map of Africa. UNESCO, Paris.

World Resources Institute, Department of Resource Surveys and Remote Sensing, Ministry of Environment and Natural Resources, Kenya, Central Bureau of Statistics, Ministry of Planning and National Development, Kenya, International Livestock Research Institute, 2007. Nature's Benefits in Kenya: An Atlas of Ecosystems and Human Well-Being. World Resources Institute, Washington, DC and Nairobi.

Chapter 13

Agricultural Expansion and Its Consequences in the Taita Hills, Kenya

Petri K.E. Pellikka*, Barnaby J.F. Clark*, Alemu Gonsamo Gosa†, Nina Himberg‡, Pekka Hurskainen*, Eduardo Maeda*, James Mwang'ombe§, Loice M.A. Omoro‡ and Mika Siljander*

*Department of Geosciences and Geography, University of Helsinki, Helsinki, Finland

†Department of Geography and Planning, University of Toronto, Toronto, Canada

‡Faculty of Agriculture and Forestry, University of Helsinki, Helsinki, Finland

§Kenya Forest Service, Kenya

1 INTRODUCTION: LAND-COVER CHANGE IN SUB-SAHARAN AFRICA

Africa had lost 16% of its forests and 5% of its woodlands and grasslands over the period of 1975–2000, equating to over 50,000 km^2 per year of natural vegetation. Most of the land change is associated with agriculture: between the years 1975 and 2000, the agricultural areas increased by 57% in sub-Saharan Africa (Brink and Eva, 2009). While the increase of agricultural production on the other continents in the world is based on improving technology as intensified use fertilizers and machinery, in sub-Saharan Africa, the increase of food production is still highly dependent on clearing more land for agriculture. In the future, land grabbing is expected to cause even more intensified increase of croplands as African land is also demanded by non-African nations in order to secure their food security. Also, natural land-cover change may exist due to changes in precipitation pattern or fires (McMichael and Hope, 2007), which may foster human populations to seek arable land elsewhere or to change their farming practices or crops (Maro, 1988; Sunderlin et al., 2000).

The forest loss is especially hard on areas with high biodiversity and in the Afromontane areas, where the decrease is estimated to be 3.8% annually according to Eva et al. (2006). A significant part of Afromontane Africa is in Kenya. Besides the continental scale studies of land changes (Brink and Eva, 2009), many forest or land-cover-change studies in East Africa indicate

Developments in Earth Surface Processes, Vol. 16. http://dx.doi.org/10.1016/B978-0-444-59559-1.00013-X

intensified land use pressure, evidenced by the loss of forests or bushlands after the 1950s (Imbernon, 1999; Lung and Schaab, 2006; Baldyga et al., 2007; Clark and Pellikka, 2009; Pellikka et al., 2009). This study highlights the results of 10-year-long study of land-cover changes in the Taita Hills in southeastern Kenya.

2 STUDY AREA: THE TAITA HILLS ARE ISLANDS IN THE SEA OF SAVANNA

The verdant and humid Taita Hills offer a unique venue for land change studies as it is located on the dry Tsavo plains between the Tsavo East and West National Parks, constraining the human population and agricultural lands to expand naturally to the plains. The hills cover an area of 1000 km^2 and are surrounded by semiarid *Acacia*/*Commiphora* shrubland and dry savanna. While the surrounding plains are at an elevation of 600–900 masl, the Taita Hills rise abruptly in a series of ridges with the highest peak of Vuria at 2208 m, although the average elevation of the hills is 1500 m (Figure 1). The climate is influenced by the Intertropical Convergence Zone, which leads to a bimodal rainfall pattern, with long rains during March–June and short rains in October–December, although the annual variability of precipitation is high. Located 150 km from the Indian Ocean, orographic rainfall plays an important part in the local climate as the hills form the first significant barrier, which moisture-laden air encounters. Mist and cloud precipitation occurs throughout the year; consequently, while the annual rainfall is circa 600 mm on the plains, it is over 1200 mm in the hills (Beentje and Ndiang'ui, 1988). A rain shadow effect is discernable on the northwestern side of the hills, with the distinctive *Euphorbia candelabrum* and more commonly *Euphorbia bussei* var. *kibwezensis* growing in the drier conditions.

FIGURE 1 The plains in the front of the Taita Hills are at 950 masl in the southwest, while the highest peak Vuria in the middle reaches 2208 masl (Photograph: Petri Pellikka, 2012). (For colour version of this figure, the reader is referred to the online version of this chapter.)

In prehistorical times, the Taita Hills may have been covered by hundreds of square kilometres of indigenous rain forests, but due to its favourable climatic and edaphic conditions, the forests have been cleared for agriculture from long ago. Many authors take it for granted that 1% of the forests remain, but none seem to know how this 1% is derived or by who. The degradation and fragmentation of the forests was noted early (Hildebrandt, 1877), with observations that the slopes were cleared for agriculture up to the head of all streams. Large-scale decreases in forest cover were caused by railway construction and traffic in the early 20th century, and before the introduction of the black wattle (*Acacia mearnsii*) for the tannin industry, the overall tree cover may have been at its lowest. Exotic trees such as cypress (*Cupressus lusitanica*), blue gum (*Eucalyptus* spp.) and pines (*Pinus* ssp.) were introduced between the 1960s and 1980s and *Maesopsis eminii* and silver oak (*Grevillea robusta*) during the late 1970s and 1980s. While eucalyptus, pine and cypress were planted for wood production, silver oak was planted within the fields to support farmers with building material and to decrease the pressure on the indigenous forests. The conservation of indigenous forests began in 1977 by the presidential decree banning the cutting of indigenous forests without a licence (Himberg et al., 2009). Based on the visual analysis of the airborne remote sensing data of the 1950s and presently, as well as landscape photographs taken in 1924, 1957 and 1992 (Turner et al., 1998), the Taita Hills are characterized by small exotic plantations among intensive agriculture with terraced fields in the 1990s and 2000s, while during the 1950s, terracing was almost absent and a significant part of the land in the hills was used as pastures. Even though there were more indigenous bushland and small forest patches during the 1950s, there were practically no trees on farms.

The Taita Hills are also part of a very important biodiversity hotspot. It is the northernmost part of Africa's Eastern Arc Mountains, which have been identified as one of the top 10 biodiversity hotspots in the world, and with over 90% forest loss during the last 200 years, the hills rank among the most threatened sites of it (Lovett and Wasser, 1993). The Taita Hills continue to boast an outstanding diversity and endemism of flora and fauna (Beentje and Ndiang'ui, 1988). Of particular scientific and conservation interest are the indigenous forest patches. Today, only a small amount of native forest remains, occurring in a scatter of four larger hilltop remnants and further much smaller fragments embedded in a mosaic of human settlements, smallholder cultivation plots and plantations of exotic tree species. The indigenous forest cover not only has been termed upland moist or mist forest by Beentje and Ndiang'ui (1988) but also is referred to as montane forest or cloud forest by other workers. The characteristic tree species include *Newtonia buchananii*, *Tabernaemontana stapfiana*, *Macaranga conglomerata*, *Albizia gummifera*, *Phoenix reclinata*, *Strombosia scheffleri*, *Cola greenwayi*, *Podocarpus* spp., *Ochna holstii* and *Millettia oblate* (Rogers et al., 2008; Omoro et al., 2010). These indigenous forest patches also play an important role in both capturing additional moisture and storing the precipitation on the hilltops.

The population of the whole Taita Taveta District has grown from 90,000 persons in 1962 to over 300,000; consequently, this has been a driving factor behind the land use change in the Taita Hills. There has been an increase in the area under cultivation, and due to poor agricultural management, erodible soils and rough topography in the hills, the foothills are subject to land degradation and soil erosion (Muya & Gicheru, 2005). The remnant forests are threatened by encroachment (for settlement, agriculture and livestock grazing), extraction of firewood and building materials, poor enforcement of government policies and regulations and lack of awareness among the communities living adjacent to forests, fires and colonization by suppressive and fast-growing exotic tree species (EAWLS, 2005). The situation in the Taita Hills, however, is not hopeless compared to the neighbouring Sagala Hills some 20 km apart.

Because of its unique biogeographic history, the Taita Hills have been under research during the last decade concentrating on the impact of forest fragmentation and degradation on different components of biodiversity (Bytebier, 2001), forest health monitoring (Rogers et al., 2008) and population ecological and genetic studies on bird, mammal, amphibian and insect species (Lens et al., 2002).

The University of Helsinki has been studying the land-cover changes of the Taita Hills since the late 1980s (Pellikka, 1990) but more intensively since 2003. The main objectives have been to develop methodologies for land-cover-change studies applying remote sensing and geographic information systems; to assess the land-cover changes in the Taita Hills and its surrounding lowlands; and, finally, to research the consequences of land-cover changes to climate, water resources, biodiversity and livelihoods. The research has resulted in more than 23 M.Sc. theses and seven Ph.D. theses (Soini, 2006; Gonsamo Gosa, 2009; Clark, 2010; Himberg, 2011; Maeda, 2011; Siljander, 2010; Omoro, 2012). Located in the middle of the Tsavo plains and having its small water towers (Ngangao, Vuria, Chawia, Mbololo and Yale), the Taita Hills are like a miniature of Kenya and serve as a model for land-cover change in Kenya.

3 MATERIAL AND METHODS: LAND-COVER-CHANGE STUDIES USING TWO SCALES

Land-cover-change studies have been carried out in two scales. In the fine-scale study, airborne remote sensing data were applied to study the forest cover changes between 1955 and today in the Taita Hills, while in the coarse-scale study, multispectral SPOT XS satellite imagery has been applied to study land changes between 1989 and today in the surroundings of the Taita Hills.

Black and white aerial photography used in the fine-scale study was acquired during January–February 1955 by the Survey of Kenya at a scale of 1:30,000. The prints were scanned using a resolution of 14 mm resulting to ground resolution of 46 cm. The colour airborne digital camera data were acquired during January 2004 using NIKON D1X colour digital camera equipped with a 14 mm lens producing approximately 40 cm ground resolution. The individual frames were mosaicked to a seamless orthorectified image mosaic using EnsoMOSAIC software in which the aerial photographs of 1955 were co-registered (Pellikka et al., 2009). The land cover of Ngangao, Chawia, Fururu, Macha, Mwachora and Yale forest fragments of the Taita Hills was classified visually on a computer screen using 2004 data producing eight forest classes (indigenous, eucalyptus, pine, cypress, silver oak, black wattle, mixed exotic and bushland) and eight other classes (rock, built-up, water, agriculture, wetland, barren land, agroforestry and heathland dominated by giant heather (*Erica mannii*)). The class structure was recorded using the Land Cover Classification System (LCCS), which is a standardized classification system developed by the Food and Agriculture Organization (FAO) of the United Nations and the United Nations Environment Programme (Di Gregorio, 2005). In 1955, no exotic plantations except black wattle existed in the area, and therefore, the 1955 data were classified into indigenous forest, rock, bushland, heathland and barren land classes only. The land-cover model was created in ArcGIS from which the area for each land-cover class was summed. The area of the indigenous forest of each fragment and total forest area including indigenous forest, exotic forests and bushlands were recorded. The change detection was analysed by the areal change, by percentage change and also by change trajectory analysis (Pellikka et al., 2009).

The SPOT XS satellite data used were from 1 July 1987, 25 March 1992 and 15 October 2003, with a 20 m pixel resolution and green, red and near-infrared (NIR) spectral bands. All images represent dry conditions, despite the non-anniversary dates, and share very similar off-nadir sensor view angles. Besides orthorectification and coregistration, atmospheric and topographic effects affecting reflectance characteristics were removed (Clark and Pellikka, 2009). LCCS was applied to record the following classes: cropland, shrubland, thicket, woodland, exotic forest, broad-leaved closed-canopy forest (indigenous rain forest), grassland with scattered shrubs and trees, bare soil and other unconsolidated material, built-up area, bare rock and water. For the initial classification, object-based segmentation algorithms of Definiens were applied. The accuracy for the 2003 data was assessed in the field work and with the airborne remote sensing data of 2004. An overall accuracy of 89% was reached after manual editing (Clark and Pellikka, 2009).

A geospatial database containing layers such as road network, hydrography, administrative borders and digital elevation model (DEM) was generated from a scanned topographic map of a 1:50,000 scale of the Survey of Kenya. The resulting DEM had a 20 m raster in a transverse Mercator projection with a Clarke 1880 spheroid and Arc 1960 datum and a planimetric accuracy of 50 m with an altimetric accuracy of 8 m (Clark and Pellikka, 2009). Mean monthly precipitation surfaces were produced from monthly rainfall records of 11 stations within the Taita Hills from 1987 to 2005 obtained from the Kenya Meteorological Department (Erdogan et al., 2011), and soil information was derived from the Exploratory Soil Map of Kenya of scale a 1:1,000,000 (Sombroek et al., 1980).

Using the results from the land-cover classification of satellite imagery and additional datasets, the soil erosion potential of the Taita Hills was assessed using the Universal Soil Loss Equation (USLE) (Wischmeier et al., 1958). The parameters required by the USLE are rainfall erosivity factor (*R*), slope-length and slope-steepness factor (LS), soil-erodibility factor (*K*), cover–management factor (*C*) and support-practice factor (*P*) for its application. Rainfall erosivity factor, slope factor and soil-erodibility factor are estimated using rainfall data, DEMs and soil maps, respectively. A support-practice factor is applied if man-made soil-control infrastructure is used in the study area. In this study, support-practice factor was used as a constant value and cover–management factor was derived from land-cover classification using the SPOT XS imagery (Erdogan et al., 2011).

Using the remote sensing data, the forest cover changes of the indigenous forest fragments between 1955 and 2004 were studied, the land-cover changes in the Taita Hills and its surroundings were studied between 1989 and 2003. The results were used also for simulating agricultural expansion scenarios up to 2030 in the study area using a spatially explicit model of landscape dynamics (Maeda et al., 2010a) and also to model soil erosion potential changes from 1989 until 2030 (Maeda et al., 2010b).

4 RESULTS: LOSS OF INDIGENOUS FORESTS IN THE HILLS AND AGRICULTURAL EXPANSION IN THE LOWLANDS

The results of the fine-scale assessment indicate that the indigenous forest in the Taita Hills decreased by 50% (260 ha) between 1955 and 2004, but the total forest cover area decreased only by 2% due to the establishment of exotic plantations (Pellikka et al., 2009). Each individual forest fragment, however, was managed differently during this time period. The indigenous forest decreased in Chawia, Fururu, Macha, Mwachora, Ngangao and Yale by 28%, 76%, 78%, 68%, 20% and 70%, respectively (Table 1). Based on the change detection results, the indigenous forest was changed not only to agricultural areas but also to some extent to exotic plantations. The exotic plantations were typically established not only in barren and dry areas but also in forests and on agricultural areas. In total, 264 ha of non-forested areas was changed to exotic plantations, while 33 ha was converted to indigenous forest. The conversion to indigenous forest is mostly caused by fast growth of succession species such as *P. reclinata* on abandoned open areas, but in places, such as at the top of Ngangao, primary indigenous forests have developed in areas that were non-forested in 1955. The change of the Ngangao Forest is presented in Figure 2.

TABLE 1 The Change of Non-Forested Area to Forests in Hectares Between 1955 and 2004 in Selected Study Areas in the Taita Hills

Change from 1955 to 2004	To Indigenous Forest	To Exotic Forest
Chawia	3.2	9.2
Fururu	4.1	45.1
Macha	1.7	23.8
Mwachora	1.0	10.1
Ngangao	12.0	81.5
Yale	11.1	94.2
Total	33.1	263.9

Adapted from Pellikka et al. (2009).

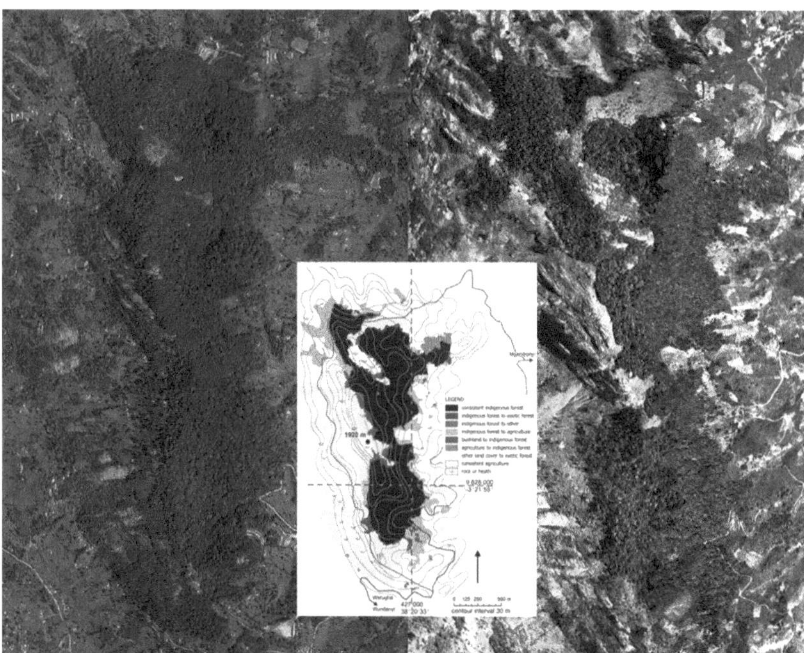

FIGURE 2 Airborne digital camera image mosaic of 2004 on the left and aerial photograph of 1955 on the right. In the middle is analysed change in which the black colour indicates constant indigenous forest from 1955 to 2004. (For colour version of this figure, the reader is referred to the online version of this chapter.)

As a result of a coarse-scale study, the croplands covered 239 km^2 in 1987 in the Taita Hills and its surroundings, but the area had increased to 244 km^2 by 1992 and to 331 km^2 by 2003 (Clark and Pellikka, 2009). This increase was caused mainly by conversion of thicket and shrublands to croplands in the lowlands and foothills. Grasslands increased by 4 km^2 by conversion from shrublands, especially between the years 1992 and 2003. Plantation forests increased in area, which may be caused by a maturing of the stands, increased biomass and consequent improved identification in the classification process. There was no important change in the area of broad-leaved closed-canopy forests (indigenous forests) between 1987 and 2003 in the main three fragments of Chawia, Ngangao and Mbololo. The large increase in bare soil may be partially caused by phenological effects, since the 1987 image was taken in July after the rainy season, while the 2003 image was taken at the end of the dry season (Figure 3). However, by applying an object-oriented classification method, misclassifications were reduced to as low a level as possible. The decrease in water is mainly caused by the decrease of the water surface in the Mwatate reservoir due to silting of the reservoir.

The land-cover-change scenario is presented with the historical land use maps for 1987 and 2003 in Figure 4 (Maeda et al., 2010a). The land use for 2030 resulted from two land-cover-change scenarios: unrestricted business-as-usual (BAU) scenario and restricted governance (GOV) scenario. It is observed that, in 1987, croplands were already clearly established in the highlands. The protected areas, situated in the northeast (grassland area) and southwest parts of the study area, were effective in containing the agricultural expansion between 1987 and 2003. The influence of this variable was reflected in both simulated scenarios, where a clear boundary constraining

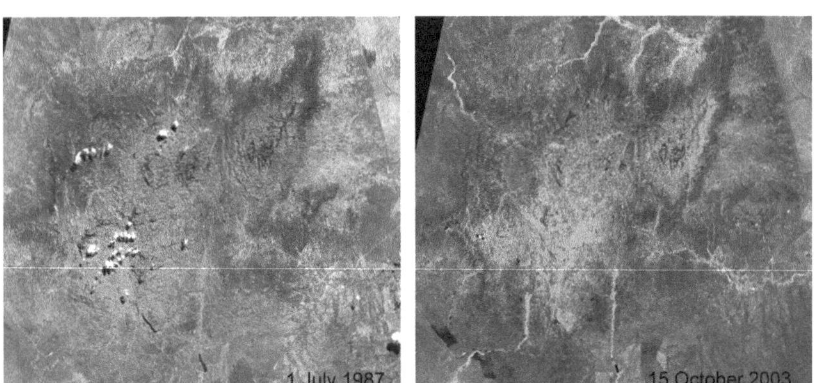

FIGURE 3 The SPOT XS image of 1 July 1987 taken at the end of the long rains represents completely different greenness compared to the image taken on 15 October 2003 after the dry season. The evergreen indigenous forest patches are much more visible from the landscape during the dry season when the fields are bare. (For colour version of this figure, the reader is referred to the online version of this chapter.)

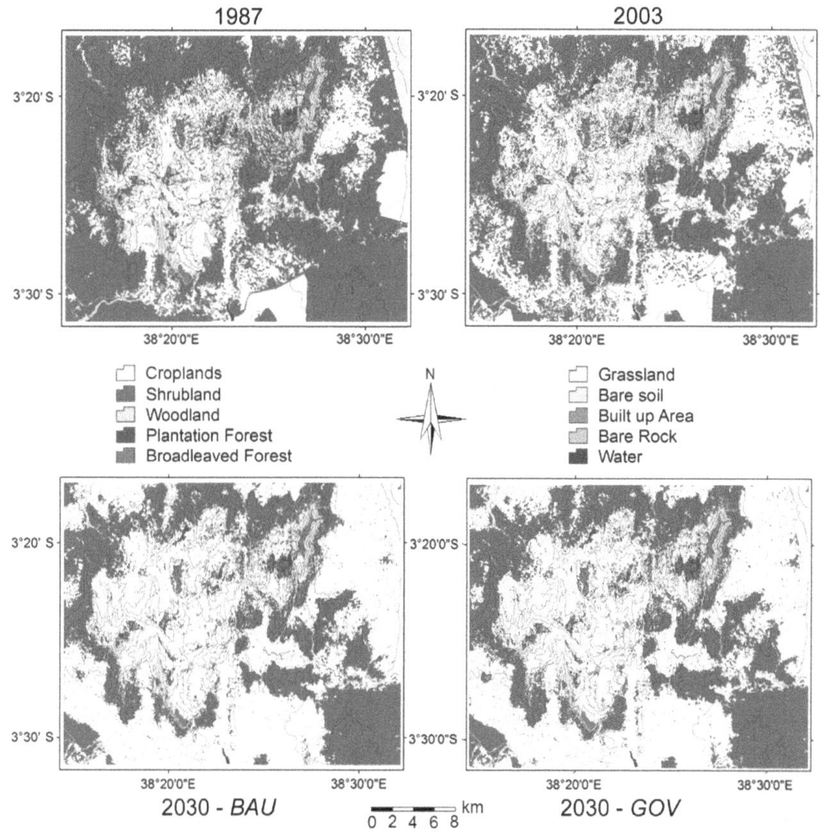

FIGURE 4 Land use maps for 1987 and 2003 (upper left and upper right) and simulated scenarios for 2030 (lower left and lower right). *Adapted from Maeda et al. (2010a).* (For colour version of this figure, the reader is referred to the online version of this chapter.)

the expansion of croplands is observed in the limits of the protected areas. Hence, although a high environmental pressure is present along the fringes of the conservation areas, such as Tsavo East National Park, the areas inside the park are likely to be preserved.

Between 1987 and 2003, croplands started to be implemented with higher intensity in the lowlands, given that suitable areas for agriculture activities in the highlands were already almost entirely taken. Although each simulated scenario was created independently, using different transition rates, the spatial distribution of new cropland patches followed the same patterns in all simulations. As suitable agricultural areas in the highlands disappeared, the expansion of new patches was distributed in the foothills. Based on the explanatory factors (Maeda et al., 2010a), among the main driving forces of such distribution were the distance to markets (here represented by villages or towns), distance to roads, and distance to rivers. Distance to markets and roads played an interesting role in croplands distribution, in the sense that the effects of these two variables in the landscape dynamic were closely related. Towns and villages acted as core points, which are interconnected by roads creating axes in which new cropland patches were settled. Such patterns were observed mainly in the southern and southwestern parts of the study area. It is also important to notice the enhanced importance attributed to rivers in the land use dynamic. Given that in this region, the foothills typically have higher average temperatures and lower precipitation volumes, the proximity to water bodies is essential to the establishment of agricultural activities. The numerical results of the simulations are presented in Figure 5. In the Business As Usual (BAU) scenario, the cropland areas expanded to around 515 km^2 in 2030, corresponding to about 60% of the study area, representing an increase of 40% in comparison to the year 2003, when croplands occupied around 365 km^2. Although the effects of the governance (GOV) scenario cannot be easily identified in Figure 4, the simulated land use policies resulted in a significant reduction in agricultural expansion. The total area used for agriculture in 2030 for the GOV scenario was approximately 485 km^2.

The land-cover changes have direct impact on the cover/management factor *C* of the USLE model. Previous studies have shown that croplands increase soil exposure to weathering more than original vegetation such as forests or shrublands. For this reason, the *C* factor of agricultural areas is usually higher, resulting in increased soil erosion rates. The *C* factor values by Erdogan et al. (2011) indicate that average *C* increased 10% from year 1987 to 2003. In the LUCC simulation for the year 2030, *C* increased by 9% from 2003. The areas with higher slope values are likely to undergo higher soil erosion rates, and when new fields are located in steep slopes, the soil erosion

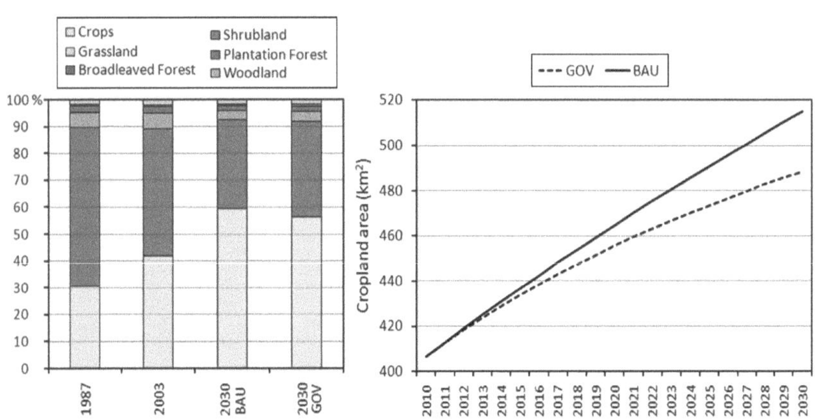

FIGURE 5 (a) Percentage of the main land use/land-cover classes during the years 1987 and 2003 and in both scenarios simulated for 2030 and (b) LUCC simulation results from 2010 to 2030. *Adapted from Maeda et al. (2010a).* (For colour version of this figure, the reader is referred to the online version of this chapter.)

potentiality is even greater. Fortunately, this is not the case in the Taita Hills. Agricultural patches established in the last decades were mainly settled in areas with favourable topography. This pattern was reinforced after 1987, when the availability of space in the hills was scarce and the agriculture started expanding to flat areas along the foothills. In the agricultural expansion simulated for 2030, a slight increase in cropland patches has occurred in steeper slopes; however, the most significant increase has occurred again in areas with gentle slopes or even flatlands.

5 DISCUSSION: ENDANGERED ECOSYSTEM SERVICES

Land-cover change has numerous ecological, physical and socioeconomic consequences. On the positive side, agricultural expansion may increase food production for a growing population, although it is unsure how productive the last exploited lands will be as they are typically the least favourable. There are numerous negative consequences with both known and unknown links and feedback mechanisms.

Converting the natural vegetation to agricultural land is likely to change the radiation balance of the given unit of area. In principle, the albedo increases as land is without vegetation at least part of the year causing more solar energy to reflect back to the space. Other environmental impacts include the decrease in soil water-holding capacity. As natural vegetation is replaced by agriculture, soil porosity may be reduced by soil compaction, decreasing infiltration capacity and increasing the risks of soil erosion. In mountainous areas, the conversion of the forests to agricultural lands decreases as does the occult precipitation as croplands capture less atmospheric moisture than multilayered indigenous forest or forest of any kind (Holder, 2004). Cloud formation over the land unit also decreases as the evapotranspiration rate is less from fields than from forests causing evidently reduced precipitation.

Further studies project that, by 2030, agricultural land is likely to expand to regions with higher evapotranspiration potential (Figure 6). This expansion will increase by approximately 40% the annual volume of water necessary for irrigation (Maeda et al., 2011).

As soil water-holding capacity is reduced, the risk of hydrologic droughts during dry seasons is increased, while during the rainy seasons, soils are more susceptible to erosion. These soil loss and sediment-deposition processes may have a significant impact on agriculture, local economies and ecosystems (Alcantara-Ayala et al., 2006). Although increasing evidences indicate that anthropogenic changes in the landscape are likely to lead to regional and global climate change, the levels and scale of this relationship remain largely unknown. However, it is clear that converting forestland to agricultural land causes changes in local climate via the changes in radiation and water balance. Changes in precipitation and temperature patterns will likely have important impacts on the sustainability of agricultural systems.

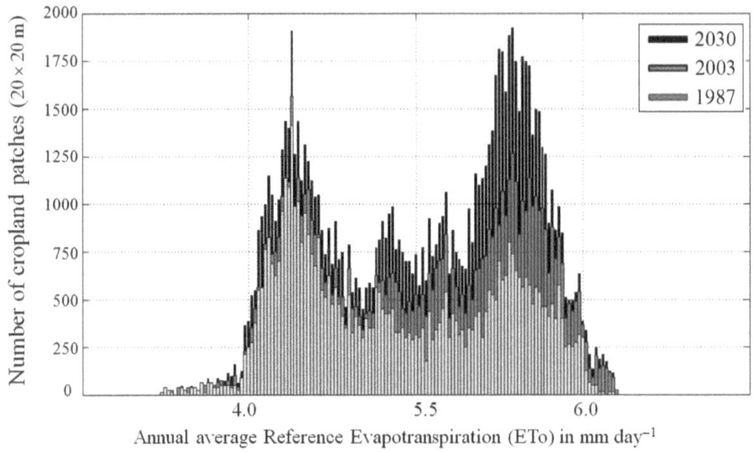

FIGURE 6 Histogram showing the cropland patches distribution during 1987, 2003 and 2030, in relation to the historical average potential evapotranspiration in the Taita Hills. *Adapted from Maeda et al. (2011).*

The land-cover change taking place in the Taita Hills and its surrounding has been continuing since human settlers arrived in the area and started to convert native vegetation to agricultural land. With ever-growing population and demand for land for cultivation of food crops and other crop types in addition to increase of reserved and protected areas in the study area, land is evidently becoming a valuable natural resource. Land use conflicts have already taken place in the area between farmers, conservationists, settlers and sisal plantation managers (Vanonen, 2008).

Converting natural vegetation, forest or grassland to agricultural areas decreases biodiversity, reduces the capability of vegetation to capture atmospheric moisture and retain water in the vegetation cover, exposes land to be subject to water and wind erosion and changes the radiation balance of the land surface as land is exposed and barren part of the year. These all have still unknown impacts on regional climate.

ACKNOWLEDGEMENTS

The University of Helsinki is involved in a number of international projects targeting to understand the consequences of the land-cover change in the Taita Hills and in Africa in general. CHIESA project financed by the Ministry of Foreign Affairs of Finland studies the impact of climate and land-cover change to ecosystem services, such as pests and pollinators of avocado, coffee, crucifers and maize, to water resources and to habitats. BIODEV project financed by the Ministry of Foreign Affairs of Finland concentrates to monitor forest and cropland area biomass and aboveground and soil carbon in the Taita Hills and its surroundings and simultaneously develops methods for increasing timber on the farms in order to sequestrate carbon. TAITAWATER project funded by the Academy of Finland researches the occult precipitation captured by the forest, water-holding capacity of the forests and human water use in the Taita Hills. For the long-term research programme, the University of Helsinki established Taita Research Station in Wundanyi in 2011 (http://blogs.helsinki.fi/taita-research-station/).

REFERENCES

Alcantara-Ayala, I., Esteban-Chavez, O., Parrot, J.F., 2006. Landsliding related to land-cover change: a diachronic analysis of hillslope instability distribution in the Sierra Norte, Puebla, Mexico. Catena 65, 152–165.

Baldyga, T.J., Miller, S.N., Driese, K.L., Gichaba, C.N., 2007. Assessing land cover change in Kenya's Mau Forest region using remotely sensed data. Afr. J. Ecol. 46, 46–54.

Beentje, H.J., Ndiang'ui, N., 1988. An ecological and floristical study of the forests of the Taita Hills, Kenya. Utafiti 1 (2), 23–66.

Brink, A.B., Eva, H.D., 2009. Monitoring 25 years of land cover change dynamics in Africa: a sample based remote sensing approach. Appl. Geogr. 29 (4), 501–512.

Bytebier, B., 2001. Taita hills biodiversity project report. National Museums of Kenya, Nairobi, 121 pp.

Clark, B.J.F., 2010. Enhanced Processing of SPOT Multispectral Satellite Imagery for Environmental Monitoring and Modelling, Department of Geosciences and Geography A5, University Print, Helsinki. http://urn.fi/URN:ISBN:978-952-10-6306-0.

Clark, B.J.F., Pellikka, P.K.E., 2009. Landscape analysis using multi-scale segmentation and object-oriented classification. In: Röder, A., Hill, J. (Eds.), Recent Advances in Remote Sensing and Geoinformation Processing for Land Degradation Assessment. Taylor & Francis, Leiden, pp. 323–342.

Di Gregorio, A., 2005. Land Cover Classification System (LCCS), Version 2: classification Concepts and User Manual. In: Food and Agriculture Organization (FAO) Environmental and Natural Resources Series8FAO, Rome.

EAWLS, 2005. East African Wildlife Society: Stakeholders Workshop on the Conservation and Management of Taita Hills Forests—Workshop Summary Report.

Erdogan, E.H., Pellikka, P., Clark, B., 2011. Impact of land cover change on soil loss in the Taita Hills, Kenya between 1987 and 2003. Int. J. Remote Sens. 32 (21), 5919–5945.

Eva, H.D., Brink, A., Simonetti, D., 2006. Monitoring Land Cover Dynamics in Sub-Saharan Africa. A Pilot Study Using Earth Observing Satellite Data from 1975 and 2000. Institute for Environment and Sustainability, Joint Research Centre of the European Commission, 44 pp.

Gonsamo Gosa, A., 2009. Remote Sensing of Leaf Area Index: Enhanced Retrieval from Close-Range and Remotely Sensed Optical Observations. In: Helsingin yliopiston maantieteen laitoksen julkaisuja A147Hansaprint, Helsinki. http://urn.fi/URN:ISBN:978-952-10-5873-8.

Hildebrandt, J.M., 1877. Von Mombassa nach Kitui. Zeitsch. Gesellsch. Erdkunde 14, 321–350.

Himberg, N., 2011. Traditionally Protected Forests-Justifications or Hindrances to Ethnodevelopment and Participatory Forestry in Taita Hills, Kenya. Department of Geosciences and Geography, University Print, Helsinki. http://urn.fi/URN:ISBN:978-952-10-7283-3.

Himberg, N., Omoro, L., Pellikka, P., Luukkanen, O., 2009. The benefits and constraints of participation in forest management. The case of Taita Hills, Kenya. Fennia 187 (1), 61–76.

Holder, C.D., 2004. Rainfall interception and fog precipitation in a tropical montane cloud forest of Guatemala. Forest Ecol. Manage. 190, 373–384.

Imbernon, J., 1999. Pattern and development of land use changes in Kenyan highland since 1950s. Agric. Ecosyst. Environ. 76, 67–73.

Lens, L., Van Dongen, S., Norris, K., Githiru, M., Matthysen, E., 2002. Avian persistence in fragmented rainforest. Science 298, 1236–1238.

Lovett, J.C., Wasser, S.K., 1993. Biogeography and Ecology of the Rainforests of Eastern Africa. Cambridge University Press, Cambridge.

Lung, T., Schaab, G., 2006. Assessing fragmentation and disturbance of west Kenyan rainforests by means of remotely sensed time series data and landscape metrics. Afric. J. Ecol. 44, 491–506.

Maeda, E.E., 2011. Agricultural expansion and climate change in the Taita Hills, Kenya: an assessment of potential environmental impacts. Department of Geosciences and Geography A7, University Print, Helsinki. https://helda.helsinki.fi/handle/10138/24585.

Maeda, E.E., Clark, B.J.F., Pellikka, P.K.E., Siljander, M., 2010a. Modelling agricultural expansion in Kenya's eastern arc mountains biodiversity hotspot. Agric. Syst. 103, 609–620.

Maeda, E.E., Pellikka, P.K.E., Siljander, M., Clark, B.J.F., 2010b. Potential impacts of agricultural expansion and climate change on soil erosion in the Eastern Arc Mountains of Kenya. Geomorphology 123, 279–289.

Maeda, E.E., Pellikka, P.K.E., Clark, B.J.F., Siljander, M., 2011. Prospective changes in irrigation water requirements caused by agricultural expansion and climate changes in the eastern arc mountains of Kenya. J. Environ. Manage. 92, 982–993.

Maro, P.S., 1988. Agricultural land management under population pressure: the Kilimanjaro experience, Tanzania. Mt. Res. Dev. 8, 273–282.

McMichael, C.E., Hope, A.S., 2007. Predicting streamflow response to fire-induced landcover change: implications of parameter uncertainty in the MIKE SHE model. J. Environ. Manage. 84, 245–256.

Muya, E.M. & Gicheru, P.T. (2005). Assessment of land degradation and its impacts on land use sustainability in Taita Taveta catchment. Miscellaneous Paper No. 63, Kenya Agricultural Research Institute, Kenya Soil Survey, Nairobi.

Omoro, L.M.A., 2012. Impacts of Indigenous and Exotic Tree Species on Ecosystem Services—Case Study on the Mountain Cloud Forests of Taita Hills, Kenya. In: Tropical Forestry Reports41Viikki Tropical Resources Institute, University of Helsinki. http://urn.fi/URN:ISBN:978-952-10-8449-2.

Omoro, L.M.A., Pellikka, P., Rogers, P., 2010. Tree species diversity, richness, and similarity between exotic and indigenous forests in the cloud forests of Eastern Arc Mountains, Taita Hills, Kenya. J. Forest. Res. 21 (3), 255–264.

Pellikka, P., 1990. Land Use and Its Classification Using a Multispectral SPOT XS Satellite Image in the Taita Hills, SE Kenya. Unpublished M.Sc. Thesis. University of Helsinki, Faculty of Science, Department of Geography, Helsinki 1990.

Pellikka, P., Lötjönen, M., Siljander, M., Lens, L., 2009. Airborne remote sensing of spatiotemporal change (1955–2004) in indigenous and exotic forest cover in the Taita Hills, Kenya. Int. J. Appl. Earth Obs. Geoinf. 11, 221–232.

Rogers, P.C., O'Connell, B., Mwangombe, J., Madoffe, S., Hertel, G., 2008. Forest health monitoring in the Ngangao forest, Taita Hills, Kenya: a five year assessment of change. J. East Afric. Nat. Hist. 97 (1), 3–17.

Siljander, M., 2010. Geospatial environmental data modelling applications using remote sensing, GIS and spatial statistics. Department of Geosciences and Geography A1, Yliopistopaino, Helsinki. http://urn.fi/URN:ISBN:978-952-10-6110-3.

Soini, E., 2006. Livelihood, land use and environment interactions in the highlands of East Africa. https://helda.helsinki.fi/handle/10138/21188.

Sombroek, W.G., Brown, H.M.N., van der Pouw, B.J.A.,1980. The Exploratory Soil Map and Agro-climatic Zone Map of Kenya. Report No. E1 (Nairobi: Kenya Soil Survey).

Sunderlin, W.D., Ndoye, O., Bikié, H., Laporte, N., Mertens, B., Pokam, J., 2000. Economic crisis, small-scale agriculture, and forest cover change in southern Cameroon. Environ. Conserv. 27, 284–290.

Turner, R.M., Ochung, H.A., Turner, J.B., 1998. Kenya's Changing Landscape. The University of Arizona Press, Tucson, 177 pp.

Vanonen, J., 2008. The conflicts of land in Taita Taveta District, Kenya. Unpublished M.Sc. Thesis. University of Helsinki, Faculty of Science, Department of Geography, Helsinki 1990.

Wischmeier, W.H., Smith, D.D., Uhland, R.E., 1958. Evaluation of factors in the soil loss equation. Agric. Eng. 39, 458–462.

Chapter 14

The Importance of Fire Ecology in Protected Areas Management

Ilaria Palumbo
Joint Research Centre—European Commission, Via E. Fermi, 2749. 21027- Ispra, Italy

1 INTRODUCTION

In this chapter, we introduce the ecological role of fire on the ecosystems and biodiversity. This does not aim to be an exhaustive review of fire ecology but rather a synthesis of the most important effects of fire in tropical ecosystems. We complete this introduction highlighting the benefits of controlled burning for land management, ecosystems and biodiversity.

The core of this chapter is then dedicated to the assessment of fire in Kenyan ecosystems and protected areas. The results are based on 10-year fire observations from space, which are crossed with the major vegetation types. Finally, we discuss the relation between the fire activity observed between 2002 and 2012 and the land-cover change assessment from 1990 to 2010 presented in chapter 12.

1.1 Fires in the Ecosystems

Fire occurs in most ecosystems of the world affecting about 3 Ml km^2 every year. According to the fossil charcoal findings, its presence on Earth dates back to the Silurian period, 420 million years ago, when the first terrestrial plants appeared (Scott and Glasspool, 2006).

With such a long presence in the terrestrial biomes, fires have contributed to shape and create some of the present ecosystems (Keeley and Rundel, 2005; Beerling and Osborne, 2006). In addition to this, fire occurrence is also fundamental for the well-being and the biodiversity of many habitats with several positive effects:

- The stimulation of new sprouts in trees and grass: the fast regrowth of grass after fire allows fresh forage to become available to herbivores or domestic animals (grazers) during the dry season.

Developments in Earth Surface Processes, Vol. 16. http://dx.doi.org/10.1016/B978-0-444-59559-1.00014-1

- The removal of dead vegetation: this favours new plant growth and reduces the risk of uncontrolled fires.
- The release of nutrients back to the soil.

These positive effects are observed in ecosystems adapted to fire, whereas, of course, if they have little or no adaptation, fires can be detrimental and compromise their survival.

According to the 'Preliminary Assessment of Fire as a Conservation Issue' (Global Fire Initiative, Hardesty et al., 2005), ecosystems can be divided into four major categories according to their predominant response to fire:

1. Fire-independent ecosystems: these ecosystems are normally not affected by fire because they do not have enough fuel or ignition sources; some examples are deserts, tundra and rain forests.
2. Fire-dependent ecosystems: these ecosystems have evolved with fire and their structure and composition are maintained by recurrent burnings. Fires occurring in these ecosystems have different frequency and intensity, and this variability is reflected in the variety of ecosystems that belong to this class: savannas, woodlands and shrublands and also some of the Mediterranean, temperate and boreal forests. Plants and animals from these environments have adaptations to fire and high resilience to burning. In comparison to the other categories, these ecosystems cover the largest area in the world, about 53% of the global ecoregions assessed in the 'Preliminary Assessment'.
3. Fire-sensitive ecosystems: these ecosystems experience fires rarely because they are not very flammable; however, if they are subject to disturbances and get degraded, they can experience more fires and be seriously damaged because they are not adapted to it. Some examples are the tropical and subtropical humid forests.
4. Fire-influenced ecosystems: they are in the transition areas between fire-dependent and fire-sensitive ecosystems, for example, the tropical dry forests.

Besides its ecological role, fire is also an important tool largely used in many human activities. For example, in the tropics, it is used in agriculture, farming, hunting and for fuel control.

1.2 Fires and Biodiversity

Fires are characterized by their intensity, frequency and seasonality and also by the part of vegetation structure they affect more, so depending on the case, they can be *crown fires* or *surface fires*. The combination of all these aspects determines the *fire regime*. Understanding the effects of different fire regimes on the ecosystems is crucial to understand the ecosystem itself and also to improve the land management practices applied to it; these can go from domestic uses (agriculture and farming) to conservation programmes.

According to the Nature Conservancy, altered fire regimes are key threats to biodiversity: 'too much, too little or the wrong type of fire are a major threat to biodiversity conservation' (Hardesty et al., 2005). The 'Preliminary Assessment' reports that around 84% of places identified by scientists as critical for global conservation have altered fire regimes. This alteration can cause biodiversity loss and habitat degradation.

For these reasons, in many protected areas, fire is recognized a key role for the well-being of the biological community (plants and animals), and it is included in the park management plans to achieve conservation objectives.

In order to apply successful fire programmes in the protected areas, conservation plans should also consider the socioeconomic context of the park and its relation to fire, for example, understanding the burning practices applied outside the protected areas and their possible effects on the park. This approach is called 'Integrated Fire Management' and it aims at designing fire plans that incorporate the social and economic needs related to burning. This allows not only to choose appropriate conservation plans, to restore or maintain the natural fire regimes, but also to promote the sustainable use of the natural resources and the ecosystem services needed for people's livelihoods (Myers, 2006; Santoso et al., 2007).

2 FIRES IN AFRICA

Africa has been often referred to as the 'fire continent' (Komarek, 1965) because of the regular burning that affects large areas every year, about 64% of the global burned area extent (Tansey et al., 2004). Around two-thirds of the continent is dominated by savanna ecosystems that have been shaped by natural fires that occurred in the last millions years (Scott, 2000). Fires occur so regularly in the tropical and subtropical regions of Africa because of the strong alternation of wet and dry seasons; this creates a build-up of fine fuel, which becomes very flammable during the dry season. In the savannas, fires were originally caused by lightings, but nowadays, they have mainly a human origin (Komarek, 1971). Burning is used for many purposes: prepare an area for domestic livestock, hunt, firewood collection, agriculture and many more (Phillips, 1965). Nowadays, fire is also recognized as a useful tool for the conservation of natural habitats, and for this reason, specific fire programmes are designed and applied in many protected areas (Govender et al., 2006; van Wilgen et al., 2011; Walters, 2012).

In the savannas, plants and animals can benefit from the recurrent burnings (Bond et al., 2005) because these stimulate the fast growth of palatable grass for herbivores or cattles during the dry season (Figure 1); in addition to this, fire helps maintain a variety of habitats in the same area by controlling the proportion of grass, shrubs and trees. The absence of fire in these ecosystems would alter their natural functionality and composition, ultimately compromising their own survival (Trollope, 1990; Thomson, 1992; Bothma, 1996).

FIGURE 1 Vegetation regrowth few weeks after the burning (Pendjari National Park, Benin). (For colour version of this figure, the reader is referred to the online version of this chapter.)

2.1 Fire in Kenya and Its Protected Areas

In this section, we present statistics about the fire activity of the last 10 years in Kenya. We based our analysis on two satellite products distributed by NASA (http://earthdata.nasa.gov/firms) and the University of Maryland (http://modis-fire.umd.edu/). These products are derived from the Moderate Resolution Imaging Spectroradiometer (MODIS) imagery sensor on board the platforms AQUA and TERRA: the active fires and the burned areas (Giglio et al., 2003; Roy et al., 2005). They provide different types of information on the fire activity: the first reports the location and timing of fires and the second product informs on the extent of the area affected by the fire. They have different spatial resolution, the active fires being available at 1 km resolution whereas the burned areas has 500 m spatial resolution.

Although their spatial resolution is not very high, they allow the retrieval of information on fire activity systematically and over the protected areas (Justice and Korontzi, 2001).

We used these time series to derive the fire occurrence over periods of 10 days and describe the fire seasonality over the country and in the protected areas. Although the dataset is available since the late 2000, we used data from 2002 so as to have both AQUA and TERRA datasets, since the AQUA satellite became available only in 2002.

2.1.1 Fire Season

Kenya has two rainy seasons alternated with a short and a long dry season, when fires usually occur. The typical fire occurrence is shown in Figure 2

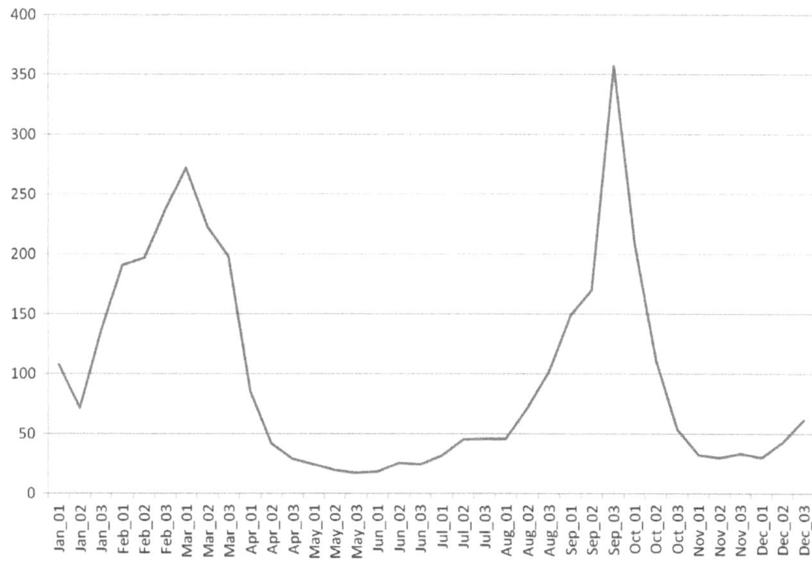

FIGURE 2 The average distribution of the fire counts along the year. The time step is 10 days indicated as '01, 02, 03' in each month. We show the average fire counts for the period 2002–2011. (For colour version of this figure, the reader is referred to the online version of this chapter.)

using the average fire counts, per 10-day periods, from year 2002 to 2011. We observe fire activity from late January to early April and from August until the end of October with peaks in February–March and September–October.

Although rainfall has similar seasonality across the country, with peak precipitations in April and November, the amount of rain varies quite substantially among different regions with effects on the fire occurrence. Some examples of regional distributions of rain and fire are shown in Figure 3 for the year 2012.

In the northern part of the country, rainfall is about half of what we observe in the south and not many fires are detected: fire activity tends to be very low in areas with poor precipitation because the lack of water limits the availability of grass and other fine fuels needed to ignite and sustain fire. The western region (plots 1 and 2) shows the highest precipitation values; here, the Afromontane ecoregion, dominated by closed trees, is mixed with the Somalia–Masai region where sparse vegetation is more common; in both examples, fire occurrence is quite high.

Besides rain, another factor driving fire occurrence is population density (Archibald et al., 2009; Lehsten et al., 2010). This can explain the fire activity observed in the Zanzibar–Inhambane area along the eastern coast (plot 4); here, rainfall is less abundant than in the examples 1 and 2, but fire counts are similar to those in plots 1 and 2. These three examples (plots 1, 2 and 4)

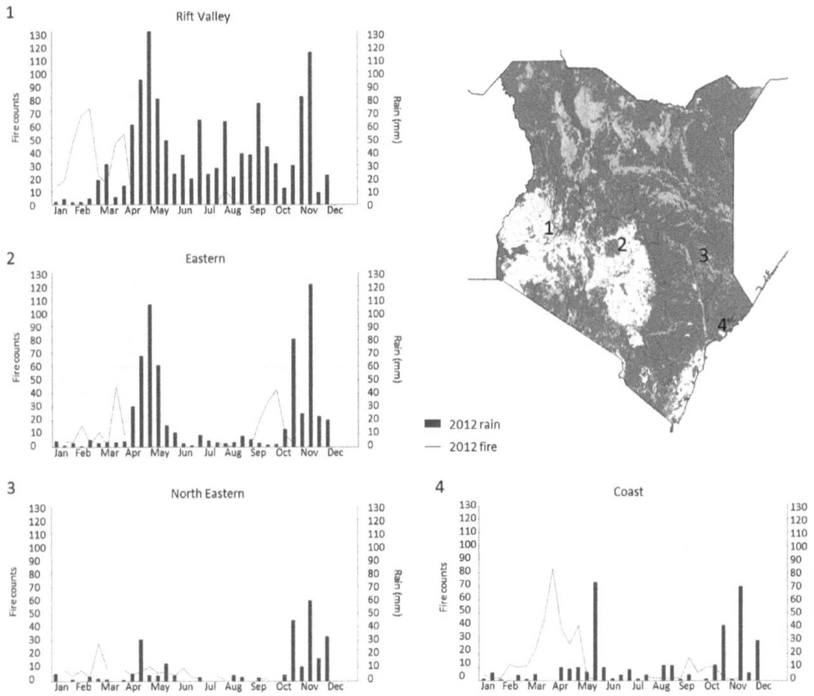

FIGURE 3 Fire occurrence and rainfall patterns in 2012, values refer to 10-day periods; rainfall and fire counts range between 0 and 130. (For colour version of this figure, the reader is referred to the online version of this chapter.)

have in common high population densities that may explain, together with rain, their fire occurrence patterns. The last example (plot 3) is from the eastern region and shows low fire activity; its precipitation values are similar to those in plot 4; however, the very low population density of this region could be the reason for such a low fire occurrence (the population density map for Kenya is shown in Chapter 12, Figure 14).

2.1.2 *Fire Occurrence in the Major Vegetation Classes*

We also analyzed the fire events and the burned areas extent in the major vegetation classes. The vegetation classes are based on the Africover land-cover map for Kenya (FAO, 2000) and re-grouped in the following categories: closed trees, wood and shrubland, aquatic and temporal flooded vegetation, grasslands, agriculture, mangroves, water and bare.

The distribution of the fire counts is shown in Figure 4a and b, annually and by vegetation class. Values range from few hundreds to more than 2500 fire events in a year showing the highest counts in the 'wood and shrubland'

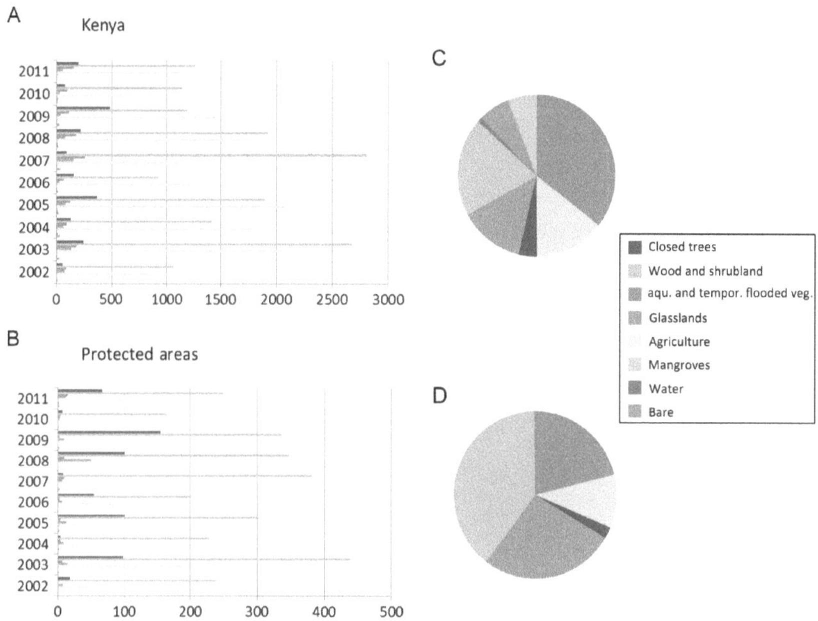

FIGURE 4 Fire activity by vegetation class in Kenya and the protected areas (period 2002–2011). (a,b) Fire counts in the major land-cover classes by year; (c,d) percentage of land cover burned as average value (2002–2011) in Kenya and the protected areas, respectively. (For colour version of this figure, the reader is referred to the online version of this chapter.)

and 'agriculture' classes in Kenya; whereas in the protected areas, the highest values are found in the 'wood and shrubland' class.

In terms of burned area, the average annual burned area in Kenya is around 175,000 ha, with the largest areas in the 'wood and shrubland' class. However, if we consider the values in terms of percentage of the vegetation class burned, the 'aquatic and temporal flooded vegetation' class is the most affected (Figure 4c). This vegetation class is covered in abundant grassy biomass during the dry season, which supports large burnings; these fires are probably related to agriculture and farming practices and lead to large burned areas also in other parts of Africa with the same vegetation type, for example, in South Sudan.

In the protected areas, the highest percentage of vegetation burned is found in the 'wood and shrubland' and the 'grassland' vegetation classes (Figure 4d). In these vegetation types, prescribed burnings can be applied for conservation purposes to control bush encroachment and maintain the open grassland for herbivores. In the 'aquatic and temporal flooded vegetation', the percentage burned is about half of what is found in the rest of the country, confirming that a lot of the burning occurring in this vegetation type is related to human activities.

2.2 Fire and Land-Cover Change

In this final section, we discuss the relation between fire activity and the change of land cover over 20 years. The information on the land-cover change is derived from the assessment presented in the Chapter 12 about vegetation of Kenya.

The land-cover assessment used high-resolution satellite images to analyze the vegetation change between the years 1990 and 2010. The change is

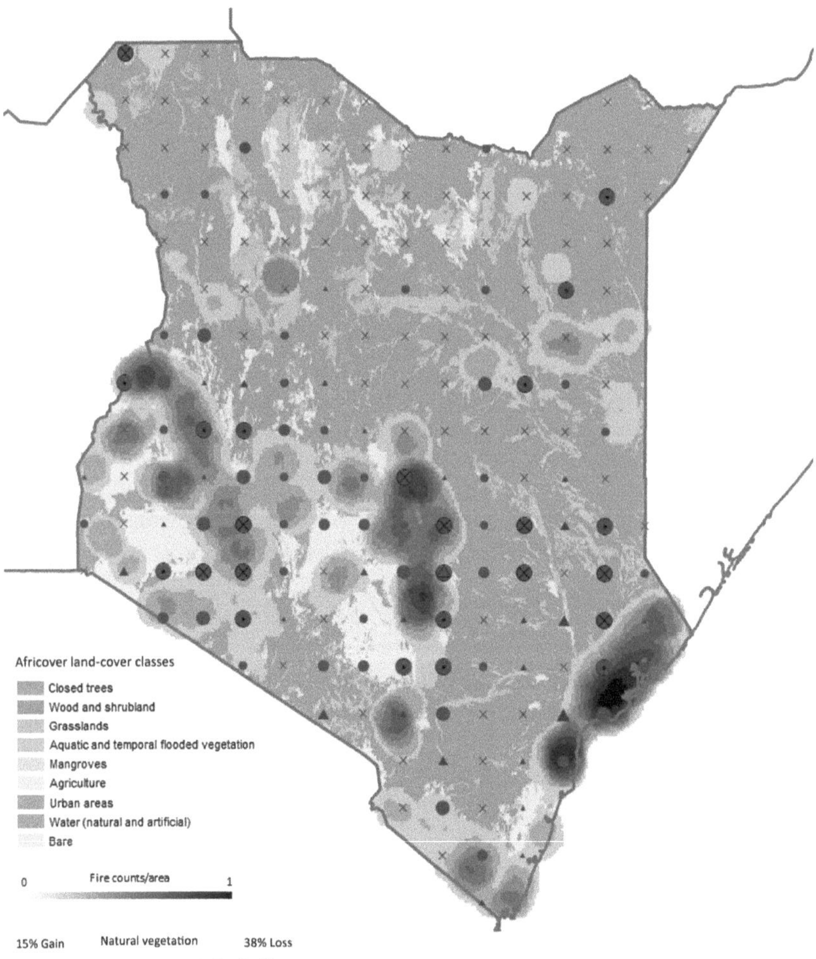

FIGURE 5 Map of the natural vegetation loss/gain with the active fire density and the Africover land-cover map as a background. The natural vegetation information refers to the period 1990–2010, whereas the fire density covers the period 2002–2012. (For interpretation of the references to colour in this figure legend, the reader is referred to the online version of this chapter.)

indicated as a 'loss' or a 'gain' of natural vegetation for each sample window (20 × 20 km) placed on a regular grid every 0.5°.

The fire activity is based on the fire counts from the MODIS product for the period 2002–2012 (available until early December). Previously in this chapter, we limited the time series until year 2011 because we presented results annually or as average values, but for this assessment, we include also year 2012 to show the most up-to-date information on fire.

Fire activity is shown as the density of fire counts by area; values go from 0 to 1. The areas are circles with incremental 25 km radius. The results from the land-cover change analysis are combined with the information on the fire occurrence in Figure 5.

We observe high fire activity in correspondence to areas with high natural vegetation loss (red circles in Figure 5). In the southwestern and central Kenya, the highest natural vegetation loss is found along the limit between 'agriculture' and 'wood and shrubland' land covers; this area also shows the highest fire occurrence of the last 10 years. This can indicate that fire is used not only for land management, for example, to maintain already established areas for agriculture, but also for land conversion, along the interface between the natural vegetation (woodland and shrubland) and the agriculture areas (an example of this is shown in Figure 6).

FIGURE 6 Area burned along the border between woodland/shrubland vegetation and agriculture. (For colour version of this figure, the reader is referred to the online version of this chapter.)

Another area with high loss of natural vegetation is in the southeastern part of the country; here, we find two different situations in terms of fire activity: (i) little fire activity in the wood and shrubland class of the Somalia–Masai ecoregion and (ii) very high fire activity along the coast where wood and shrublands are still dominant but belong to the Zanzibar–Inhambane region. This net change of fire occurrence is probably due to the different rainfall patterns in the two regions, the Zanzibar–Inhambane having higher precipitations and therefore richer in fine flammable fuel than the other. These two regions have also very different population densities, with the coastal area being much more populated than the internal region. It is important to continue monitoring these areas of change because the high fire activity in these areas may lead to further land conversion in the future.

REFERENCES

Archibald, S., Roy, D.P., van Wilgen, B.W., Scholes, R.J., 2009. What limits fire? An examination of drivers of burnt area in Southern Africa. Glob. Chang. Biol. 15 (3), 613–630.

Beerling, D.J., Osborne, C.P., 2006. The origin of the savanna biome. Glob. Chang. Biol. 12, 2023–2031.

Bond, W.J., Woodward, F.I., Midgley, G.F., 2005. The global distribution of ecosystems in a world without fire. New Phytol. 165, 525–538.

Bothma, J., du, P., 1996. Game ranch management. Van Schaik Publishers, Pretoria.

FAO, 2000. Africover multipurpose land cover databases for Kenya. FAO, Rome. www.africover.org.

Giglio, L., Descloitres, J., Justice, C.O., Kaufman, Y., 2003. An enhanced contextual fire detection algorithm for MODIS. Remote. Sens. Environ. 87, 273–282.

Govender, N., Trollope, W.S.W., van Wilgen, B.W., 2006. The effect of fire season, fire frequency, rainfall and management on fire intensity in savanna vegetation in South Africa. J. Appl. Ecol. 43, 748–758.

Hardesty, J., Myers, R., Fulks, W., 2005. Fire, ecosystems and people: a preliminary assessment of fire as a global conservation issue. George Wright Forum 22 (4), 78–87, The Nature Conservancy.

Justice, C.O., Korontzi, S., 2001. A review of the status of satellite fire monitoring and the requirements for global environmental change research. In: Ahern, F.J., Goldammer, J., Justice, C.O. (Eds.), Global and Regional Wildland Fire Monitoring from Space: Planning a Coordinated International Effort. SPB Academic Publishing, The Hague, pp. 1–18.

Keeley, J.E., Rundel, P.W., 2005. Fire and the Miocene expansion of C4 grasslands. Ecol. Lett. 8, 683–690.

Komarek, E.V., 1965. Fire ecology: grasslands and man. In: Proc. Tall Timbers Fire Ecol. Conf.4, pp. 169–220.

Komarek, E.V., 1971. Lightning and fire ecology in Africa. In: Proc. Tall Timbers Fire Ecol. Conf.11, pp. 473–511.

Lehsten, V., Harmand, P., Palumbo, I., Arneth, A., 2010. Modelling burned area in Africa. Biogeosciences 7 (10), 3199–3214.

Myers, R.L., 2006. Living with fire: sustaining ecosystems and livelihoods through integrated fire management. Global Fire Initiative. The Nature Conservancy.

NASA-FIRMS, 2012. MODIS Active Fire Detections. Data set. http://earthdata.nasa.gov/firms.

Phillips, J.F.V., 1965. Fire as a master and a servant: its influence in the bioclimatic regions of Trans-Saharan Africa. In: Proc. Tall Timbers Fire Ecol. Conf.4, pp. 7–10.

Roy, D.P., Jin, Y., Lewis, P.E., Justice, C.O., 2005. Prototyping a global algorithm for systematic fire-affected area mapping using MODIS time series data. Remote. Sens. Environ. 97 (2), 137–162.

Santoso, H., Alvarado, E., Ainuddin Nuruddin, A., Rodríguez-Trejo, D.A., Swaty, R., Schmidt, D., Kaufmann, M., Myers, R., Alencar, A., Kearns, F., Johnson, D., Smith, J., Zollner, D., Fulks, W., 2007. Fire, ecosystems and people: threats and strategies for global biodiversity conservation GFI Technical Report 2007-2. The Nature Conservancy, Arlington, VA.

Scott, A.C., 2000. The pre-Quaternary history of fire. Palaeogeogr. Palaeoclimatol. Palaeoecol. 164 (1–4), 281–329.

Scott, A.C., Glasspool, I.J., 2006. The diversification of Paleozoic fire systems and fluctuations in atmospheric oxygen concentration. Proc. Natl. Acad. Sci. U.S.A. 103, 10861–10865.

Tansey, K., Grégoire, J.-M., Stroppiana, D., Sousa, A., Silva, J., Pereira, J.M.C., Boschetti, L., Maggi, M., Brivio, P.A., Fraser, R., Flasse, S., Ershov, D., Binaghi, E., Graetz, D., Peduzzi, P., 2004. Vegetation burning in the year 2000: global burned area estimates from SPOT vegetation data. J. Geophys. Res. 109, D14S03. http://dx.doi.org/10.1029/2003JD003598.

Thomson, R., 1992. The Wildlife Game. Nyala Wildlife Publications Trust, Westville, South Africa.

Trollope, W.S.W., 1990. Veld management with specific reference to game ranching in the grassland and savanna areas of South Africa. Koedoe 33 (2), 77–86.

University of Maryland. MODIS Burned Area Product. http://modis-fire.umd.edu/.

van Wilgen, B.W., Govender, N., Forsyth, G.G., Kraaij, T., 2011. Towards adaptive fire management for biodiversity conservation: experience in South African National Parks. Koedoe, 53 (2), 9, 982.

Walters, G., 2012. Customary fire regimes and vegetation structure in Gabon's Bateke Plateaux. Human Ecol. Interdiscip. J. 40 (5), 661–810.

Chapter 15

Importance and Dynamics of the Mangroves in Kenya

Bernard K. Kirui

Department of Natural Resources, Egerton University. P. O. Box 536, 20115 Egerton, Kenya, and Joint Research Centre—European Commission, Via E. Fermi, 2749 Ispra, Italy

1 INTRODUCTION

Mangroves are the dominant vegetation along many tropical and subtropical coastlines with an estimated global coverage of 137,760 km^2 (Giri et al., 2011). They provide a wide range of ecosystem goods (fuelwood, medicine, food and construction materials) and services (fisheries nursery grounds, sediment trapping, coastal protection from storm surges and tsunamis, sewage phytoremediation and carbon sequestration) of immense value to local, national and global communities (Mumby et al., 2004; Barbier et al., 2008).

Despite this importance, these unique coastal forests are among the most threatened habitats in the world (FAO, 2007). Local communities living adjacent to mangrove ecosystems have traditionally collected fuel wood, fish and other natural resources from them at rates that were sustainable (Dahdouh-Guebas et al., 2000). However, in recent decades, many coastal areas have come under intense pressure from rapid urban and industrial development, compounded by a lack of effective governance and/or power among responsible government institutions (Field, 1995; Abuodha and Kairo, 2001). Mangroves have been overexploited or converted to various other forms of land use, including agriculture, aquaculture, salt ponds, urban and industrial development and coastal roads and embankments. Analyses of the true economic value of mangroves indicate that their destruction for short-term profit is usually economically irrational; in fact, the discrepancy between their value as intact systems and their value after destruction is one of the greatest for all habitats (Balmford et al., 2004).

In Kenya, estimates of mangrove cover vary between 46,000 and 52,980 ha (FAO, 2007; Kirui et al., 2013). Mangroves are found in creeks, estuaries, bays and river deltas along the 536 km coastline, which extends over 3° latitude from 1°42′ south to 4°40′ south. The largest formations occur in the north coast around the Lamu area and at the Tana River delta, which consist of 70% of mangrove formations (Kirui et al., 2013) in Kenya. The mangroves in Lamu

Developments in Earth Surface Processes, Vol. 16. http://dx.doi.org/10.1016/B978-0-444-59559-1.00015-3

are the most productive due to their proximity to the upwelling conditions up north (Kairo et al., 2002). Smaller mangrove formations are found in the mouths of semiperennial and seasonal coastal rivers on the south coast, at Shimoni–Vanga, Funzi and Gazi bays, and Port Reitz, Tudor, Mtwapa, Kilifi and Mida Creeks. All the nine East African mangrove species are represented in Kenya; *Rhizophora mucronata* (locally called mkoko) and *Ceriops tagal* (mkandaa) are very common, while *Heritiera littoralis* and *Xylocarpus mekongensis* (syn *X. moluccensis*) almost rarely found. Two basic formations have been described: fringe communities that do not show patterns of zonation and creek mangrove formations found in low-energy, more sheltered environments, which are often larger in aerial extent and may show zonation patterns. This is generally highly influenced by the level of inundation and the salinity of soil and water; a typical (FAO, 2003) pattern from seaward to landward is *Sonneratia alba*, *R. mucronata*, *Bruguiera gymnorrhiza*, *C. tagal*, *Avicennia marina*, *Xylocarpus granatum*, *Lumnitzera racemosa* and *H. littoralis*.

Mangrove forests in Kenya provide many direct products—both timber and nontimber. Timber products include firewood, building poles and charcoal used in urban and rural areas. Poles used in construction are normally graded into different utilization classes depending on their uses (Table 1).

TABLE 1 Mangrove Poles Sizes and Their Uses

Local Name for Utilization Classes	Diameter Class (cm)	Uses	Preferred Species
Fito	2.5–3.5	Used to fill walls	Usually young *R. mucronata* and *C. tagal*
Pau	4.0–7.5	Used as roof frames	Usually young *R. mucronata* and *C. tagal*
Mazio	7.5–11.5	Used to build main frame of house walls	Usually young *R. mucronata* and *C. tagal*
Boriti	11.5–13.5	Used to build fences, mainly for tourist developments	*R. mucronata*
Nguzo	14.0–20.0	Used at the corner of houses to support the roofs	Mature *R. mucronata* and *B. gymnorrhiza*
Vigingi	20.5–35.0	For support of the main roof of larger tourist hotels	Mature *B. gymnorrhiza*

Source: Kairo et al. (2001)

Historical records show that as early as 200 BC, mangrove poles were exported to Arabian countries (Rawlins, 1957). By 1970, the country was exporting about 34,000 scores (1 score = 20 poles) of mangrove poles to Somalia and countries in the Middle East (FAO, 1981). A presidential ban that was introduced in 1982 and meant to reduce mangrove overexploitation effectively ended mangrove exports (Government of Kenya, 2009).

Other uses of mangrove poles include boat masts and fish traps/stakes. Larger logs of mangroves, especially of *A. marina*, are used in traditional boat construction. Aerial roots of *S. alba* are also used as floaters for fishing nets. Mangrove wood is also utilized by the local communities for furniture. Among the nontimber products derived from mangrove forests include honey harvesting, medicinal values, crabs and fish caught inside the forest (Abuodha and Kairo, 2001).

Mangrove also provides an array of ecosystem services. Mangrove areas are primary nursery areas for commercially important species of fish (Kimani et al., 1996) and contribute to offshore productivity via the outwelling of detritus (UNEP, 2011). The root systems retard water flow and create a quiet environment that encourages sedimentation and inhibits resuspension (Field, 1995). This encourages sediment stabilization, which leads to shoreline protection (Kairo et al., 2009). Mangroves also perform waste disposal services by removing fertilizer and pesticide surpluses, industrial wastes and sewage (PUMPSEA, 2008). Fixation of carbon dioxide is also identified as a service by mangroves (Bouillon, 2011). Indirect interest in species found in mangrove ecosystems stems from an appreciation of the diversity of life and/or individual species (ecotourism). Mangroves also provide services that are associated with knowledge, for example, many mangroves are visited by students and scientists (Kairo et al., 2009).

2 THREATS TO KENYAN MANGROVES

As is the case in many parts of the world, the mangrove ecosystems in Kenya face a number of challenges that threaten their existence. Overexploitation for wood products is the main factor responsible for their degradation (Abuodha and Kairo, 2001). Lack of cutting plans escalates problems of mangrove management in Kenya. The government agency vested with the responsibility of managing mangroves and other forests in Kenya lacks adequate resources for implementing management guidelines. In most cases, therefore, selective removal of quality poles of suitable species has tended to leave out inferior species unsuitable for the market. Quality poles have been wiped out in most mangrove areas of Mombasa, Kwale and Kilifi districts where population density is highest along the coast (Abuodha and Kairo, 2001).

Salt extraction has also led to loss of mangroves. Currently, there are more than six salt works in Ngomeni where most of extraction is carried in Kenya, landing 71,400 tonnes of salt per year (Yap and Landoy, 1986). Environmental impacts associated with salt extraction include hypersalinity in areas close to mangroves, leading to their deaths. Poor land use practices in the hinterland have increased sediment loads into mangroves, leading to siltation of breathing roots of the trees and eventual death of the system. The situation was worsened during the 1997/1998 El Niño rains that hit the country causing massive death of mangroves in many areas along the coast, most of which have experienced no recovery to date (M'rabu et al., 2012). Another threat facing Kenyan mangroves is oil pollution. For instance, between 1983 and 1993, Mombasa port and surrounding waters experienced tonnes of oil spills that affected the mangroves of Port Ritz and Makupa creeks (Abuodha and Kairo, 2001).

A new threat to mangroves in Eastern Africa is the projected sea-level rise due to climate change. Climate change impacts are also associated with increased flooding/sedimentation and aridity (IPCC, 2007). Since a coastal area where mangroves occur is low-lying land, a small increase in sea level will mean that mangroves will be submerged unless they can migrate to new areas in the mainland (Huxham et al., 2010). At the Kenyan coast, most areas where mangroves could migrate to have already been occupied by human settlement and/or infrastructure. Evidence of death of mangroves due to climate change-related impacts has been observed in several areas along the coast such as Gazi bay, Mwache creek, Ngomeni, Tana River and Dodori creek.

3 CONSERVATION MEASURES

With the realization of the loss of mangrove goods and services and the potential catastrophic effects both on local and national economies, the government agencies together with the communities living adjacent to the mangrove ecosystems have put in place efforts to ensure sustained utilization of mangroves. In 2005, a new forest act was enacted; among the provisions in this act was the creation of an agency with the mandate of managing forests in Kenya including mangroves. Through its licencing system, the agency regulates the harvesting of mangroves. The act also provides for comprehensive community involvement in forest management through creation of community forests associations. For a long time, because of lack of comprehensive management plans, coupled with lack of resources to monitor harvesting system used by mangrove cutters, mangrove cutting is unregulated in all areas. However, there are recent ongoing efforts aimed at development of mangrove management plans.

The restoration of mangroves will benefit local communities, and these habitats are rich in natural resources and also are effective filters for pollution. Mangroves are the first marine ecosystem in Kenya where attempts at restoration have been implemented. In Gazi bay, located in the south coast of Kenya, a research–community partnership has led to the restoration of over 100 ha over a period of 10 years (Government of Kenya, 2009). An economic valuation of a restored 12-year plantation has estimated the net income to be approximately US$ 3000/ha/year (Kairo et al., 2009). Such information can be used to persuade the government to invest in mangrove reforestation. Also, there is a need to scale up the mangrove reforestation activities to other parts of the Kenyan coast if it is to have a significant impact at national level.

Mangrove forests are highly productive and efficient at capturing carbon, much of which ends up buried below ground and can therefore be permanently stored away from the atmosphere. This opens up possibilities for using payments for carbon credits to help mangrove conservation and to bring revenue for local people. Currently, efforts are underway at Gazi bay in the south coast of Kenya aimed at establishing the first mangrove Payment of Ecosystem Services (PES) project in Kenya. Once in operation, the mangrove-adjacent communities are expected to receive benefits in return for retaining carbon stocks in the conserved forest.

REFERENCES

Abuodha, P.A.W., Kairo, J.G., 2001. Human-induced stresses on mangrove swamps along the Kenyan coast. Hydrobiologia 458, 255–265.

Balmford, A., Gravestock, P., Hockley, N., McClean, C.J., Roberts, C.M., 2004. The worldwide costs of marine protected areas. Proc. Natl. Acad. Sci. U.S.A. 101, 9694–9697.

Barbier, E.B., Koch, E.W., Silliman, B.R., Hacker, S.D., Wolanski, E., Primavera, J., Granek, E.F., Polasky, S., Aswani, S., Cramer, L.A., Stoms, D.M., Kennedy, C.J., Bael, D., Kappel, C.V., Perillo, G.M.E., Reed, D.J., 2008. Coastal ecosystem-based management with nonlinear ecological functions and values. Science 319, 321–323.

Bouillon, S., 2011. Carbon cycle: storage beneath mangroves. Nat. Geosci. 4, 282–283.

Dahdouh-Guebas, F., Mathenge, C., Kairo, J.G., Koedam, N., 2000. Utilization of mangrove wood products around Mida Creek (Kenya) amongst subsistence and commercial users. Econ. Bot. 54, 513–527.

FAO, 1981. Tropical forest resources assessment project. Forest resources of tropical Africa. Part II: country briefs. In: FAO, U. (Ed.), p. 586.

FAO, 2003. Status and trends in mangrove area extent worldwide. In: Wilkie, L.M., Fortuna, S. (Eds.), Forest Resources Assessment Working Paper 63. Forestry Department, Food and Agriculture Organization of the United Nations, Rome.

FAO, 2007. The world's mangroves 1980-2005. A thematic study prepared in the framework of the Global Forest Resources Assessment 2005, FAO Forestry Paper 153.

Field, C.D., 1995. Impact of expected climate-change on mangroves. Hydrobiologia 295, 75–81.

Giri, C., Ochieng, E., Tieszen, L.L., Zhu, Z., Singh, A., Loveland, T., Masek, J., Duke, N., 2011. Status and distribution of mangrove forests of the world using earth observation satellite data. Glob. Ecol. Biogeogr. 20, 154–159.

Government of Kenya, 2009. State of the Coast Report: Towards Integrated Management of Coastal and Marine Resources in Kenya. National Environment Management Authority (NEMA), Nairobi, 88 pp.

Huxham, M., Kumara, M., Jayatissa, L.P., Krauss, K.W., Kairo, J.G., Langat, J., Mencuccini, M., Skov, M.W., Kirui, K.B., 2010. Intra and inter-specific facilitation in mangroves may increase resilience to climate change threats. Philos. Trans. R. Soc. 365, 2127–2135.

IPCC, 2007. Climate Change 2007: The Physical Science Basis, Summary for Policy Makers. Contribution of Working Group I to the Fourth Assessment Report of the Intergovernmental Panel on Climate Change. WMO & UNEP.

Kairo, J.G., Dahdouh-Guebas, F., Bosire, J., Koedam, N., 2001. Restoration and management of mangrove systems—a lesson for and from the East African region. S. Afr. J. Bot. 67, 383–389.

Kairo, J.G., Kivyatu, B., Koedam, N., 2002. Application of remote sensing and GIS in the management of Mangrove Forests within and adjacent to Kiunga marine protected area, Lamu, Kenya environment. Dev. Sustain. 4, 153–166.

Kairo, J.G., Wanjiru, C., Ochiewo, J., 2009. Net pay: economic analysis of a replanted mangrove plantation in Kenya. J. Sustain. For. 28, 395–414.

Kimani, E.N., Mwatha, G.K., Wakwabi, E.O., Ntiba, J.M., Okoth, B.K., 1996. Fishes of a shallow tropical mangrove estuary, Gazi, Kenya. Mar. Freshw. Res. 47, 857–868.

Kirui, K.B., Kairo, J.G., Bosire, J., Viergever, K.M., Rudra, S., Huxham, M., Briers, R., 2013. Mapping of mangrove forest land cover change along the Kenya Coastline using Landsat imagery. Ocean Coast. Manag. 83, 19–24.

M'rabu, E., Bosire, J.O., Cannicci, S., Koedam, N., Dahdouh-Guebas, F., 2012. Mangrove die-back due to massive sedimentation and its impact on associated biodiversity. In: Dahdouh-Guebas, F., Satyanarayanam B. (Eds.), Proceedings of the International Conference 'Meeting on Mangrove ecology, functioning and Management - MMM3', Galle, Sri Lanka, 2-6 July 2012. VLIZ Special Publication 57, pp. 104–105.

Mumby, P.J., Edwards, A.J., Arias-Gonzalez, J.E., Lindeman, K.C., Blackwell, P.G., Gall, A., Gorczynska, M.I., Harborne, A.R., Pescod, C.L., Renken, H., Wabnitz, C.C.C., Llewellyn, G., 2004. Mangroves enhance the biomass of coral reef fish communities in the Caribbean. Nature 427, 533–536.

PUMPSEA, 2008. Peri-urban mangrove forests as filters and potential phytoremediators of domestic sewage in East Africa. Final activity report. European Commission: FP6, INCO-CT2004-510863, Instituto de Ciências Aplicadas e Tecnologia, Lisbon, Portugal, 447 pp.

Rawlins, S. 1957. The East African mangrove trade. East Africa Herbarium, National Museum of Kenya, Nairobi, Kenya.

UNEP, 2011. Economic Analysis of Mangrove Forests: A Case Study in Gazi Bay, Kenya. UNEP, iii + 42 pp.

Yap, W.G., Landoy, R.J., 1986. Report on a Survey of the Coastal Areas of Kenya for Shrimp Farm Development, Rome.

Chapter 16

Freshwater Ecology of Kenyan Highlands and Lowlands

Dorothy Wanja Nyingi*, Nathan Gichuki[†] and Mordecai O. Ogada[‡]
*Ichthyology Section and KENWEB, National Museums of Kenya, P.O. Box 40658, Nairobi, Kenya
[†]University of Nairobi, P.O. Box 30197, Nairobi, Kenya
[‡]Laikipia Wildlife Forum, P.O. Box 764, Nanyuki, Kenya

1 INTRODUCTION

Kenya lies astride the equator, but the country experiences wide variations in climate. Freshwater ecosystems occur mainly in the narrow coastal plain, the central highlands and western parts of Kenya. These cover about 25%, while the rest of the country (75%) experiences arid or semiarid climate. The annual rainfall in the country averages 621 mm per year, but it varies seasonally and spatially. The amount of annual rainfall available to maintain freshwater ecosystems varies from a maximum of 750 mm in the dry areas to a maximum of 1700 mm in the coastal plain, the central highlands and Lake Victoria basin. Surface runoff and natural river flows as well as the rate of seepage of rainwater are influenced by variations in intensity of rainfall, density of vegetation, type of soil and the presence or absence of wetlands.

The headwaters of Kenya's five primary catchment areas all arise in five indigenous montane forest areas commonly referred to as Kenya's 'water towers' (Figure 1). The water towers, namely, Aberdare Range, Mount Kenya, Mount Elgon, Mau Complex and Cherangani Hills, have an average annual rainfall of 2300 mm and a potential rainfall–runoff ratio of 65%, which is more than 75% of the renewable surface water resources of Kenya (19,691 million m^3 per year). They occur in a wide range of rainfall regimes and at varying levels of altitudes, mainly between 1500 and 2500 m above sea level. These towers provide water for domestic uses, industrial and irrigation developments, hydropower generation and other needs.

The five drainage systems in Kenya are determined mainly by the Great Rift Valley, which runs in an almost north–south direction, and the Lake Victoria system that drains the western flanks and eastwards-flowing rivers to Indian Ocean. These drainage basins are Lake Victoria, covering 8.0% of the country; Rift Valley

Developments in Earth Surface Processes, Vol. 16. http://dx.doi.org/10.1016/B978-0-444-59559-1.00016-5

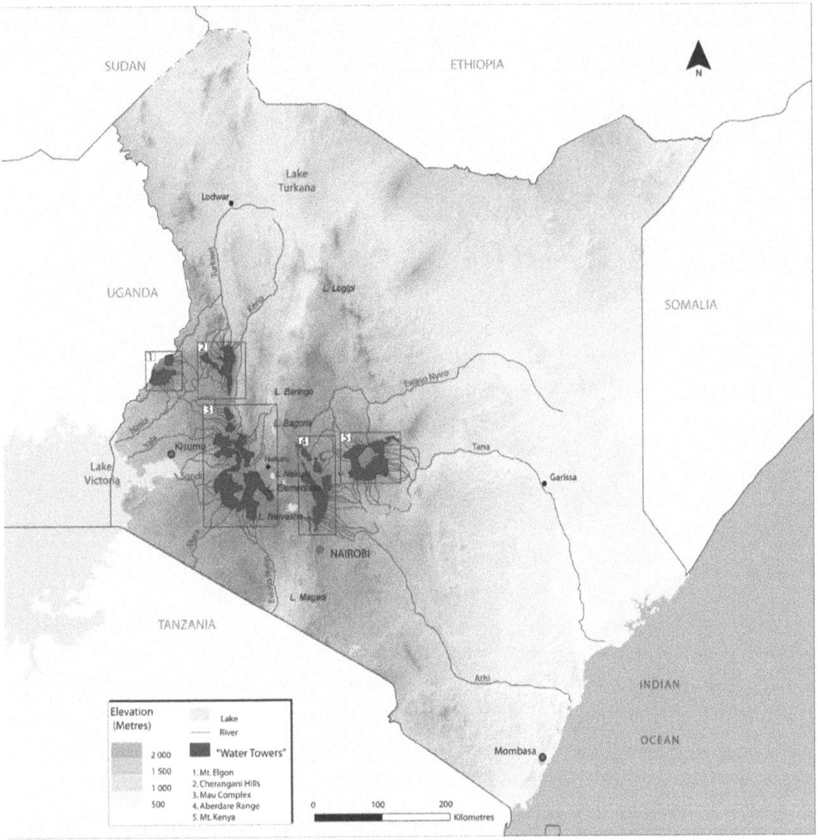

FIGURE 1 Map of Kenya highlighting the five water towers or headwaters of the main river catchments.

and inland lakes, covering 22.5% of the country; Athi River and coast, covering 11.5% of the country; Tana River, covering 21.7% of the country and Ewaso Ng'iro, covering 36.3% of the country (Seegers et al., 2003). All drainage or river basins are defined by the watershed, which begins in the high-altitude areas or highlands from which snowmelt and precipitation feed into streams and rivers. These rivers in turn feed into a middle reach that is the mainstream of the river and then flow to the lowlands, which empty into the ocean or into swamps or lakes; in some cases, it may dry out. This chapter describes the present status, threats and management challenges of Kenya water towers and the main drainage systems.

2 KENYA WATER TOWERS

2.1 Mount Kenya

Mount Kenya is the highest mountain in Kenya and the second highest in Africa, after Kilimanjaro. The highest peaks of the mountain are Batian

(5199 m (17,057 ft.)), Nelion (5188 m (17,021 ft.)) and Point Lenana (4985 m (16,355 ft.). Mount Kenya is located in central Kenya, just south of the equator, around 150 km (93 miles) north-northeast of the capital Nairobi. Mount Kenya is a stratovolcano created approximately 3 million years after the opening of the East African rift. Before glaciation, it was 7000 m (23,000 ft.) high. It was covered by an ice cap for thousands of years. The erosion from the gradual ice melt has resulted in numerous valleys radiating from the centre, which form the sources of several streams and tributaries that join to and from major rivers further downstream.

There are currently 11 small glaciers and the forested slopes are an important source of water for part of the Ewaso Ng'iro basin and the Tana River basin. The Mount Kenya ecosystem and unique natural resources, however, are being degraded and in many cases destroyed by uncontrolled use and exploitation. The forests especially are under extreme threat from human-induced illegal activities. In 1999, an aerial survey of the then forest reserve conducted by the Kenya Wildlife Service highlighted the ongoing destruction of the forest, leading to the decision to regazette the forest reserve as a natural reserve (KWS, 2003). The report gives an excellent summary of the threats on Mount Kenya's forests. Since the 1999 report and following the subsequent regazettement of the reserve and the formation of the Mount Kenya Trust, a follow-up report was released. This report was compiled by the Kenya Wildlife Service, UNEP, the Kenya Forest Working Group and the Durrell Institute of Conservation and Ecology, University of Kent. The report shows encouraging improvements; however, extensive destruction still continues to threaten the future of Mount Kenya's precious resources.

The major threats to Mount Kenya's ecosystems currently include the following: illegal logging—the KWS survey recorded fresh and old logging of indigenous trees in natural forest. Some of the most targeted species were identified separately. They include camphor (*Ocotea usambarensis*), cedar (*Juniperus procera*), wild olive (*Olea europaea*) and E.A. rosewood (*Hagenia abyssinica*). The major drivers of this activity are the widespread use of woodfuel in the lowland areas and the demand for cedar fencing posts in the farms and ranches of the Laikipia plateau. Intensive charcoal making, in various stages of production, was also observed in the lower parts of the Upper Imenti Forest Reserve, whereby significant areas of indigenous forests have been destroyed.

Secondly, there is the grazing of livestock. In total, 4258 heads of livestock were counted in the reserved forests during the survey. This number rises to over 10,000 during the periodical droughts in northern Kenya when pastoralist communities drive their livestock into the forest in search of pasture. In the rainy seasons, goats are a particularly acute problem because their browsing has a significant negative impact on the growth of seedlings and tree recruitment. The third major threat to this forest is fire. Fires occur in the forest as a result of illegal charcoal-burning activities, honey gatherers, and carelessness on the part of guides and tourists climbing the mountain and camping in the forest. In March–April 2012, over 60,000 ha of the forest was destroyed in a fire that was estimated to have cost the country over Ksh. Eight billion (US $95,000,000). The risk level is particularly high in the dry season when the deep peat layer in the moorlands is easily combustible and very difficult to extinguish without the use of heavy equipment.

Significant policy weaknesses underlie several of the threats to Mount Kenya and other forests in Kenya, notably the plantation establishment livelihood improvement scheme. This was established by the Kenya Forest Service (KFS) to engage communities in establishing plantation forests as they practise agriculture in the forests.

2.2 Mount Elgon

Mount Elgon is an extinct shield volcano on the border of Uganda and Kenya, north of Kisumu and west of Kitale. The mountain's highest point, named 'Wagagai', is located within the country of Uganda. Notable physical features on Mount Elgon include the caldera, which is one of the largest intact calderas in the world. It also has a large network of caves including Ngwarisha, Makingeny, Chepnyalil and Kitum. Kitum Cave is over 60 m wide and penetrates 200 m into the mountain. It was formed over the centuries by the activities of wild elephants that lick the salt exposed by gouging the walls with their tusks. The mountain soils are fertile red laterite, which have resulted in intensive cultivation on the slopes. The mountain is the catchment for the several rivers such as the Suam River, which becomes the Turkwel downstream and which drains into Lake Turkana, the Nzoia River and the Lwakhakha, which flow to Lake Victoria. The town of Kitale is on the foothills of the mountain. The forest around the mountain is protected by two Mount Elgon National Parks, one on each side of the international border. However, the effect of the protection is limited by the small size of these national parks. They have been effective in protecting the rare flora found on the mountain, including *Ardisiandra wettsteinii*, *Carduus afromontanus*, *Echinops hoehnelii*, *Ranunculus keniensis* and *Romulea keniensis*, but there is not enough ground cover to prevent siltation of the rivers.

Policy weaknesses are a challenge to the conservation of Mount Elgon forest, and this is further complicated by its situation on an international border. The reserve on the Ugandan side was upgraded to a 'national park' in 1993, and its boundaries were expanded from the reserve's previous *de facto* endowment—a process that entailed, by some estimates, the eviction of more than one hundred thousand people from within the new boundaries. The people were forcibly evicted from their homes by the Uganda Wildlife Authority (UWA) and the Uganda People's Defence Forces (UPDF, the Ugandan military). Incidences of displaced villagers confronting rangers have become commonplace as they fight to reclaim their lost land. In Kenya, the forest has also been the scene of fighting between security forces and local militias. This social instability has compromised environmental conservation efforts in Mount Elgon over the years.

Further pressure on forest and land resources results from population explosion of approximately 2 million people living on the slopes of Mount Elgon and growing at a rate of 2.3–4.3% per annum. Limited livelihood options pose a threat to the environmental integrity in the Mount Elgon ecosystem because a large proportion of the population survive on resources obtained directly from the forest, including mushrooms, bamboo shoots, vegetables, firewood and timber. Predominant subsistence farming on the slopes where the rivers flow from has resulted in poor farming practices, for example, cultivation of riparian zones. Illegal settlements in the forest have led to the felling of trees in order to create space for construction and cultivation. This has also compromised the quality of wildlife and biodiversity habitats in the area. Poaching has substantially reduced the biodiversity as animals are hunted down for game meat. Records at the Kenya Wildlife Service and UWA show that the targeted species include duikers, buffaloes, leopards, red-tailed monkeys, various antelopes and birds. Hitherto, potential tourist attractions have all escaped to the Kenya side of the mountain and many of the species are now at the brink of local extinction. Perhaps the most robust challenge to law enforcement in this particular water catchment is the 'no man's land' that exists between international boundaries. Perpetrators of environmental crimes in Kenya regularly escape across the border to Uganda and vice versa.

2.3 Aberdare Range

The Aberdare Range is located in central Kenya and straddles the equator. The Range stretches over 125 km from Nyahururu in the north to Limuru in the south. It is the third highest mountain in Kenya, with two main peaks, Oldonyo Lesatima (also known as Sattima) and Kinangop, which reach, respectively, altitudes of 4001 and 3906 m. The Range presents a deeply dissected topography sloping gradually to the east. In contrast, the western side drops along impressive fault escarpments towards the Rift Valley. The climate of the Aberdare Range is generally characterized by two rainy seasons—from April to May and October to November. Rainfall varies with altitude and exposure to the dominant wind from the Indian Ocean but reaches a maximum of around 2600 mm annually on the southeastern slopes and drops to less than 900 mm a year on the northern and southwestern lee slopes. Various vegetation zones can be distinguished on the Aberdare Range, including the closed-canopy forest belt, the bamboo zone and the subalpine and alpine vegetation. The forest belt covers a major part of the Range. Most of the forest is gazetted as forest reserves. However, parts of the upper forest zone fall within the Aberdare National Park. The forest belt of the Aberdare Range is characterized by a high diversity of forest types, because of the wide altitudinal range (from 1800 to 3600 m) and the climatic differences between the slopes.

The Aberdare highlands form one of the most economically important water catchments in Kenya, supplying over 70% of the water needs for Nairobi city, which in turn produces approximately 40% of Kenya's GDP. It is also the source of the Ewaso Ng'iro River, which is the lifeline of the northern Kenya rangelands, including the Laikipia plateau.

The major threat to the ecological integrity of the Aberdare mountains is human population growth and related habitat transformation. It is estimated that 6100 ha has already been largely converted to agriculture. This destruction and conversion into settlements includes 4475 ha around Kipipiri and 1104 ha in Ragia forest. Unattended charcoal kilns and illegal logging of indigenous trees occurs extensively in the northern, western and southern slopes. In the southern slope for example the Rhino Ark, a charitable trust supporting conservation in Kenya, counted over 10,000 charcoal kilns in a recent aerial survey. The latter is an important refuge for the endangered mountain bongo (*Tragelaphus eurycerus*) in Kenya. In addition to the common threats posed by livestock grazing, quarrying is a major threat that is particular to this forest complex. Some 23 quarries were recorded in the Aberdare Range forests. Most of them were located in Nyeri Forest Reserve and Kiganjo Forest Reserve on the eastern slopes and on the northern part of the Aberdare Forest Reserve. All the quarries were outside fenced forest areas. This threat is largely driven by the construction industry in Nyeri town and its demand for the murram and rock that lies at an average depth of 4 m below the forest floor. This pressure from quarrying activities in the lower Aberdare forest underscores the importance of the fence erected around the protected area.

2.4 Cherangani Hills

The Cherangani Hills is an expansive area traversing three districts. It is situated in parts of Trans-Nzoia, Elgeyo-Marakwet and West Pokot districts. Of the water towers in Kenya, it is arguably the most heavily impacted by human activity and agriculture, since it occurs in Trans-Nzoia County, which is part of the country's 'bread basket' or the most productive maize-growing area. The destruction was largely driven by the politically based forest excisions that occurred in Kenya during the 1980s and 1990s. The actual forest cover is diminished to a level where it is not listed as a forest area in the current KFS strategic plan other than the Embobut area. The Cherangani Hills still functions as a catchment, drawing relief rainfall, but the poor condition of the forest is evident in the silt load carried by the Nzoia and Birbiriet rivers.

Increasing density of settlements and intensification of land-use activities driven by rapid population growth and increasing poverty continue to exert pressure on the Cherangani Hills forests. Absolute poverty in the area stands at 54%, largely driven by the weaknesses in agricultural policy and the resultant instability of maize prices in local markets. This circumstance has triggered community encroachment into the forests for timber, firewood and charcoal or in an effort to increase the land area under maize.

Cherangani Hills forest has witnessed conflicts between downstream and upstream communities over the dwindling water resources because of watershed degradation by upstream communities. This conflict often manifests as cattle rustling and ethnic hostilities. In addition, a declining natural resource base especially pastures and water has seen pastoralist communities venture into crop farming on the hillsides to diversify their livelihoods. As a result, the fertile but steep forestland has suffered serious degradation. Occasional fires are a threat to the Cherangani Hills. This is caused by immigrant labourers engaged in charcoal burning to supplement their dwindling seasonal incomes from the maize farms and clearing forest to access land, which they can occupy as squatters.

3 FRESHWATER ECOSYSTEMS OF ATHI RIVER BASIN

The Athi River Basin has three names at different parts of its flow and is thus named Athi–Galana–Sabaki river system. It is the second largest of the eastwards-flowing rivers of Kenya and drains a comparatively small catchment of 66,837 km^2 flowing across southeast Kenya into the Indian Ocean. The river provides water for urban and rural areas including Nairobi, Mombasa and Malindi towns. The Athi River flows through semiarid land with an average annual rainfall of 739 mm and a low groundwater recharge rate of 296 million m^3 per year. When compared with the other four drainage basins, Athi River Basin has the largest number of small dams (90), shallow wells (6720) and boreholes. That information clearly points to an area with limited surface water and consequentially limited freshwater ecosystems.

In the upper catchment, Athi River is fed by numerous streams originating on the densely populated Kikuyu Highland and Ngong Hills. The main permanent streams are Gatharaini, Mathare, Nairobi, Mpakasi and Kiserian. These streams originate from springs and small wetlands in the upper catchment area. There are also numerous water storage dams, such as Tigoni, Nairobi, Jamuhuri and Ngong forest dams, that were constructed across some of the streams to enhance water supply for domestic and farm use. In many places, however, the dams have become polluted with domestic or farm waste and their water is no longer suitable for human use.

Athi–Galana–Sabaki river system: Athi River at 14 falls, Ol Donyo Sabuk—Thika.

The middle part of Athi Basin is largely semiarid and most of the freshwater ecosystems are closely linked with seasonal rainfall. Most rivers, marshes, ponds and water pans are seasonal. The main rivers that supply water to Athi River for most of the year are the rivers Kiboko, Tsavo and Voi. The hydrology of the middle Athi Basin permits formation of numerous springs, water pools and shallow lakes. Rain and mist water seeps on Chyulu Hills and appears at Mzima Springs in Tsavo West National Park and at Umani Springs near Kibwezi. These water systems are extremely important because they support the water needs of the people, agriculture, livestock and wildlife.

The Kiboko wetland system has been formed by surface runoff and springs arising from seepage of rain and mist on the surrounding hills. Water seeping on Mount Kilimanjaro has permitted the formation of permanent swamps and a lake in Amboseli. The lake basin is flat and has fine alkaline sediments deposited over the years by seasonal runoff from the watershed, including sediments transported downstream by Namanga River. The basin holds shallow water during periods of heavy rainfall, but for most of the year, it remains an unfriendly dust bowl. Evaporation rate is high and soda ash is deposited on the soil surface as the water evaporates.

14 falls Below the Lugard Falls (near the R. Tsavo), the river is called Galana, while the lowermost course is the Sabaki, which flows to its mouth 11.3 km north of Malindi town. Flood plain lakes exist towards the mouth of the river and are often maintained by rainfall and runoff. These lakes include: Chem Chem, Jilore, Merikano, Mbirikano and Mekimba (Okeyo, 1998).

Athi River at Lugard Falls after which the river becomes Galana River.

Nairobi is the principal urban centre but there are other rapidly growing municipalities in the upper catchment of Athi River. These emerging municipalities include Kikuyu, Ngong, Kiserian, Athi River and Kitengela. Pollution with municipal, industrial and farm waste is a fundamental problem affecting freshwater ecosystems. Natural wetlands have also been under pressure from excessive water abstraction for domestic and farm use, encroachment by subsistence farmers and dumping of waste material. The available surface water is inadequate to meet the growing demand for water. Therefore, there is increasing abstraction of water from underground resources through wells and boreholes throughout the upper Athi catchment.

Malindi District as the most downstream riparian district of the Athi–Galana–Sabaki river system is, due to its hydrological position, dependent on any modification in water flow occurring further upstream. During the last severe drought in 1998–2000, there was inadequate quantity of water to meet the requirements of all the water users in Malindi District. Upstream water users overabstracted water, leaving none for users downstream. In the context of such water use competitions, Malindi District is vulnerable to every additional water abstraction further upstream. In the last decades, protection of river water users in Malindi District against overexploitation by co-riparians further upstream was not necessary, since most of the time, there has been enough water for all. Although Athi River as the main tributary to the Sabaki River tends to dry up in times of drought, Sabaki River in Malindi District has never dried up due to permanent flow of the Mzima Springs, which drains its water into Sabaki. A certain part of the clear Mzima Springs water is transferred by the so-called 1 pipeline to the urban region Mombasa. This water transfer (within the Athi River Basin) reduces water levels in Sabaki River. Due to sufficient amounts of Mzima Springs water, water use conflicts between water users in Mombasa and Malindi District arise only in times of drought, which occurs every 3–4 years.

4 FRESHWATER ECOSYSTEMS OF TANA RIVER BASIN

Tana River is the largest river in Kenya and flows for nearly 1000 km to the Indian Ocean. The river drains a catchment of 126,026 km^2, which encompasses the southern and eastern slopes of Mount Kenya (5100 m asl) and the eastern slopes of the Aberdare Mountain Range (4030 m asl). The upper catchment of the river receives 1000–1500 mm of annual rainfall, but the whole of Tana River Basin receives an average of 697 mm of rainfall per year. Most of the river flows through a semiarid region and its basin varies from 6 km wide in the upper part to 42 km wide in the lower part.

The upper catchment of Tana River has a large network of primary and secondary streams. Among the major secondary rivers feeding into Tana River are Sagana from the Aberdare catchment and Nyamindi and Thiba from the massive Mount Kenya catchment. Because of the high gradient of the land, most of the rivers are fast-flowing and their valleys have little marginal vegetation. However, where the land gradient in the valley is low, there are patches of riverine forest, marshes and swamps. Remnants of valley bottom wetlands occur widely in Nyeri, Kirinyaga, Embu and Meru Counties. The wetlands play an important hydrological role as they trap silt, store floodwater and regulate water flow in the valleys.

Apart from the natural freshwater ecosystems associated with streams and rivers, the upper catchment of Tana River has numerous artificial water ecosystems. These include stream impoundments for water storage, aquaculture ponds for fisheries research and production and irrigated small-scale and large crop fields. There are small fish farms at Kiganjo and Sagana, which are dedicated to cold stream fisheries research and fish fry production. These facilities produce and supply fish fingerlings to small-scale fish farmers in central and eastern Kenya. The waters of the permanent rivers Thiba and Nyamindi support 4000 ha of paddy fields at Mwea in Kirinyaga County. A few of the tributaries supplying water to the Tana, such as Sagana River, have been dammed to generate hydroelectricity, which is used locally or transmitted to the national power grid. Overall, the rivers of the Upper Tana Basin are a valuable source of water for human use, crops, livestock and wildlife, not only within the vicinity of the mountains but also further downstream.

The low-elevation sections of Tana River Basin cover at least 30% of the basin and fall in a semiarid environment. The dominant economic activities are dry land agriculture and livestock production in the semiarid rangelands.

Apart from Tana River and its numerous seasonal tributaries, the principal freshwater ecosystems in this part of the basin are the seven major impoundments constructed for water storage and generation of hydropower. These hydropower dams are Masinga, Kamburu, Gitaru, Kiambere, Kindaruma, Sasumaru and Bathi. Masinga dam is large (150 km^2), with 1560 million m^3 of water storage capacity. Most of the other dams are relatively small with storage capacities ranging from 2 to 10 million m^3 of water. The power generation capacities of these dams are gradually diminishing because of siltation and fluctuations in water supply from the Aberdare and Mount Kenya catchments.

Apart from hydropower generation, the dams store water, which is used by local people, livestock and wildlife. They support economically important fisheries resources, especially of tilapia, catfish and common carp. These resources have immense potential for sport fishing and other tourism activities. Availability of adequate water of high quality is fundamental to the development of hydropower and fisheries in the Tana Basin. Wildlife conservation and tourism as well as livestock production are currently high-value uses of the rangelands and natural wetlands in the middle and lower Tana Basin. However, expansion of cultivation in the highlands and increased levels of urbanization is contributing to diminishing water supply and increasing water pollution levels.

5 FRESHWATER ECOSYSTEMS OF EWASO NG'IRO BASIN

Ewaso Ng'iro is the largest drainage basin covering 210,226 km^2 or 36% of the total area of Kenya. However, the basin has low mean annual rainfall (411 mm), high evaporation rate and the low mean annual runoff. Nevertheless, because of its large surface area and highly permeable geologic formations, the basin has a high groundwater recharge rate and a huge groundwater reserve. Surface water is scarce and hence most freshwater ecosystems are small and largely seasonal or periodic.

The major freshwater ecosystems comprise Ewaso Ng'iro River and its tributaries, and the Lorian Marula, Rumuruti and Pesi swamps. There are also natural springs, numerous man-made dams and pans and seasonal marshes. These wetlands serve as sources of water for people, livestock and wildlife. The swamps and seasonal marshes provide valuable dry season grazing areas for livestock and wildlife. The occurrence and distribution of freshwater ecosystems in Ewaso Ng'iro Basin are influenced by a number of factors. These include the amounts of rainfall and prevailing evaporation rate as well as local topography.

The ultimate effect of the variable structure and distribution of freshwater ecosystems are consequences of changes in biotic communities.

Freshwater ecosystems cover approximately 1% of Kenya's rangelands. However, the structural and functional dynamics of the ecosystems have had profound implications on the availability of water, pasture and fisheries resources and sustainable livestock production and wildlife conservation. The freshwater ecosystems in these dry lands have served like 'magnets' that attract concentrations of people, livestock and wildlife. They are therefore highly vulnerable to disturbance and degradation through overexploitation of natural resources, such as water extraction, overgrazing by livestock, and drainage of wetlands for agriculture, settlements and development of infrastructure.

6 FRESHWATER ECOSYSTEMS OF LAKE VICTORIA BASIN

Lake Victoria is the largest freshwater body in Africa. It has a catchment area of 180,950 km^2, 21.5% of which lies in Kenya. The lake is a vital water reservoir with an average depth of 40 m and an expansive surface area of 68,870 km^2. Only a small portion (6%) of the water surface occurs in Kenya. The lake's shore line is 3450 km long, 17% of which is in Kenya. Lake Victoria receives 85% of its water from direct precipitation, which averages 1350 mm per year and the balance of 15% derives from rivers, which drain the surrounding catchment. Most of the water leaving the lake (85%) is through direct evaporation from the surface, while 15% is lost via Nile River outflow.

The eastern catchment is wholly in Kenya and comprises Mount Elgon, Cherangani, Mau Mountain ranges and other highlands in western and southwestern Kenya. The average annual rainfall in the catchment ranges from 1300 mm in the Nzoia–Yala Basin in the north through 1415 mm in the Sondu Miriu Basin in the Kisii highlands to 1040 mm in the Mara River basin southern segment of Mau ranges. The eastern catchment is drained by six main rivers: Nzoia, Yala, Nyando, Sondu Miriu, Gucha and Mara. Rivers Nzoia and Yala have the largest subbasin of 15,143 km^2 followed by Mara River (13,915 km^2) and Gucha River (6612 km^2). The other two rivers, Nyando and Sondu Miriu, have relatively small subbasins of about 3600 km^2. There are also numerous other streams and springs that originate in the lake basin and drain into the shores of Lake Victoria.

Most of the Kenyan portion of Lake Victoria lies in Winam Gulf, which lies in a 46,230 km^2 drainage basin. Apart from the gulf, there are other freshwater ecosystems, such as small lakes, dams, marshes and swamps, fish farms, sewage ponds and rice paddies, that occur on the surrounding land. The main satellite lakes are Kanyaboli, Sare and Namboyo, which occur on the Nzoia–Yala flood plain in the northern basin of Lake Victoria. In the southern basin, there is a small crater lake, known as Simbi Nyaima. The lake has saline water, though it is only about 100 m from the lake. There are also numerous water storage dams and fish ponds found in all parts of the eastern catchment.

The rivers draining the eastern catchment are short and tend to flow rapidly, transporting large loads of sediment and other pollutants from their catchment. They tend to have little aquatic vegetation, but they have linear blocks of woodland along their valleys. Their flood plains are flooded during periods of high rainfall in the catchment. The affluent rivers enter the lake through extensive swampy deltas, dominated by herbaceous vegetation, especially papyrus (*Cyperus papyrus*) and elephant grass (*Vossia pyramidalis*). The northern basin is covered by the extensive Yala Swamp with numerous open channels that drain the river and surface runoff slowly into the lake. There are also extensive delta wetlands at the mouths of Nyando River and Sondu Miriu.

The main freshwater issues in Lake Victoria basin are exploitation of the fishery, water quality and supply, rampant waterborne diseases and introduction of invasive species. The lake and marginal wetlands receive factory effluent, oil and grease, and sewage and storm water from the 18 towns in the basin and catchment area. The loading of untreated effluent discharge from urban centres has contributed to loss of oxygen and build-up of total phosphorus and nitrogen content in the lake. Much of the water in Winam Gulf is therefore murky, smelly and choking with algae and water hyacinth. This deteriorating water quality has serious implication on the shrinking aquatic biodiversity, the diminishing fisheries resources and limited access to clean and potable water, which is a fundamental livelihood and public health issue for the residents of Lake Victoria basin.

7 FRESHWATER ECOSYSTEMS OF THE RIFT VALLEY BASIN

Kenya's Rift Valley extends for 900 km from north to south and its altitude ranges from 400 m in the north to 1880 m in the central part and 600 m in the southern part. Except for the elevated central part of the valley, this internal drainage basin has semiarid to arid climate. Annual rainfall varies from 250 mm in Lake Turkana Basin through 900 mm around Lake Nakuru to 400 mm around Lake Magadi in the south. As a result of the low rainfall received in the basin, surface water is relatively scarce. The basin contributes 1% of national surface water yield but a significant proportion (20.3%) of the national groundwater yield. The uplands flanking the Rift Valley receive 1000–2000 mm of rainfall per year and serve as the main rainfall catchment for numerous springs, streams and rivers that drain into the valley.

The rivers and streams flowing into the Rift Valley are short and flow rapidly, especially as they flow down the escarpments. The main rivers are Turkwel, Kerio, Suam and Suguta, which flow into Lake Turkana; Molo, Ol Arabel and Perkerra, which flow into Lake Baringo; Subukia, which flows into Lake Bogoria; Malewa and Gilgil, which flow into Lake Naivasha; Njoro and Nderit, which flow into Lake Nakuru; and Mara and Ewaso Ng'iro (South), which flow into Lake Victoria and Lake Natron, respectively. There are also many seasonal streams that supply water to the Rift Valley lakes and wetlands. The rivers valleys and other areas with shallow underground water table support characteristic woodland vegetation dominated by *Acacia xanthophloea* and fig trees, such as *Ficus natalensis*. The rivers enter the lakes via small deltas with characteristic wetland reeds and sedges. The riverine woodlands provide vital habitat corridors for movement of birds and primates between the valley floor and the adjacent forested escarpments. The rivers are important for water supply and fishes, which support subsistence fishing, sport fishing and camping sites.

The Rift Valley basin contains a chain of freshwater and saline lakes, which occupy closed basins with no surface outflows. There are also permanent swamps and seasonal marshes as well as numerous streams that drain into the valley from the adjoining highlands. The Kenya Rift Valley contains seven major lakes, one of which is fresh (Naivasha), two are moderately saline (Turkana and Baringo), three are saline (Bogoria, Nakuru and Elementaita) and one is hypersaline (Magadi). Most of the Rift Valley lakes tend to be shallow, long and narrow. The saline and hypersaline lakes have narrow shores with large boulders, gravel, sand or expansive fine silt and crystalline salt. There are also seven minor lakes that are scattered along the valley floor from the north to the south. These minor lakes are Logipi, Kamnarok, Kijirtit, Solai, Ol Bolossat, Kwenia and Kabogo. The basins of the lakes are under semiarid climatic conditions, where the evaporation rate far exceeds the prevailing rainfall. In the short term, lake water levels fluctuate with surface water flow and supply from seepage in the watershed.

The water chemistry and associated sediment types influence the community structure of microorganisms, animals and plants found in each lake.

Freshwater ecosystems associated with the Rift Valley support a wide range of aquatic and terrestrial organisms. Most of the primary productivity in the saline lakes can be attributed to the floating photosynthetic algae (*Spirulina platensis*) and the variety of phytoplankton community. The freshwater lakes, especially Lakes Naivasha and Baringo, have abundant floating, submerged and marginal vegetation and freshwater algae and phytoplankton communities. All these primary producers contribute the organic energy needed to maintain short animal food chains and food webs in the lakes. The freshwater Lake Naivasha and the moderately saline Lakes Baringo and Turkana support moderate diversity of fishes and regionally important fisheries resources.

Lake Turkana occupies an area of 7560 km^2 and it is the largest single lake in Kenya. This long but narrow lake has three volcanic islands: north, central and south islands. The lake is located in a hot desert environment, but receives 90% of its water from the Omo River which drains the humid central highlands of Ethiopia. The river enters the lake via a swampy delta at the northern end. Rivers Turkwel and Kerio have increasingly become seasonal, reaching the lake during the extremely wet seasons via a sandy and gravel delta on the southwestern shores. The lake supports 48 known fish species, seven of which are endemic to the lake. The central island has three saline crater lakes, which support some endemic cichlids (Tilapia spp.). There are 12 commercially exploited fish species with littoral tilapia fishery, which is dominated by *Oreochromis niloticus*, comprising 52% of the fishery. The demersal fishery, which is dominated by *Labeo horie* and *Bagrus bayad*, contributes 21%, while the pelagic fishery consists of seven species dominated by the Nile perch (*Lates niloticus*, 10% and *Alestes*, 7%). This fishery is characterized by 10 year boom and bust cycles, which are associated with water fluctuations arising from climatic variability in the catchment of Omo River.

The Naivasha basin contains the freshwater Lake Naivasha (160 km^2) and two saline lakes Oloidien and Sonachi. Lake Naivasha is one of the most studied tropical lakes and has been shown to support a high diversity of aquatic plants, birds and invertebrates. Following the extinction of its only native fish species, *Aplocheilichthys antinori*, in the early 1970s, Lake Naivasha has no native fish species but has had 10 fish introductions since 1925. Out of the 10 species introduced, only three successfully established in the lake: blue spotted tilapia (*Oreochromis leucostictus*), redbelly tilapia (*Tilapia zilli*) and black bass (*Micropterus salmoides*). In the late 1990s, the common carp (*Cyprinus carpio*) invaded the lake from fish ponds in Kinangop Plateau and Dundori Hills. By the year 2010, the common carp was contributing nearly 70% of the total fishery of Lake Naivasha. The occurrence of abundant food in the form of fish, invertebrates and aquatic vegetation has supported nearly 420 species of birds in the lake and the surrounding littoral zones and woodlands. Lake Naivasha is a wetland of international importance (Ramsar Site) because of the high diversity and large concentrations of resident and migratory species of birds.

The most conspicuous wildlife is birds, which occur in large concentrations in both the saline and freshwater lakes. The saline lakes and the associated hot springs support rich communities of microorganisms, especially algae and bacteria. They also support large concentrations of birds, especially the lesser flamingos, greater flamingos, pink-backed pelicans and aggregations of migrant and resident ducks and shore birds. The birds inhabiting the lakes and wetlands move periodically among the lakes of the Rift Valley, including Rift Valley lakes in Ethiopia and Tanzania. The freshwater lakes also support significant groups of hippopotami, crocodiles, terrapins and other wildlife, such as waterbuck, giraffe, buffalo and eland on the surrounding terrestrial habitats.

Lesser flamingos form large concentrations in the alkaline Rift Valley lakes of Kenya.

The scenic beauty of the Rift Valley and the large concentrations of birds associated with the lakes, wetlands and woodlands in the valley constitute an important tourist attraction. The freshwater ecosystems in the Rift Valley basin, such as Lake Naivasha, Nakuru, Bogoria and Baringo, are major destinations of international tourists in Kenya. The rivers and freshwater wetlands and the freshwater lakes, such as Lake Naivasha, are important sources of water needed to support agriculture, horticulture and livestock, industrial and household needs. Fertile soil in the wetlands is attractive to farmers, while abundant foliage in the permanent wetlands is attractive to livestock herders. The principal threats to the freshwater ecosystems in the Rift Valley basin are high rates of siltation and pollution from catchments of affluent streams, excessive abstraction of water from affluent rivers and from the lake itself, drainage of wetlands for farming and settlement and overexploitation of fisheries and plant resources.

Human settlements and human impacts are the greatest threat to freshwater ecosystems in the Rift Valley basin. Effective and sustainable management of these ecosystems and their globally important bird populations and endemic fish species is necessary. The first priority is to ensure that the water catchment areas of streams supplying water to the lakes and wetlands on the valley floor are protected and, where necessary, rehabilitated. Thus, soil erosion and water abstraction from affluent streams are strictly controlled. This will ensure reduced sediment load entering into the lakes and continuous water supply. Land and water pollution in the lake's watershed should be controlled and exploitation of fish and plant resources from the lakes and wetlands regulated. In addition, concerted efforts should be made to minimize negative impacts of climate change on the lakes and the biodiversity that they support.

8 DISCUSSION

The Green Revolution in the 1940s aimed at increasing cultivated areas and water use for irrigation of high-yielding plants has in the global context been the cause for reduction in freshwater resources (Finger et al., 2011). In addition, developments in the industrial sectors have increased the economic use of water, accounting for about 20% of the global freshwater withdrawal. Water for domestic use is a very small fraction in comparison to both agricultural and industrial uses.

In Kenya, the water towers are under serious threat from population growth and poor agricultural deforestation and lack of implementation of policy decisions. The deforestation of Kenya's water towers over a 10 year period, 2000–2010, was approximately 50,000 ha, which yielded timber and fuelwood volume of 2010 cubic metres per hectare, with a cash value of 272,000 KES/ha. However, the indirect costs of deforestation are borne by sectors and households elsewhere in the economy through the reduction in value of regulating services of the ecosystem. The negative effect of deforestation on the economy by 2010 was approximately KES 3.8 million, which outweighs the immediate cash benefit from timber and fuelwood sales by 2.8 times.

The protection of these biomes is crucial for two major reasons. Firstly, they are important water catchments, on which millions of people depend, in both urban and rural settings. The destruction of these water towers would therefore be a potential disaster affecting the quality and quantity of water available over a wide geographical area. Secondly, the threat posed by invasive species also underscores the importance of the upper reaches of rivers because they sometimes remain unreached by the invasive species due to various physicochemical barriers, and when they are invaded, it is not possible to protect the lower reaches due to the flow direction and periodical flooding. A case in point is the Louisiana red swamp crayfish *Procambarus clarkii* in Kenya. In the Ewaso Ng'iro River, this species was introduced by settlers into the lower reaches but is restricted to ambient water temperatures above 14°C, even in contiguous habitats (Ogada et al., 2009). The upper reaches of the Ewaso Ng'iro like the Burguret and Naro Moru Rivers were therefore found to be pristine habitats still inhabited by freshwater crab *Potamonautes neumannii*, unlike the lower reaches where these had been supplanted by the invasive crayfish. This habitat quality was reflected by the density of otters *Aonyx capensis*, with approximately 6.5 individuals per km of river is over three times that of the invaded area at 1.8 otters per km (Ogada, 2007). In another case, the steep Kinangop escarpment has protected the upper Naivasha catchment and the Aberdares from invasion by the common carp *Cyprinus carpio* from Lake Naivasha in the Rift Valley.

The importance of conservation of these biomes is further underscored in reference to point-source chemical pollution. Crustaceans cannot survive in acidic waters because the low pH interferes with their calcium uptake and shell formation (Lodge and Hill, 1994) and by this token, the negative impacts of the Athi River pollution from Nairobi are still prevalent in the apparent absence of crustaceans in the Galana River hundreds of kilometres downstream of the source in Tsavo East National Park. Kenya's water catchment areas are something of a 'confluence' of aquatic and terrestrial ecosystems, and are, therefore, crucial to our understanding of both biomes. It is well known that transfers of material and energy between ecosystems shape the ecology of entire landscapes. In the study of cross-boundary connections, much attention has been paid to linkages between aquatic and terrestrial ecosystems. It has become apparent that despite the physical differences between the two systems, they impact each other directly. Rivers in particular have been shown to be especially well linked to terrestrial systems, with significant exchange of nutrients from river to land, and vice versa either by gravity or through primary or secondary consumers. In resource-poor countries like Kenya where we are unable to protect entire rivers, there is a case for priority to be given to the catchment areas for the reasons previously mentioned.

8.1 Significance of Freshwater Ecosystems

Freshwater ecosystems play a fundamental ecological role and provide economically important products and services. They provide critical habitats for a large number of aquatic plants, fishes, reptiles, birds and mammals. They host many migratory and threatened species of birds, reptiles and fish. The freshwater ecosystems are areas of tourist attraction by providing recreation sites for game and bird watching.

Freshwater ecosystems, especially vegetated wetlands, play an important role in mitigation against climate variability. They do so through a number of ecosystem functions including flood control, water purification, shoreline stabilization and sequestration of carbon dioxide. At landscape level, wetlands control soil erosion and retain sediments and in so doing concentrate nutrients in the wetland soil. They also provide economic benefits such as fresh water, fisheries, fuelwood, building material, medicinal products, honey and foliage for livestock and wildlife. Wetlands provide fertile land for agricultural, mineral salts, sand and soil for making pottery and building bricks. Wetlands are central to rural subsistence economies and livelihood activities of many rural communities in Kenya. Freshwater ecosystems in general are critical to poverty alleviation and creation of employment and wealth.

Freshwater ecosystems provide water for home and farm use and for power generation and industrial development. Their biodiversity supports recreational activities and tourism. The valley bottom wetlands regulate stream flow and improve water quality through sediment filtration and absorption of heavy metals and other toxic pollutants. Wetlands absorb and store large quantities of rainwater and floodwater, thereby facilitating recharging of underground water. By recharging underground aquifers, the water table is raised, hence making groundwater available in springs and wells to support domestic, industrial and agricultural activities.

8.2 Threats and Management Challenges

In spite of having several freshwater sources, Kenya has been classified as a water-scarce country, which in a global context refers to a country whose renewable freshwater is between 1000 and 1700 cubic metres per capita. The Water Resources Availability in Kenya in 2010 was estimated at 1172 cubic metres per capita up from 647 cubic metres per capita estimated in 1992. The main challenge is that water in Kenya is unevenly distributed in time and space, which is a factor of the variability of rainfall patterns. In addition, extreme weather events during droughts and floods have become more frequent and affect water availability. Several other natural and anthropogenic factors affect water supply and quality.

The natural processes that threaten freshwater ecosystems and their biodiversity are variability in water supply, increasing soil erosion and sedimentation rates, ecological succession processes and introduction of invasive species. Intensification of land use and urbanization in the catchment pose serious threats to all freshwater ecosystems in general. The rapid increase of human population and its growing demand for land, shelter and food production has led to rapid loss of wetlands due to conversion into agricultural land. Increased demand for water and other wetland resources has initiated unsustainable exploitation with serious alterations in ecosystem structure and functions.

Introduction and dispersal of alien invasive species in freshwater systems have threatened the ecosystem balance and regeneration of diverse resources. The introduction of the Nile perch (*Lates niloticus*) into Lake Victoria, for instance, had increase the economic value of the lake's fishery, but the species caused damage to the ecosystem, resulting in the loss of some native species with significant economic potential. The introduction and spread of Louisiana crayfish in freshwater ecosystems in central and eastern Kenya has had disastrous effects on fish communities and their productivity. Ornamental aquatic plants, such as water hyacinth, Nile cabbage and water fern, have invaded dams, lakes and rivers, thereby hindering fishing and transportation activities. They have contributed to unfavourable changes in native plant and animal communities. The impacts of invasive species are immense, damaging to the native species and usually irreversible.

Pollution as a result of urbanization and industrialization poses a big threat to the well-being of freshwater ecosystems. Rivers, streams and swamps are often used as receptacles of liquid effluents from municipalities and waste dumpsites for solid waste. Damming of rivers to produce hydroelectric energy helps in trapping sediments from flowing water, but they result in reduced water flow downstream, thereby affecting ecological functions and economic activities downstream. Lack of ecosystem approaches to development of hydropower and water use, especially for irrigation and municipal supply, has serious consequences on the freshwater ecosystem and human well-being in Kenya. Concerted efforts should be made to monitor threats to the country's freshwater ecosystems and adopt ecosystem approaches to their management.

Water resource degradation was estimated to cost the country at least Ksh 3.3 billion annually (Mogaka et al., 2006). In spite of this, the government reduced spending for water resource management in Kenya from KES 4249 million in 1994 to KES 1765 million in 2001 (Mogaka et al., 2006), thus further reducing the capacity of the Ministry of Water and Irrigation to control negative anthropogenic activities in the water sector and to protect the water catchments.

Water resource assessments have deteriorated greatly in Kenya with debilitated and vandalized river gauging stations, making it impossible to carry out in-depth water resource analyses and planning. Water allocation has thus been made based on inadequate and in some cases no information on the flow patterns and quantities, leading to failure of development projects and also to frequent disruptions of water supply and rationing common in big cities such as Nairobi, Mombasa, Kisumu, Nakuru and Malindi.

REFERENCES

Finger, R., El Benni, N., Kaphengst, T., Evans, C., Herbert, S., Lehmann, B., Morse, S., Stupak, N., 2011. A meta-analysis on farm-level costs and benefits of GM crops. Sustainability 3 (5), 743–762.

Kenya Wildlife Service, 2003. Changes in the state of Conservation of Mt. Kenya forests: 1999–2003: Aerial survey report. KWS, Nairobi.

Lodge, D.M., Hill, A.M., 1994. Factors governing species composition, population size, and productivity of cool-water crayfishes. Nord. J. Freshw. Res. 69, 111–136.

Mogaka, H., Gachere, S., Davis, R., Hirji, R., 2006. Climate Variability and Water Resources Degradation in Kenya: Improving Water Resources Development and Management. World Bank working paper, No. 69, Washington, DC, 128pp.

Ogada, M.O., 2007. Effects of the Louisiana crayfish invasion on the food and territorial ecology of the African clawless otter in the Ewaso Ng'iro ecosystem. PhD thesis, Kenyatta University, Kenya.

Ogada, M.O., Aloo, P.A., Muruthi, P.M., 2009. The African clawless otter (*Aonyx capensis* Schinz, 1821) and its diet as an indicator of crayfish invasion dynamics in aquatic systems. Afr. J. Ecol. 47 (1), 119–120.

Okeyo, D.O., 1998. Updating names, distribution and ecology of riverine fish of Kenya in the Athi-Galana-Sabaki river drainage system. NAGA ICLARM. 21 (1), 44–53.

Seegers, L., De Vos, L., Okeyo, D.O., 2003. Annotated checklist of the freshwater fishes of Kenya (excluding the lacustrine *Haplochromines* from Lake Victoria). J. East Afr. Nat. Hist. 92, 11–47.

Chapter 17

The Management of Wildlife and Fisheries Resources in Kenya: Origins, Present Challenges and Future Perspectives

Mordecai O. Ogada* and Dorothy Wanja Nyingi†
*Laikipia Wildlife Forum P.O. Box 764-10400, Nanyuki, Kenya
†Ichthyology Section and KENWEB, National Museums of Kenya, P.O. Box 40658, Nairobi, Kenya

1 INTRODUCTION

The prevailing view about conservation of wildlife and management of fisheries in Kenya is that its origins lie within the structures that are in place to manage it. This line of thinking is largely responsible for the difficulties currently faced in the effective and sustainable management of these resources. It presumes that there was no thought or philosophy that guided the way in which precolonial African societies lived and interacted with the wild fauna around them. One of the results of this thinking is that wildlife research in Africa is largely designed, implemented and funded by external agents, who are widely believed to have 'brought conservation' to Kenya. This chapter examines the inherent failures of this conservation paradigm and suggests the directions of required shift in Kenyan conservation thinking and practice. In Africa, there has always been and there still is a large proportion of resources obtained directly from the environment. These include grazing, woodfuel, fish, game meat and water. It therefore stands to logic that these communities possess some level of knowledge on how to live among and exploit these resources in a sustainable manner. The basis on which the forest, wildlife and fisheries management structures in Kenya were developed is the Victorian gamekeeper model (Plate 1).

This was the prevalent thinking in England, when British colonial administration was first established in Kenya in 1895 during the latter years of Queen Victoria's rule. Generally, this model involved prevention of subsistence use of

Developments in Earth Surface Processes, Vol. 16. http://dx.doi.org/10.1016/B978-0-444-59559-1.00017-7

PLATE 1 Early 20th century gamekeepers at Pale Hall, one of the big country estates of Victorian Britain. Photo credit: Daviek.

natural resources by the proletariat in order to ensure availability of the same for recreational use by the nobility or elite. There was no technical or social qualifications required or used in the management of wildlife and fisheries by the Kenya Game Department, which was established in 1935. This was in response to the proliferation of arms in the hands of bored demobilized troops who were hunting haphazardly since the end of the First World War in 1918 (Steinhart, 2006). Coincidentally, the history of fisheries management in Kenya also dates back to the early 1900s when the colonial government decided to manage sport fishing, an activity popular among the settlers and constantly bored army troops and to manage the harvest of inter alia, pearls, sea cucumbers (also known as beche-de-mer) and ambergris. Consequently, there were no laid down policies or plans and these activities were diligently pursued by these colonial officers on largely subjective bases with detailed reports being published on a regular basis. An examination of these reports reveals the genesis of several challenges still confronting conservation practitioners in Kenya today. The following is an excerpt from this document:

...In August I proceeded to the Kerio Valley, where agricultural activity was hindered by large numbers of wild pigs, baboons, monkeys, and porcupines. Mtama and wimbi** were the only foodstuffs obtainable, and ripe heads of these were soaked in a solution of arsenite of soda, dried in the sun and distributed. This proved to be entirely effective, and the Marakwet and Elgeyo natives were loud in their expressions of appreciation. One case of injury by a rhino came to my notice. The injured man's friends had carried him to the railway line, stopped the train and put him aboard. I motored him to the dispensary and learnt his story. He and a friend were passing along a path when a rhino with a young calf dashed out of a thicket, tossed him on*

its horn, gored him in three places, and then made off. As this pugnacity was clearly in defence of the calf, and unlikely to be habitual, I took no steps. In July I was at Konza endeavouring to deal with lions stated to be living in the Reserve and frequently killing European-owned cattle on the neighbouring farms. A week's search failed to locate the marauders, and then I shot wildebeests, etc., baited with strychnine, and distributed. The depredations then ceased. Whilst there I arrested several Wakamba poachers armed with bows and poisoned arrows, whom I detected stalking a herd of wildebeests in the reserve. Suitable punishment followed...

(Kenya Colony Game Report, 1936).

The key challenges to conservation practice referred to earlier that show up in this excerpt include the obvious ethnic/racial prejudice in the consumptive use of wildlife and management responses to human–wildlife conflict. Another problem that emerges is the use of ethically and environmentally unsustainable problem animal control. This indicates the low capacity of the practitioners mandated by the colonial government to manage wildlife in Kenya.

The exclusion of local communities has been a stumbling block to conservation success in Kenya until the deliberate change of policy driven by the theme of the World Parks Congress in Durban 2003 'Benefits beyond boundaries'. On the second issue of unethical and unsustainable management actions, poisoning is no longer used in wildlife management, but communities picked this up from the earlier practices and poisoning is now the greatest threat to carnivore and vulture populations in much of East Africa (Ogada et al., 2011).

2 TRANSITIONS

There have been two major transitions in Wildlife Management in Kenya, namely, the ban on trophy hunting in 1977 and the transformation of the Game Department into Kenya Wildlife Service (KWS) in 1989. The ban on hunting in some ways was an 'equalizer' because hunting in Kenya (and much of Africa) has always been a pursuit of the foreign elite (Plate 2). Suddenly, all 'users' of wildlife were on the same 'nonconsumptive' plane. This sudden 'democratization' of wildlife conservation and management has compelled adjustments that have taken a number of stakeholder decades to fully accept. Many proponents of sport hunting or 'consumptive management' of wildlife in Kenya and other parts of Africa tout it as being a source of revenue and other benefits to local communities. It is also supported on a scientific basis through reference to its 'sustainability', based on the populations of various species. However, the social human dimension of sport hunting has largely been avoided, despite being a crucial ingredient in any conversation on the subject. There is repeated reference by public and private sector conservation practitioners in Kenya to Southern African models. This tendency is oblivious

PLATE 2 Taxidermy shop. Nairobi 1961. © Life Magazine.

to the fact that these models are the result of socially and politically undemocratic governance systems. The only meaningful way in which the locals participated in conservation was as wildlife rangers, guides and trackers, avenues that do not include any intellectual participation in any management decisions or the results thereof. An example is illustrated by this excerpt: "Although the economic value of game meat is low, it provides benefits beyond its cash value. Distributed to low-income families unable to afford domestic meat, or in areas where trypanosomiasis and tsetse fly have reduced alternative domestic meat supply, game meat can be a welcomed source of protein otherwise not available. Formal compensation for wildlife-inflicted damage to property or life in all cases consists of game meat derived from culled problem animals. Further, devolving greater proportions of financial revenues from lucrative use options such as safari hunting to rural inhabitants has been problematic in some community-based programme areas. In such cases, meat supply from directed cropping, or as a by-product from licensed hunting or problem animal control, represents the bulk of wildlife benefits accrued". From Food For Thought: The Utilization Of Wild Meat In Eastern And Southern Africa published by TRAFFIC (1999). It actually suggests that meat from safari-hunted animals can be used to appease local communities and even to compensate for loss of life or property to wildlife damage. Equally telling is the assertion that devolution of financial revenues from hunting to communities can be 'problematic'. This document is a classic example of the theory presented from 1936 earlier on the origins of wildlife management in Kenya, still persisting into the twenty-first century. There is no attempt here (or anywhere else in Africa) to involve local communities intellectually or philosophically in the management of wildlife or the consumptive use thereof. The main

PLATE 3 Subsistence poachers under arrest. Tsavo, 1961. © Life Magazine.

challenge that persists to the necessary transitions in Kenyan wildlife conservation is the simplistic view taken by scientists and practitioners that limits conservation down to 'science', that is natural resource use or ecology. That is why the arguments for and against consumptive use of wildlife still refer exclusively to studies on values like 'sustainable offtake' of key species while reducing the human beings who share their habitats to nonentities (Plate 3).

3 MANAGEMENT AND MITIGATION OF HUMAN–WILDLIFE CONFLICT IN KENYA

Human–wildlife conflict remains the greatest challenge to conservation efforts in Kenya. In the absence of consumptive use, it does remain the major duty of wildlife managers in Kenya. When examined as the negative interactions between people and wildlife, several different manifestations occur including human death and injury caused by various reptile and mammal species, livestock depredation by carnivores and crop damage by primates, ungulates, birds and elephants. These modes of conflict have all occurred for millennia since the beginnings of agriculture and livestock keeping in East Africa. However, two have come to the fore since the beginning of structured wildlife conservation in Kenya, namely, crop raiding by elephants and livestock depredation by the large carnivores (lion, leopard, hyena, cheetah and, to a lesser extent, wild dog). There are two major reasons for this prominence: the magnitude of the resource needs and the consequent damage by these animals, and the psychological effect of their significance as tourist attractions. Elephants are known to cause extreme damage to crops not only from the

PLATE 4 Cow killed and eaten by lions, Laikipia. © Laurence Frank. (For colour version of this plate, the reader is referred to the online version of this chapter.)

quantities they eat but also from trampling fields as they move. They are highly intelligent animals that can find and eat food in storage while causing extensive damage to stores and granaries. The factors in the preceding text and the physical danger posed by elephants to the farmers defending their crops make them a significant threat to food security in the affected areas. With reference to livestock depredation, the reaction of livestock ranchers and pastoralists is beyond what is merited by the monetary loss and exacerbated by the position that lions hold in our psyche. The death of a cow from a lion attack can elicit violent reactions from a livestock owner who may lose several animals to tick-borne diseases within the same time frame (Plate 4). Human–wildlife conflict management has a strong human dimension to it that is often overlooked by practitioners, occasionally resulting in the application of noneffective mitigation attempts. Two case studies presented here illustrate this point.

4 CASE STUDY 1: HUMAN–ELEPHANT CONFLICT AND ATTEMPTED MITIGATION MEASURES IN LAIKIPIA COUNTY

Laikipia is the mostly dry plateau covering 10,000 square kilometres to the northwest of Mt. Kenya in the central region of the country. This area is rich in wildlife, including a population of over 4000 elephants and approximately 700,000 people, many of who are involved in subsistence farming. Consequently, conflict between humans and elephants has been one of the major challenges to wildlife conservation efforts and food security in Laikipia for several generations. In 2010, an electric fence to separate the regions where elephants are tolerated from the farmlands was initiated by the Laikipia Wildlife Forum as a mitigation strategy. The impact of this particular intervention

TABLE 1 Fence Breakages by Fence Sections

Fence Section	Location of Breakage (s)	No. of Breakages	Total For Each Section
Ol Maisor	OLM30–OLM35	1	1
Mathira	Total	0	0
Ngorare Ranch	NG5–NG10	1	1
	NG10–NG15	1	1
Lombala Ranch	Total	0	0
Kifuko Ranch	K5–K10	2	2
Matigari	Total	1	1
Mutara	M15–M20	3	42
	M20–M25	1	
	M25–M30	38	
OPC	OPC25–OPC30	1	2
	OPC45–OPC50	1	
		Grand total	50

From Ochieng et al. (2011).

has differed vastly between geographic locations along the fence, with 84% of breakages occurring in one section. This is in spite of the fact that the entire range is used by largely the same elephant population (Table 1). This implies that the variation is likely due to human factors that differ along the fence line. The practitioners implementing the project identified regular fence-breaking elephants and determined that they were males using their tusks to break the live wires and breach the fence. A decision was then made to capture and detusk the two male problem elephants in order to curb this behaviour.

The capture and detusking of elephants did not have any significant impact, an example of the cost that practitioners are willing to incur rather than explore the human dimensions of conflict with wildlife. Capture and detusking of elephants also involves considerable risk to the people and animals involved, not to mention animal welfare considerations. The culling of 'problem elephants', though not attempted, has been suggested as a solution to deal with individual 'problem elephants'. It is not likely to have long-term impact unless carried out on a massive scale, which would defeat the conservation objective of the exercise. The human factors at play here include communities compromising the fence in order to access various resources like firewood, water, pasture and markets on the other side. The Laikipia elephant fence has therefore functioned as a *de facto* 'people fence' in various places with the resultant difficulties. Another source of variation in fence function is the management input from community members in certain areas, maintaining the voltage of the fence. There is need for conservation practitioners and researchers in Kenya to move away from the strictly ecological perspective in order to properly address the human-based challenges to conservation. Placement of fences can be a source of conflict in cases where they are seen as the only solution. This is clearly illustrated by this case where the exact same elephants are consistently breaching the fence in some areas while adhering to its limits in other areas.

5 CASE STUDY 2: LIVESTOCK DEPREDATION AND THE USE OF MONETARY COMPENSATION IN AMBOSELI, KENYA

This section refers to the area covered by Mbirikani Predator Compensation Fund. The Mbirikani Group Ranch (MGR) is an area of community-owned land that borders the Amboseli National Park in southern Kenya. This scheme was initiated by the Maasailand Preservation Trust and is funded by private donations. Its aim is to provide incentives to the MGR community residents not to kill predators in retaliation to livestock losses. This scheme was later extended to the community-owned group ranches surrounding the Amboseli National Park.

MGR no. of livestock killed per species

No. killed	Cow	Shoats	Donkey	Total
	40	61	2	103
Percentage (%)	38.83	59.22	1.94	100

MGR no. of predator compensation claims per loss type

No. of livestock killed per species	Lost	Bad boma	No penalty	Total
	43	4	14	61
Percentage (%)	70.5	6.56	23	100

MGR compensation value (Kshs) per livestock species

Total payment per stock type	Cow	Shoat	Donkey	Total
	200,800	100,000	5425	306,225
Percentage (%)	65.57	32.65	1.77	100

MGR compensation value (Kshs) per loss type

Total payment per claim type	Lost	Bad boma	No penalty	Total
	214,500	7725	84,000	306,225
Percentage (%)	70.04	2.52	27.43	100

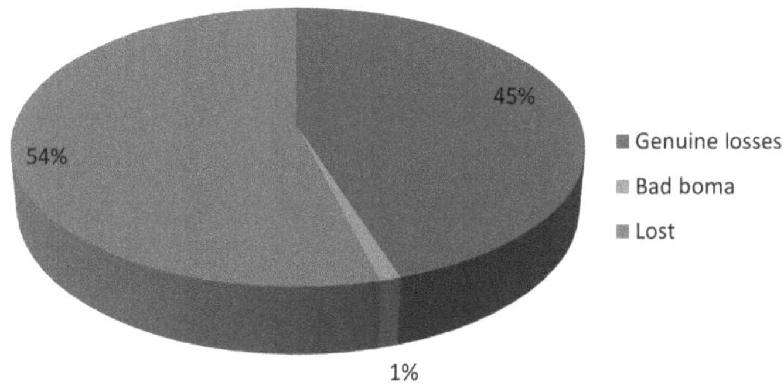

FIGURE 1 Categories of claims in the Mbirikani predator compensation scheme. (For colour version of this figure, the reader is referred to the online version of this chapter.)

The dominance of the 'lost' claims implies that there is a lacuna in livestock management, because livestock are highly valued by the Maasai people and generally well protected from wild animals and theft by humans. The indication here is that the practitioners did not thoroughly examine the human factors in this scheme (Figure 1).

6 LAIKIPIA

This is a comparison of Mbirikani to lion mortality in the Laikipia Plateau and the adjoining districts, into which wildlife population regularly moves. Laikipia District is a key wildlife habitat in Kenya and is unique in that it supports high densities of wildlife, but is not a protected area. The expansive livestock ranches, pastoralist grazing areas and agricultural holdings create an extensive human–wildlife interface and can be taken as a microcosm of the conflict between wild carnivores and livestock producers on a national scale. There is no compensation scheme in place here, and the losses have to be fully borne by the livestock owners. Lions have historically posed the major threat to livestock here, accounting for 72% of stock lost to wild predators (Ogada et al., 2003). It is also the one area where this study accessed relatively complete data on human-caused lion mortality. This circumstance forms the basis of the assumption that lion mortality per unit area/time in Laikipia is the maximum sustainable by a viable lion population in Kenya. This implies a highly robust population in Laikipia, relative to others in Kenya. Awareness of carnivore conservation issues is relatively high in this area due to the impact of the Laikipia predator project, which has built a very strong rapport with livestock ranchers and pastoralists over the last 12 years. The data indicate that livestock depredation is a serious problem in the district, with 96% of the human-caused lion mortality related to stock raiding. The majority of killings (79%) appear to be on commercial ranches relative to pastoralist areas (18%), but this could be more a reflection of information availability and

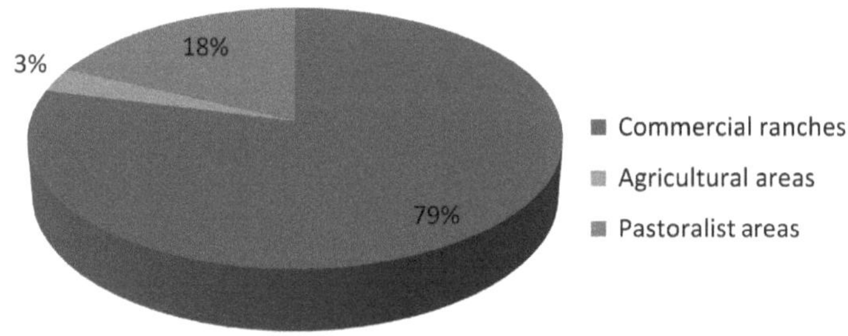

FIGURE 2 Locations of reported lion mortalities in Laikipia (2000–2010). (For colour version of this figure, the reader is referred to the online version of this chapter.)

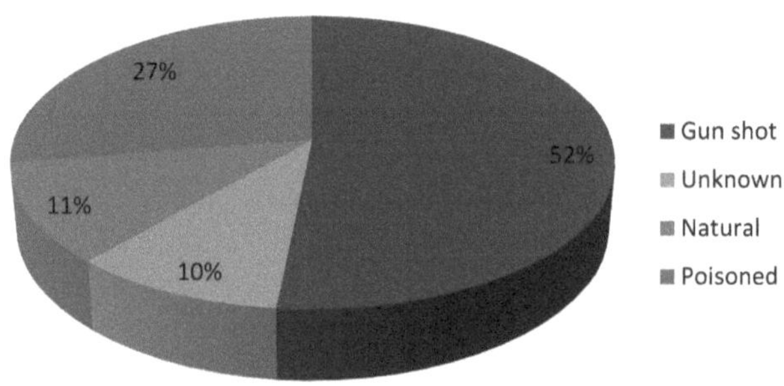

FIGURE 3 Causes of reported lion mortalities in Laikipia (2000–2010). (For colour version of this figure, the reader is referred to the online version of this chapter.)

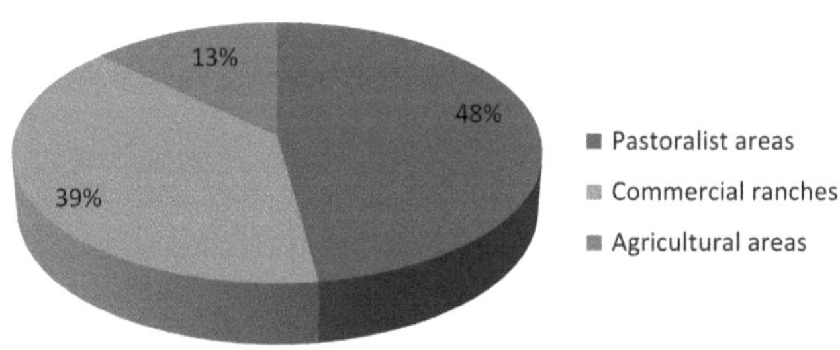

FIGURE 4 Locations of known lion poisonings in Laikipia. (For colour version of this figure, the reader is referred to the online version of this chapter.)

density of lions therein rather than difference in attitudes (Figures 2 and 3) towards lions. Poisoning (Figure 4) is also a notable problem accounting for 27% of mortality, with 39% of occurrences recorded in commercial ranches versus 48% in pastoralist areas. It should be noted, though, that this reflects the locations of poisoned carcasses, which may not necessarily be the place where the poisoning occurred. Overall, there has been a known mortality rate of 9.1 lions per annum in the Laikipia area over the last 10 years from 1999 to 2009.

Evidence from literature indicates that compensation is neither an effective nor sustainable conservation tool (Nyhus et al., 2003). It is extremely difficult to investigate the veracity and circumstances surrounding claims. Data from the schemes themselves do not show any significant reduction in lion mortality as a result of the monetary compensation being handed out. The trend is picking up, and Olgulului–Ololarashi Group Ranch, a 240,000 acre area to the west and south of Amboseli National Park, started one in June 2008. They are paying 14,500 shillings for every cow or donkey killed by lions and 2500 shillings for every sheep or goat. What is the driving force behind this compensation trend?

A compensation scheme, by definition, is a pool of unplanned and unaccounted financial resources. Nobody is immune to the attractions of such funds, whether they are to be used to fund research community projects or other noble objectives. The financial motive is also evident in the scheme designs. The Olgulului scheme only compensates depredation caused by lions, despite the fact that there are losses caused by hyenas, leopards and cheetah as well. The reason given for this is that leopards and cheetah 'aren't much of a problem', and the communities know how to 'deal with' hyenas. Lions, however, tend to kill the highest numbers of livestock, being responsible for 74% and 60% of losses reported in Laikipia and Isiolo districts, respectively (Ogada, 2001, 2008). The more important reason is that they (the pastoralists) consider lions the most attractive and exciting animal to tourists and the most important to the KWS. They are therefore most likely to get paid not to kill lions rather than any of the other species. Further probing reveals that the 'dealing with' hyenas is most likely poisoning, an unfortunate side effect of the compensation scheme, which needs to be addressed as well.

Conservation and field research in Kenya and Africa is expensive and largely supported by private funding. In many instances, the donor strains to see the direct impact of his/her funding on the welfare of animals on the ground. Compensation schemes present no such problem and are consequently unmatched in their 'feel-good' factor among donors. They can therefore attract very significant amounts of money from donors with relative ease. A number of these schemes are operated by up-market tourist facilities, with their clientele being the prime source of contributions and the schemes themselves becoming selling points for their business.

The people involved in these schemes are aware of their dubious legal and ethical status, so there is wide use of the term 'consolation' ostensibly because the amount being paid out is 'not equivalent to the loss'. Knowledge of these schemes spreads rapidly, and payouts in Amboseli are already causing some discontent as far as Samburu and Laikipia in the north. This creates a culture of dependency, rather than involving communities fully in the benefits, knowledge and responsibilities of conservation. In this context, it would be better to do nothing at all than hand out some benefits that cannot be sustained in the long term.

7 WILDLIFE PROTECTION AND ANTIPOACHING OPERATIONS

What was initially regarded as poaching was actually subsistence hunting practiced by communities living in and around wildlife habitats. However, these communities came into conflict with the 'Victorian gamekeeper model' to preserve the wildlife from poaching by the 'natives' so that the 'sportsmen' could shoot them for pleasure and have their trophies mounted. After the ban, there was a rise in commercial poaching for trade-in game trophies to meet market demand that had been created in the hunting era. A further challenge was corruption by senior officials in the Kenyatta government, and the then first lady Mama Ngina Kenyatta, who were suspected of colluding with poachers to flout the hunting ban (Steinhart, 2006). The seed of elitism in wildlife matters sown by the colonialists had grown and spread its tendrils even into the arena of wildlife crime. The preservationist mentality became obsolete with the hunting ban, but the paradigms that were developed then persist to this day, much to the detriment of conservation practice in Kenya. They manifest today in all sorts of ways, for example, KWS dealing with calls to cull hyenas in certain areas 'because they are killing baby rhinos' or to cull lions in certain areas 'to save grevy's zebras', the creation of 'sanctuaries' where the wealthy can keep wild animal enclosures on their land for their pleasure, the 'saving' of an elephant's calves that may have been killed by lions after separation from the herd. Kenya has yet to escape the preservationist mind-set because managers still attempt to protect one species at the expense of another and this cannot work with living, dynamic ecosystems. It is narrowness of thought that makes some practitioners think only about a rhino calf being killed and not the hyena cubs that will share in that meal or an elephant calf/Grevy's zebra being killed and not the lion cubs being raised on that meat. A prime example is the recent erection of a 'predator proof' enclosure to protect the hirola antelopes (*Damaliscus hunteri*) from predators like wild dogs and lion. This project was approved by the KWS expressly against their own carnivore conservation plans that named habitat fragmentation as one of the biggest challenges to their conservation. The 'preservationist' approach to conservation is at odds with the entire concept of 'wildlife', which implies entire ecosystems that include the cycle of life and death. The research departments of the KWS need to be strengthened in order to guard against 'preservationist' tendencies that are designed by practitioners to cultivate lucrative niche funding sources. African wildlife has fascinated Western minds for generations, and this manifests in some private grants that are given by well-meaning but ignorant donors for projects that do not serve the overall objectives of conservation.

8 THE DEVELOPMENT AND ROLE OF WILDLIFE CIVIL SOCIETY ORGANIZATIONS

Kenya has long been regarded across the world as a leader in matters concerning wildlife. This included hunting (before 1979), wildlife research, tourism development and practice. This has been due to a combination of different factors including climate, geographic location and patronage of its tourism products by internationally prominent personalities over the years. In addition to the preceding text, the making of several globally recognized wildlife films has been a cause (and result) of Kenya's high profile in matters of conservation. The most recognizable and enduring result of this circumstance has been the high level of external funding that has been invested in Kenya's wildlife conservation sector, giving rise to a vibrant conservation civil society, the oldest example being the East African Natural History Society (now Nature Kenya) founded in 1902. This, and other membership organizations, for example, the East African Wildlife Society (founded 1955), led the way in voicing concerns and opinions about the interactions between wildlife and the wider populace. In addition to these, there are two other types of CSOs. Firstly, there are the area-specific organizations, for example, Laikipia Wildlife Forum, Nakuru Wildlife Forum, and Northern Rangelands Trust. These have been formed by residents of the areas from which they take their names at the behest of KWS, which deliberately set out to involve communities in conservation. They had humble beginnings and started with individual membership contributions, but the successful ones now have broadened revenue streams including corporate sponsorships and grants from various agencies and governments. Being that their focal areas (both geographic and thematic) have been very finite, these organizations have managed to make significant impacts in natural resource management and continue to do so. Animal welfare organizations have also found Kenya's conservation sector an important rallying point because of the country's no hunting policy, which is unique in Africa. Prominent among these are the International Fund for Animal Welfare and the Born Free Foundation. These organizations are capable of mobilizing significant amounts of funding internationally and have channelled the same to KWS. Lastly, there are several smaller organizations that are species-, site-, or theme-specific. Most of these rose out of individual research projects or initiatives and are still largely run by their founders or the views and paradigms they espoused, with varying levels of success and degrees of relevance. The voice views of the wildlife civil society in Kenya are strident and far from unanimous in approach or opinion on any issue. However, their importance cannot be overstated in a country where the wildlife authority (KWS) is an armed force with all the state strictures and codes of conduct that accompany that status. The practice of conservation and wildlife management is often subject to philosophical and political arguments of several issues, but KWS and their personnel have never overtly participated in any of these discussions, leading to a perception that they do not have the capacity to handle these issues and giving rise to civil society actors positioning themselves as 'filling the gap' left by KWS. The wildlife civil society has, however, scored some significant victories in instances where government has been seen to breach conservation ethics in one way or another. A notable example is the 2005 proposed sale of various wild animals to Thailand in a deal agreed between President Mwai Kibaki of Kenya and Thaksin Shinawatra, the then Prime Minister of Thailand. This consignment included plains zebras (*Equus burchelli*), impala (*Aepyceros melampus*), buffalo (*Syncerus caffer*) and several other nonendangered species of least conservation concern. This was a watershed in Kenya's conservation history because the opposition was not based on the status of the species concerned, the welfare of the animals or the price offered, but on the principle that they were not Kibaki's animals to sell. An attempt by the elite to exclude the proletariat from the philosophy of wildlife management had failed in Kenya for the first time in postcolonial history. This is an important philosophical threshold that still stands uncrossed in much of sub-Saharan Africa leading to dependent, rather than sustainable wildlife management. The vibrant conservation civil society in Kenya has also shown some negative effects in policy discourses and has on various occasions obstructed developments in this area. A notable example of this is the new draft Wildlife Act, which has taken a total of 11 years of debate and dispute between different interest groups before it was finally submitted to the Kenya cabinet in September 2012. The foreign sources of funding for conservation civil society in Kenya often create a somewhat incestuous link with tourism that clouds the fundamental differences between conservation and tourism objectives.

9 THE KENYA FISHERIES SECTOR

The first Fish Protection Act (Cap 379 of the Laws of Kenya) was enacted in 1902 and, shortly after in 1905, trout was introduced into the country. These introductions were the first of several others and were presumed to have no negative impacts on endemic fish species since there was no overlap of species niches at an altitude of 6000 feet where the introductions were made. However, since no base studies were carried out before the introductions, there was no evidence to justify the introductions. Recent studies have shown that trout introduction has indeed affected the distribution of indigenous species such as the mountain catfishes of the genus *Amphilius*, which today are hardly present in mountain streams, even at lower altitudes. In 1948, the Trout Ordinance (Cap 380) was enacted in order to not only regulate its exploitation but also hinder local communities from competing with the settlers for trout. This colonial enforcement of segregation in fisheries sector is one of the reasons that some communities in trout fishing areas such as around the Mt. Kenya region hardly took interest in trout fishing, culture and consumption as a viable activity even in postcolonial times. In 1954, a Division of Fisheries headed by a Fisheries Department was established within the Game Department in order to enforce the various legislations. Thereafter, the Fisheries Protection Act (Cap 379) was amended in order to create the Department of Fisheries headed by a chief of fisheries officer, which later became the director of fisheries (Fisheries Policy, 2008). The late 1950s and 1960s saw the decline of several fisheries and continued introductions of fishes. The Nile perch (*Lates niloticus*) was introduced into Lake Victoria in order to improve sport fisheries in Western Kenya. In this lake region, local communities developed a considerable interest in commercial fisheries particularly of two endemic species of tilapia, the Graham's tilapia *Oreochromis esculentus* and Victoria tilapia *Oreochromis variabilis*. Between 1954 and 1962, the Nile tilapia was introduced in several phases into Lake Victoria from stocks originally from Lake Albert and cultured in Kajjansi fish ponds in Uganda, first into the Kagera River in 1954 and then into the Kenyan and Tanzanian gulfs of Lake Victoria between 1956 and 1958 (Trewavas, 1983). Massive stockings were also carried out between 1961 and 1962 from Entebbe (Uganda) with fry of unknown origin (Balirwa, 1992). This species rapidly colonized the entire lake and by the end of the 1960s was well established in inshore habitats (Ogutu-Ohwayo, 1990; Twongo, 1995). In view of the inconsistency in these introductions, the exact source of *Oreochromis niloticus* in Lake Victoria remains uncertain. Since its introduction, *O. niloticus* prospered and became the most important commercial cichlid fish of Lake Victoria. It is unfortunately held responsible for the disappearance of the two native tilapia species. Increase in commercial fisheries particularly by local communities led to the expansion of the Fisheries Department's mandate to include management of commercially exploited freshwater and marine fisheries and also aquaculture development. In 1989, the Fisheries Act (Cap 378) and the Maritime Zones Act (Cap 371) were enacted in order to provide a legal framework for the expansion of the department's mandate. Both form the principal statutes that regulate and govern fisheries of Kenya to date. The fisheries sector or department has so far been hosted in several ministries since independence including Ministries of Information and Tourism, Natural Resources, Tourism and Wildlife, Environment and Natural Resources, Regional Development, Water Irrigation and Land Reclamation, Agriculture and Rural Development, Livestock and Fisheries Development, and Fisheries Development. Even though the mandates of the Fisheries Department and the functions of the director of fisheries remained constant, the various hosting arrangements had implications on the prioritization of the fisheries sector (Fisheries Policy, 2008). In addition, the oceans and fisheries sector has operated without a policy and legal framework since 1963, leading to inconsistency in the fisheries management activities placing the fishing industry. The drafting of the policy begun in 2003 and was completed in 2008. The overall objective of Kenya's Fisheries Policy 2008 is to "enhance the ocean's and fisheries sector's contribution to wealth creation, increase employment for youth and women, food security and revenue generation through effective private, public and community partnerships". The policy focuses on the promotion of aquaculture, implementation and monitoring of sustainable management and responsible fishing practices, the promotion of fish consumption and securing the rights of vulnerable and traditional fisher communities. The Ministry of Fisheries Development in recognition of depleted wild fish stocks in both fresh and marine water, and to meet the objective of promoting aquaculture, begun the Economic Stimulus Programme (ESP) in order to highlight fish farming as a viable economic activity in the country by raising the income of farmers and other stakeholders in the fishing industry. The first phase of the ESP worth 1120M Kenya shillings (€10.67M) was launched by the Ministry of Fisheries Development to construct 200 fish ponds in 140 constituencies by June 2013. According to existing plans, each constituency is geared to receive 8 million Kenya shillings for ponds (~100,000 US$) (Nyingi and Agnese, 2011). In Kenya, the Sagana Fish Farm under the Fisheries Department provides fingerlings for warm water freshwater species. So far, the centre has been efficient in provision of seed to farmers and in research and production of suitable feed. However, a lot more effort is required in provision of suitable fish species for the various regions. Some complaints on the inability of the fish from Sagana to grow to desirable sizes would also need to be considered in ongoing initiatives (Nyingi and Agnese, 2011). With the governments supporting upcoming initiatives, the greatest challenge is to identify a suitable species that will ensure high yield while also safeguarding native species from the impacts of introducing aquacultured species. Unfortunately, in Africa, the search for a suitable species for aquaculture has often disregarded potential impacts on the native species. The most important culture species are still mainly taken from the wild and populations are often translocated into basins far beyond their native range, potentially bringing closely related but formerly isolated species of populations into contact with each other. If there has been inadequate research and planning, an introduced cultured species may directly compete with the native species or may hybridize with them, as noted for *O. niloticus* when it was introduced to Lake Victoria (Nyingi and Agnese, 2011). Today, the nation still awaits the decision of parliament in regard to the fisheries policy. When the policy transforms to law, the government will make institutional arrangement to improve coordination of research and management of fisheries including the following: create the Kenya Oceans and Fisheries Council chaired by the minister to advise and review the oceans and fisheries sector; create the Kenya Oceans and Fisheries Service (KOFS) to harmonize fisheries extension, development, management and marketing; and restructure the Kenya Marine and Fisheries Research Institute to offer effective and timely research to guide KOFS (Fisheries Policy, 2008).

10 CONCLUSIONS

The wildlife and fisheries sectors in Kenya have seen numerous changes over the years, particularly because of the changes in the respective state of the two resources. The colonial government sought to create very strong state organs to govern them but in doing so created organizations that were not agile enough to change at the same rate as the resources they were managing and the environments within which they were operating. The wildlife sector spawned a vibrant civil society that could handle the finer policy matters and somewhat filled the agility gap of the wildlife authorities. The fisheries sector, however, is still solely the preserve of the state organs and suffers from the shortcomings thereof. Both sectors have the benefit of a rich human resource base in the country. What they both lack is the ability to fully tap this pool and bring all these skills to bear on their respective policies and practices. KWS to date has not managed to separate its security and scientific arms, often leaving the final decision on scientific matters in the hands of security personnel who are incapable of handling them competently and confidently. The Fisheries Department still operates as a manager of offtake rather than a manager of ecosystems, often mixing species in the name of aquaculture and lacking scientific bases for location and durations of fishing bans and catch monitoring. It still retains the colonial government tendency to issue orders to resource users without engaging them intellectually. Natural resources in Kenya (as everywhere) do not exist in a vacuum, so Kenya would greatly benefit from stronger intra- and interagency cooperation in fisheries and wildlife management. The authorities and environment would also benefit from greater devolution of management thinking, acknowledging the fact that local communities have an intellectual contribution to offer in both fields.

REFERENCES

Balirwa, J.S., 1992. The evolution of the fishery of Oreochromis niloticus (Pisces: Cichlidae) in Lake Victoria. Hydrobiologia 232, 85–89.

Kenya Colony Game Report, 1936. Kenya Colony Game Department, Nairobi, Kenya.

National Oceans and Fisheries Policy, 2008. Republic of Kenya Ministry of Fisheries Development, Nairobi, Kenya.

Nyhus, P., Fischer, H., Madden, F.M., Osofsky, S., 2003. Taking the bite out of wildlife damage: the challenges of wildlife compensation schemes. Conserv. Prac. 4 (2), 37–43.

Nyingi, D.W., Agnese, J.F., 2011. Tilapia in eastern Africa—a friend and foe. Species in the Spotlight, Chapter 3, IUCN 2011 African Freshwater publication.

Ochieng', T., Kipkorir, A., Graham, M., 2011. Elephant Conservation and Management in the Laikipia Ecosystem. SfG annual Report.

Ogada, M.O., 2001. Relationships between livestock husbandry methods and depredation by large carnivore in Laikipia District, Kenya. MSc. Thesis, Kenyatta University, Nairobi, Kenya.

Ogada, M.O., 2008 Human-Wildlife Conflict Trends In The Buffalo Springs & Shaba Conservation Area. Unpublished report to African Wildlife Foundation.

Ogada, M., Woodroffe, R., Frank, L.G., Oguge, N.O., 2003. Limiting depredation by African carnivores: the role of livestock husbandry. Conserv. Biol. 17 (6), 1521–1530.

Ogada, D.L., Keesing, F., Virani, M.Z., 2011. Dropping dead: causes and consequences of vulture population declines worldwide. Ann. N. Y. Acad. Sci. 1249, 57–71.

Ogutu-Ohwayo, R., 1990. The decline of the native fishes of lakes Victoria and Kyoga (East Africa) and the impact of introduced species especially the Nile perch *Lates niloticus* and the Nile tilapia *Oreochromis niloticus*. Environ. Biol. Fishes 27, 81–96.

Steinhart, E.I., 2006. Black Poachers, White Hunters. A Social History of Hunting in Kenya. Oxford University Press, Nairobi, Kenya.

Trewavas, E., 1983. Tilapiine Fishes of the Genera *Sarotherodon, Oreochromis* and *Danakilia*. British Museum (Natural History), London.

Twongo, T., 1995. Impact of fish species introductions on the tilapias of Lakes Victoria and Kyoga. In: Pitcher, T.J., Hart, P.J.B. (Eds.), The impact of Species Changes in African Lakes. Chapman & Hall, London, Chapman & Hall Fish and Fisheries Series 18.

Chapter 18

Environmental Impact Assessment in Kenya

Elijah K. Biamah*, Jacqueline Kiio† and Benjamin Kogo‡
*Professor of Environmental and Water Systems Engineering, University of Nairobi, Kenya
†Agricultural Economist, Ministry of Agriculture, Kenya
‡Environmental and Biosystems Engineer, Norken Consult, Kenya

1 INTRODUCTION

1.1 Background

Environmental impact assessment (EIA) is one of the environmental assessment tools being used worldwide to provide decision-makers and the concerned public with essential information to plan for environmentally sustainable economic development. It is a systematic analysis of projects to determine their potential environmental impacts and the significance of such impacts and to propose measures to mitigate the negative impacts.

EIA is both a planning tool and a decision-making tool. As a planning tool, EIA presents methodologies and techniques for identifying, predicting, and evaluating potential environmental impacts of projects as per the project cycle. As a decision-making tool, it provides information that promotes policy-making and actions that ensure sustainability in the implemented projects. Best-practice EIA identifies environmental risks, lessens conflicts by promoting community participation, minimizes adverse environmental effects, informs decision-makers, and helps lay the base for environmentally sound projects. Benefits of integrating EIA have been observed in all stages of a project, from exploration and planning through construction, operations, decommissioning, and beyond site closure (Sinha, 1998).

EIA has a number of strengths. First, it can be a flexible process and employ a large number of evaluation methods and techniques. Second, EIA is increasingly viewed as a process, not as a mandated document. Third, EIA is becoming more commonly parallel to and part of standard prefeasibility engineering and economic studies. In general, EIA is focused on a previously selected project, and only the better EIAs consider the sector as a whole or the wider implications, such as policies.

Developments in Earth Surface Processes, Vol. 16. http://dx.doi.org/10.1016/B978-0-444-59559-1.00018-9

In Kenya, where over time the natural resource base has become severely stressed due to unsustainable use, EIA has been necessitated under the Environmental Management and Coordination Act of 1999 (EMCA, 1999), which is the statutory law. Other subsequent environmental regulations and guidelines are mainly the EIA and Audit Regulations of 2003 and EIA Guidelines and Administrative Procedures of 2002.

1.2 Goals and Principles of EIA

The United Nations Environment Programme (UNEP) (1988) set certain goals and principles of EIA and defined it as an examination, analysis, and assessment of planned activities with a view to ensure environmentally sound and sustainable development. The Rio Declaration in 1992 has given emphasis to EIA (Principle No. 17) as a national instrument, which shall be undertaken for proposed activities that are likely to have a significant and adverse impact on the environment and are subject to a decision of a competent national authority. In Agenda 21, Chapter 8 is entirely devoted to integrating environment and development in policy, planning, and decision-making processes, providing an effective legal and regulatory framework, making effective use of economic instruments and market and other incentives, and establishing a system for integrated environmental and economic accounting (Sinha, 1998).

Whereas EIA has been used primarily as a project-oriented planning tool, the evolving concept of sustainable development has highlighted the need for the role of EIA to be expanded to include other techniques such as strategic environment assessment. With this expansion there has evolved a series of core values, guiding principles and operating principles for the practice of EIA. (Banerjee and Rathore, 1994; Biswas and Geping, 1987; Bonde and Cherp, 2000; Clerk, 1996; Erickson, 1994; Glasson et al., 1994; Goodman and Tillman, 1996; Heer and Hagerty, 1977; Jain et al., 1977; Lawrence, 1994; Rees, 1988; Sadler, 1996; Sadler and Verheem, 1996; Smith and Wansem, 1995; Wandesforde-Smith et al., 1985; Wathern, 1988; Weiner, 1997; World Bank, 1999a, World Bank, 1999b).

EIA: Three Core Values

- *Sustainability*—the EIA process will result in environmental safeguards.
- *Integrity*—the EIA process will conform to agreed standards.
- *Utility*—the EIA process will provide balanced, credible information for decision-making.

EIA: Eight Guiding Principles

- *Participation*—appropriate and timely access to the process for all interested parties.
- *Transparency*—all assessment decisions, and their basis, should be open and accessible.
- *Certainty*—the process and timing of assessment should be agreed in advance and followed by all participants.
- *Accountability*—decision-makers are responsible to all parties for their actions and decisions under the assessment process.
- *Credibility*—assessments are undertaken with professionalism and objectivity.
- *Cost-effectiveness*—the assessment process and its outcomes will ensure environmental protection at the least cost to society.
- *Flexibility*—the assessment process should be able to adapt to deal efficiently and effectively with any proposal or decision-making situation.
- *Practicality*—the information and outputs provided by the assessment process are readily usable in decision-making and planning.

2 THE ENVIRONMENTAL IMPACT ASSESSMENT PROCESS

2.1 Procedures for Initiating an EIA Process

The procedures to be taken by National Environment Management Authority (NEMA) to initiate an EIA process are as follows:

- The proponent is advised to select EIA Lead Experts to undertake the screening of the project by preparing and submitting a project report to NEMA within some stipulated period.
- NEMA issues an EIA licence or advises the proponent to proceed with a full EIA study. But before doing so, the proponent is required to prepare the terms of reference for the EIA study.

The terms of reference of an EIA for a project of this magnitude would include

1. to review existing policy and legal and institutional framework on environmental management and development planning,
2. to collect and collate baseline information on the development project,
3. to conduct interviews through the community participatory process,
4. to identify and assess positive and negative impacts of the development project,
5. to identify and analyze project alternatives,
6. to identify mitigation measures and cost estimates from all the negative impacts of project,

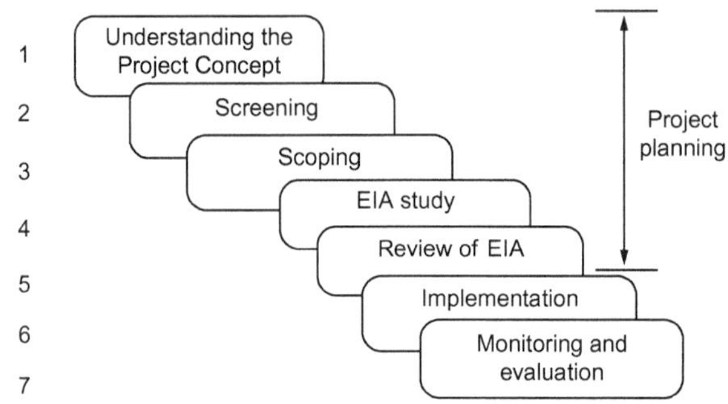

FIGURE 1 The seven stages in an environmental impact assessment process.

7. to design an environmental management plan (EMP) (including cost estimates) and a monitoring framework for the environmental/social impacts of the project.

2.1.1 Key Stages of the EIA Process

The EIA process involves seven key stages of understanding the project concept or project definition, screening, scoping, EIA study, review of EIA, implementation, and monitoring and evaluation (Figure 1).

2.1.1.1 Understanding the Project Concept

Understanding the project concept is the first stage in an EIA and the project planning phase. It involves an understanding of a proposed project within the context of its overall development strategy, viability, and sustainability.

During project identification and definition, the project proponent conducts feasibility studies, defines the usefulness of the study, considers alternatives, files a notice of intention to seek EIA clearance, and, ideally, initiates an interagency and public consultation process.

2.1.1.2 Screening

Screening is an initial assessment to decide whether a project requires further investigation in an EIA. Screening determines the extent (in terms of detail) to which an EIA is required. This process should be undertaken and agreed upon by NEMA and the project proponent. Under the European Commission's (EC) and World Bank EIA Guidelines, the project can be classified as either Category A or B or C. Within the EC EIA Guidelines, Category A projects will need an EIA, Category B will need a partial EIA, and Category C will need a full EIA. The reverse of this is the case with the World Bank Guidelines.

In the screening stage, NEMA or the EIA agency consults the project proponent and other stakeholders (agencies and public) to determine the requirements of further studies. At this stage, NEMA or the EIA agency determines whether the project may proceed as planned or if it needs to be subjected to a partial or full EIA.

According to NEMA, the screening process should be undertaken and agreed upon by NEMA and the project proponent. Thus, the proponent shall avail to the expert prefeasibility study report and any other relevant information to assist the expert conduct the screening exercise. It is recommended that the proponent work closely with the expert on the screening exercise. The screening exercise involves project and site description, collection of baseline data, data analysis, evaluation of significance of environmental impacts, evaluation of alternatives, consultation and public participation, preparation of a project report, review of project report, and an approval process.

2.1.1.3 Scoping

Scoping is undertaken to identify the key impacts requiring further investigation and prepare the terms of reference for the study. Scoping is a unique term used for EIA to identify and set the boundaries of the data required in an EIA. It identifies all the environmental impacts, who should be involved, whether external expertise is required, the input of each expert, any further information required, how the EIA fits into the project planning cycle, and the logistical and practical requirements of the study. It should ensure that all major stakeholders at all levels are contacted and their opinion sought on the expected environmental impacts of the proposed development. This exercise is done by the EIA experts and the project proponent and the proponent's project planning team. In scoping, the list of persons contacted and their viewpoints are reported.

The scoping stage, often merged with screening, deals with a more detailed plan of study for the project to identify major concerns and key impacts and to decide on assessment methods and models to be used. Stakeholder agencies and public representatives concerned with the project or the project area are consulted at this stage.

The *significance of scoping in an EIA process* is to consider reasonable and practical alternatives, inform potentially affected people, identify possible effects, understand local values, evaluate concerns expressed, define the boundaries of the EIA study, determine the analytical methods and consultation procedures, establish procedures for consultation and analysis, and establish the terms of reference.

The *nine considerations of the scoping exercise* include the major relevant environmental issues, how issues will be addressed, setting the boundaries of the baseline data collection, who should be involved, whether external expertise is required, what input is required by each person, what further information is required, how EIA fits into the project planning cycle, and the logistical and practical requirements of the study.

For a scoping report to be considered as a balanced assessment of a project, it should ensure that all major stakeholders at all levels are contacted and their opinion sought on the expected environmental impacts of the proposed development. The report therefore must include a list of people contacted with a summary of their viewpoints.

2.1.1.4 EIA Study

EIA study is the identification, analysis, and evaluation of the significance of impacts. An EIA study involves collection of data, identification of expected positive and negative impacts, causes of negative impacts, and mitigation measures to counter the negative impacts and support measures for the positive impacts. Baseline data collection and analysis are done from primary and secondary sources in the study area. The output of this study is a comprehensive EIA study report.

An EIA study carried out very early at the project concept stage or reconnaissance level takes less time than one conducted after the detailed design has been completed. This early EIA is designed to check out a project concept without worrying about a lot of detail. Such an EIA usually takes 2–3 weeks for a small project or 1–2 months for a large project. A full feasibility level EIA conducted in detail after the feasibility study has been completed would typically take 3–4 weeks for a small project or 4–6 months for a large project (Peterson et al., 1974).

The broad outline of the contents of an EIA study report would include an executive summary; a policy, legal and administrative framework; project description; baseline information; impact assessment; analysis of alternatives; an environmental management plan; a monitoring plan; and public consultation.

2.1.1.5 Review of EIA

Review of EIA is the process of assessing the adequacy of the EIA report, taking account of the points of view of stakeholders and assessing the acceptability of the proposal in terms of existing plans, policies, and standards. Review of EIA is done by NEMA through its Technical Advisory Committee (TAC) and must be done within the stipulated time frame as clearly stated by law. The output of this review is a letter giving the go ahead, a request for more environmental information or a rejection of the proposed project clearly stating the reasons. However, it should be pointed out that an EIA is meant to facilitate the project and not reject it. Why because a feasibility study must have been done prior to commissioning the EIA study. It is the responsibility of NEMA to ensure that EIAs are done for projects that have been considered as environmentally sound and viable.

During the review of an EIA, it should establish if the EIA team went to the field to collect the data and predict the impacts, establish if the team

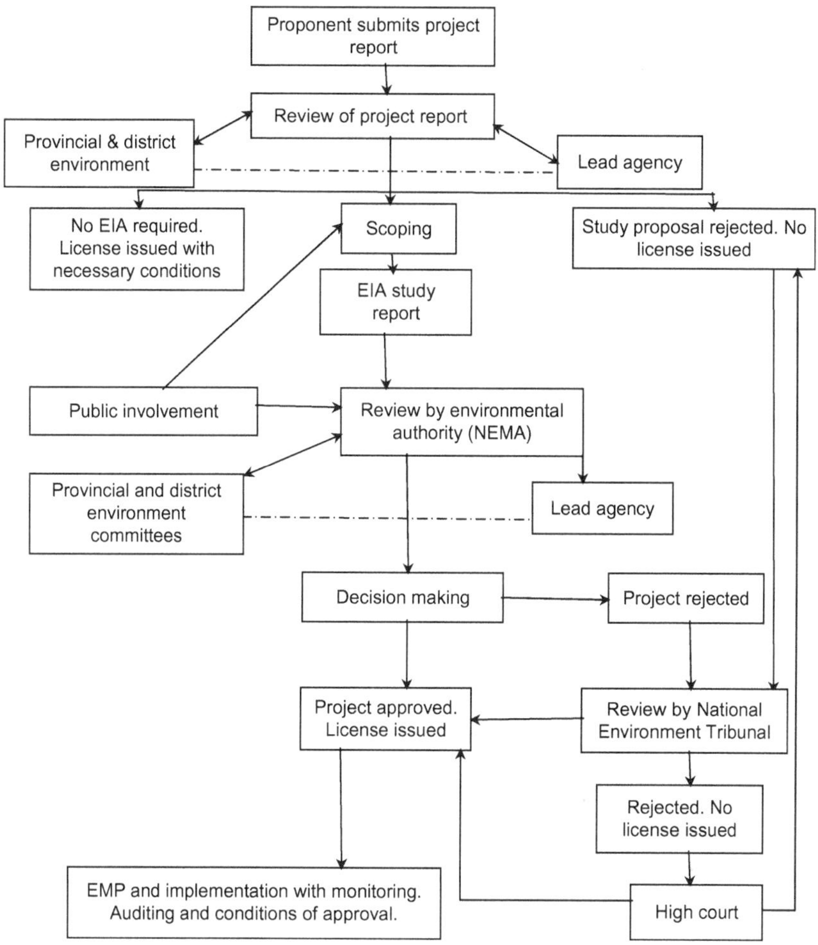

FIGURE 2 Structure of the EIA process in Kenya (EMCA, 1999).

worked together, establish the standard or quality of primary data collected, establish if community and other stakeholders were involved in the EIA process, and examine the report for completeness.

In NEMA, it is the TAC that is responsible for the EIA review process. The TAC's mandate as provided for in EMCA is to advise the authority (NEMA) on EIA-related reports. The director general is the one to prescribe the terms of reference for the TAC (Figure 2).

2.1.1.6 Environmental Monitoring and Evaluation

According to NEMA (2002), monitoring and evaluation helps to determine how effective environmental management is performing. It assesses the actual environmental impact, accuracy of prediction, the effectiveness of EIA mitigation, and enhancement measures and functioning of monitoring mechanism. Environmental monitoring and evaluation is meant to apply to ongoing projects and those commenced prior to the coming into force of the EIA regulations.

Part VII Sections 68–69 of EMCA (1999) give NEMA in conjunction with the lead agencies the responsibility of carrying out environmental audits/evaluations of all activities that are likely to have significant effect on the environment. NEMA (2007) states that "*auditing or evaluation* is used to examine the effectiveness of EIA (Leopold et al., 1971) as a decision-making tool and also enforces Environmental Management Plan (EMP)". The auditing utilizes an audit plan constructed by the proponent in consultation with NEMA.

There are two audit types in Kenya: *control audit*, undertaken by the authority, and *self-audit* undertaken by the proponent. Information generated over a period of time forms the basis of auditing of existing projects that have not undergone EIA (NEMA, 2007). In control auditing, NEMA is meant to check for compliance with the environmental parameters for the project or verify self-auditing reports. The purpose is to ensure adherence to the EMP of the project and to verify the adequacy of the EMP in mitigating negative impacts of the project.

Monitoring provides specific information on characteristics and functions of the environment and social variables in space and time (NEMA, 2007). It compares impacts predicted in an EIA with those that actually occur during and after implementation. NEMA does monitoring of ongoing projects while administering appropriate remedial action in cases of nonadherence. The three types of monitoring usually undertaken include

- baseline audit: done before project implementation,
- impact audit: done after project implementation,
- compliance audit: periodic or continuous collection of data.

3 PUBLIC CONSULTATION AND PARTICIPATION IN ENVIRONMENTAL ASSESSMENT

3.1 Rationale for Public Consultation and Participation

Public consultation and participation is an important element of the planning and implementation of project development activities. It is embedded in most donors' governance policy, sector policies, and safeguard policies on environment, involuntary resettlement (ADB 1995), and indigenous peoples. Public consultation and participation are necessary to ascertain the public's views. Providing for people's participation in project design and development is a way to improve environmental governance by providing a mechanism to influence decisions about the use and management of natural resources.

Part III Article 17 of the EIA and Audit Regulations, 2003, is explicit on public involvement in the process mainly of scoping and review of the EIA study report. The proponent is required by law to publicize the project and its anticipated impacts. Additionally, the proponent is required to hold public gatherings for consultation where a suitably qualified coordinator is appointed to receive and record both oral and written comments.

Some of the policies that require public consultation include (1) environment policy; (2) policy on indigenous peoples; (3) resettlement policy; (4) policy on confidentiality and disclosure of information; (5) sector policies on gender, water, forest, fisheries, energy, etc.; and (6) governance policy. Most multilateral and bilateral donors have prepared operational guidelines on these policies for adoption and compliance during project implementation.

Donors safeguard policies on environment and involuntary resettlement (ADB 1995); indigenous peoples recognize the importance of consulting with the project-affected individuals and groups to provide opportunities to raise community's concerns and issues. For example, a donor like Asian Development Bank (ADB) has operation manuals and guidelines for resettlement, indigenous peoples, and incorporation of social dimensions in project operations, which provide guidance on (1) identifying and assessing stakeholders, (2) assessing and evaluating the social impacts, and (3) developing and implementing appropriate resettlement and indigenous peoples' plans. These plans are designed to minimize adverse impacts on the project-affected people and provide for monitoring and evaluation of the outcomes.

3.2 Principles of Public Consultation and Participation

Public consultation and information disclosure in the EIA process is to be guided by the following general principles:

Information dissemination: Sufficient information should be provided in accessible and culturally appropriate ways. Providing information about benefits and disadvantages of the project at an early stage of the EIA process allows people time to think about the issues, consider implications, and formulate their views. An informed public will understand the trade-offs, be able to contribute meaningfully to project design, and have greater trust with the project proponent.

Information solicitation: Asking and listening to the local community, residents, and interested groups about their views and input into the EIA yield new insights and site-specific information. Past broken promises or mismanagement may have left a legacy of mistrust. Information solicitation provides the public's past experience with authorities and can initiate constructive dialogue.

Integration: Predicting likely direct and indirect impacts, short- and long-term resource use implications, evaluating their significance and risks, and developing appropriate mitigation and monitoring programmes not only require the scientific data collected by sampling and modelling but also must be based on stakeholder's input and views.

Coordination: The ability to conduct effective public consultation and participation depends on how individual team members appreciate benefits of consultation, understand their roles, and cooperate with each other. A well-integrated project team with well-defined roles and responsibility can facilitate dialogue with the executive agency to meet the donor's requirements and gain their commitment to remove any constraints to carry out public consultation throughout the project cycle.

Engaging people in dialogue: Public consultation and participation involves engaging people in dialogue—a two-way flow of information and ideas between the project proponent and the stakeholders with the opportunity for the stakeholders to express their views and concerns. Ensuring the opportunity to participate in dialogue during the early preparation stage of the EA process enables one to manage expectations and detect any potential serious conflict and help resolve issues before they lead to conflict, thus reducing financial losses due to delays.

3.3 Benefits of Public Consultation and Participation

Effective public consultation and participation can add substantial value to the EIA process. The information gained through public consultation on the stakeholders' concerns, interests, and their ability to influence decision-making helps identify key cause of environmental problems. This can be used to evaluate direct and indirect environmental impacts and assess short-term and long-term resource use implications. The input from local communities and NGOs can help evaluate alternatives and strengthen the EMP by incorporating local input and know-how. An informed public will better understand the trade-offs between project benefits and disadvantages, be able to contribute meaningfully to the project design, and have greater trust with the project proponent and support for the proposed project. Effective public consultation and participation reduces the risk of conflicts and substantial delays in project implementation. Wider public support for the project is essential for the sustainability of the project.

3.4 Stakeholder Analysis in Public Consultation and Participation

Stakeholder analysis is a tool to identify all parties that have direct and indirect interests in the project and its potential impacts on them. Failure to identify the stakeholders and consult with them could impair transparency in decision-making and that, in turn, could lead to conflicts, delaying the project process. Therefore, it is important to identify the stakeholders and the potential project impacts on them and also to evaluate their concerns and needs and their ability to understand and influence the decision-making at the project preparation stage.

There are five categories of stakeholders: local communities, civil society, government and local government bodies, private sector bodies, and other institutions. The 'primary' stakeholders are those who are directly affected, whether positively or negatively. Social development specialists will have to generate and share socioeconomic and cultural information in doing the stakeholder analysis. Social analysis techniques and methods could be used in identifying stakeholders, their needs, aspirations, and concerns regarding the project. The type of data and information required to conduct a stakeholder analysis include household level socioeconomic data, information on ethnic mix and interactions, cultural traditions, gender profile in socioeconomic activities, mechanisms for decision-making regarding their environment, and experience with similar projects. It is useful to rank the expectations, concerns, and needs of different groups of project-affected persons. It is also important to determine how their diverse interests would impact on the project outcome. One basic issue that needs analysis is how different groups and individuals will interact with each other in influencing the project outcome.

The 'other' stakeholders are those who are indirectly affected. They can be organizations through which benefits are channelled to primary stakeholders. They include local, regional, and national government, civil society organizations, individuals who live in the vicinity, private sector agencies, and multilateral and bilateral institutions. Although the potential project impact on them may be secondary in nature, they also could become primary stakeholders depending on the type of activities, the sectors involved, and the institutional capabilities to influence decision-making. Therefore, it is important that social and environmental specialists and project leaders be aware of the indirectly affected stakeholders.

Choosing stakeholders. The choice of stakeholders for direct consultation must be made on a fair and equitable basis. Meaningful public consultation requires consultation with people who represent a range of legitimate interests including the following:

1. Who will be directly or indirectly and positively or negatively affected?
2. Who are the most vulnerable groups?
3. Who might have an interest or feel that they are affected?
4. Who supports or opposes the changes that the project will deliver?
5. Whose opposition could be detrimental to the success of the project?
6. Whose cooperation, expertise, or influence would be helpful to the success of the project?

3.5 Conflict Management Through Public Consultation and Participation

Conflicts and disputes can arise over the assessment of environmental and social impacts, compensation, and decisions already made. Carrying out a careful stakeholder analysis, focusing on actual and potential conflicts, during the early stages of EIA is a useful step to avoid and manage project-related conflicts. However, if there is a potential for serious conflict or conflict has already occurred, voluntary resolution approaches such as problem-solving and consensus building should be used to mediate the situation and develop consensus among stakeholders to gain their support for the proposed project during the EIA process.

To achieve consensus on environmental issues, stakeholder representatives must meet and negotiate mutually acceptable solutions during the conflict resolution process. The four key negotiation rules include:

Assessment. Identify various stakeholder groups and their concerns and interests, assess procedural ground rules, and select a neutral party. Issues related to fundamental values or constitutional and legal rights are not amenable to consensus building, while issues of a certain way of using the resource can be mediated. Technical and scientific issues need to be dealt with as part of environmental mediation.

Areas of agreement and disagreement. Define all areas of agreement and disagreement to establish common ground and eliminate those issues from discussion. Identifying areas of disagreement with underlying reasons of each point of conflict gives negotiators an idea of the magnitude of the problem. It also ranks points of conflict in an order of importance.

Specific steps. Suggests specific steps for organizing productive meetings and decision-making processes among groups that have differences.

Issue-by-issue negotiation. Try to solve points of conflict one at a time, starting from minor issues towards major ones at the end. In this way, negotiators will address the more difficult problems after establishing a successful record of negotiation on less difficult issues.

3.6 Approaches for Public Consultation and Participation

Communication during environmental assessments involves seeking information, imparting information, or reaching agreement through dialogue. There are a number of approaches that involve public consultation and information dissemination. A critical element of preparing for public consultation is associated with the selection of appropriate approaches. Available approaches may be categorized into three types: (1) disseminating information, (2) soliciting input, and (3) getting consensus on issues. The following approaches represent best practices for carrying out public consultation.

3.7 Requirements for Public Consultation and Participation

Public consultation is mandatory as part of the EIA process for all projects requiring EIAs. The adequacy of the public consultation and participation is one of the criteria used to determine the project's compliance with the required policies.

Public consultation needs to be carried out during the early stage of EIA preparation and throughout the project implementation to address any environmental issues that affect the local communities, NGOs, governments, and other interested parties.

For all EIA compliant projects, the environment policy requires public consultation at least twice: once during the early stages of EIA field work and once when the draft EIA report is available and prior to loan appraisal by donors.

Accessibility to EIA reports. EIA reports for projects are intended to be accessible to all stakeholders including the general public. To facilitate the required consultations with all the stakeholders (including local NGOs), the information about the project's environmental issues, as well as technical data, needs to be transferred into a form and language(s) accessible to those being consulted.

Reporting the public consultation is required as part of project review by donors. The information regarding the past and future consultation is reviewed during the midterm review and annual review for all EIA compliant or sensitive projects. During the midterm review mission, the executing agency's progress and performance in EMP implementation are reviewed. Interviews with local communities should be undertaken to detect any negative environmental and social impacts that were not predicted in the EIA. The effectiveness of EMP should be assessed and any necessary corrective actions undertaken. The review of EMP and complaints feedback procedures should be described in the back-to-office report.

During the end of project evaluation, the review mission should cover issues and identified solutions if any concerning project-affected people. The executive agency needs to send the annual report to describe the progress made in the ongoing public consultation. Evaluate the consultation process and draw on the experiences for the future projects. As part of the project review, the effectiveness of the consultation process should be assessed in the project completion report. The performance audit reports should draw lessons for future projects.

The key reporting aspects listed in the succeeding text are to be included in both the EIA and IEE:

a. Relevant laws and regulation
b. Methodologies used to inform and involve the public in the EA process
c. Analysis of the data and information gathered
d. Discussion of the strategic issues related to various stakeholders and level of involvement
e. Documentation of public meetings and interviews, including dates, names, topics of discussion, and important outcomes

f. Recommendations on how the project might address concerns raised during public consultation
g. Recommended measures for continuous public consultation during the environmental management project

4 SOCIAL DIMENSIONS IN ENVIRONMENTAL IMPACT ASSESSMENT

The concept of social dimensions captures the key elements of human perspectives in development and aims at avoiding or mitigating the adverse effects of development interventions on groups that do not have the capacity to absorb such effects. The concept begins with the explicit recognition that people are the centre of development and that development is for all people. It recognizes that economic growth is a prerequisite for development, in general, and social development, in particular. Furthermore, it recognizes that (i) the poor, women, and vulnerable groups contribute to economic growth; (ii) their empowerment enhances social harmony, which is essential to economic growth; and (iii) returns from investments in women and the poor are often comparable to, and may exceed, returns from investments in infrastructure, energy, industry, and agriculture (ADB, 1993; 1998; 1999a,b and c; and 2000).

The key social dimensions considered in EIA operations are:

1. *poverty reduction*, which involves helping the poor through (a) assistance directly targeted to support productive activities that generate employment and income; (b) identification of development policies and investments that expand the employment opportunities for the poor; and (c) improvement of the access of the poor to health, family planning, education and related services, and expansion of these services (ADB 1999a,b and c);
2. *gender and development*, consisting of promoting policies and activities that help all people develop their full potential, improve their productivity, increase their contribution to the economy, and share in the rewards of development as equal partners;
3. *human resources development*, including population planning, which involves investments to help improve the skills, living standards, and quality of life of present and future generations of people and increase their contribution towards sustained and accelerated economic growth; and
4. *vulnerable groups*, including children, indigenous and tribal people, ethnic minorities, illegal settlers and squatters, disabled people, new and old immigrants, whose condition may be made worse by a policy reform, a new programme, or some form of project intervention, and for whom social safety nets and compensation mechanisms have to be provided so that they are not adversely affected by such changes.

The need to analyze social factors that influence a project continues throughout the entire life of a project, although the most crucial stage occurs during project design or the conduct of the project feasibility study when all relevant social dimensions of the proposed project are examined thoroughly and incorporated into project design. The analysis conducted as part of the feasibility study is called social analysis and may cover an assessment of (i) the groups expected to benefit from, and use the services to be provided by, the project; (ii) the needs of the groups; (iii) their demands; (iv) their absorptive capacity; (v) gender issues; and (vi) possible adverse effects on vulnerable groups and the need for measures to mitigate or compensate those adversely affected.

As the scope and content of social analysis differ among and within sectors and among countries and regions within countries, it is necessary to conduct an initial social analysis (ISA) during the project identification stage. The ISA would identify the major population groups that may be affected, beneficially and otherwise, by the proposed project and that should be the focus of the social analysis. It would also identify the specific social dimension issues that would be examined during the social analysis to be carried out as part of project preparation. A general description of the objectives, scope, and methodology for ISA is in the ADB *Guidelines for Social Analysis of Development Projects* (1991) and *Handbook for Incorporating Social Dimensions into Bank Operations* (1994). The latter document includes subsector checklists to guide the conduct of the ISA and suggests several data collection techniques and guidelines for the conduct of interviews with local groups. To supplement the information contained in these publications, the Offices of the Regional Vice Presidents issued an Inter-office Memorandum on a *Guideline for Conducting Initial Social Assessments* on 3 June 1997, which was to be incorporated in the next revision of the Handbook. In addition, ADB like other donor agencies has policies and norms of good practice that relate to the following major issues: gender, involuntary resettlement (ADB 1995), indigenous peoples, participatory processes, and involvement of NGOs.

5 LEGISLATIVE AND REGULATORY FRAMEWORK FOR ENVIRONMENTAL MANAGEMENT IN KENYA

5.1 Environmental Legislation and Legal Statutes in Kenya

Most existing environmental legislations in Kenya were originally formulated in response to specific problems. The thrust of the legislation is almost entirely negative, stressing what should not be done. It bears little relationship to environmental management, which is a concept emphasizing planning and incentives for environmentally sound choices.

A further deficiency of the past legislation was that it did not provide any specific remedies or confer any rights on private citizens, either as individuals or in groups, in the event of their interests being infringed upon by acts of environmental mismanagement. Although the law of torts provides a framework within which many environmental problems can be resolved, its use in combating environmental problems is extremely limited.

There are 57 separate legal statutes in Kenya that relate to the protection of the environment and the management of natural resources. Some of these legal statutes include Wildlife Conservation and Management (Amendment) Act of 1989, Physical Planning Act (CAP 286), Water Act 2002, Local Government Act (CAP 265), Public Health Act (CAP 242), Wildlife Conservation and Management (Amendment) Act (CAP 376), the Penal Code (CAP 63), Agriculture Act (CAP 318), and Factories Act (CAP 514), among others. There is a further 20 statutes that relate indirectly to the environment. It is because of this lack of a comprehensive act on the environment that the Environmental Management and Coordination Act (EMCA) No. 8 of 1999 was enacted. This act was on a review of the 77 statutes related to the environment. The EMCA Act of 1999 guarantees every Kenyan a clean and healthy environment.

5.2 Environmental Provisions in Kenya's Constitution

Kenya's constitution does not contain explicit environmental provisions. However, it places importance on the right to life and, by extension, encompasses the right to a clean and healthy environment. The constitution protects individual fundamental rights and freedoms, which are relevant in accessing justice in environmental matters. These include freedoms of speech, assembly, and association; the right to life; and the right to the protection of the law, which appear in Chapter 5. Section 70 of the constitution guarantees every person the right to assemble freely and associate with other persons and includes the right to form or belong to associations. The constitution also includes the right of access to the high court for redress regarding enforcement of fundamental individual rights and freedoms.

5.3 Environment Management and Coordination Act (1999)

The Environment Management and Coordination Act (EMCA) (1999) creates an overall and all-embracing agency for the management of the environment as opposed to hitherto existing legislation that set up sectoral agencies, often leading to regulatory competition. EMCA confers *locus standi* on individuals to enforce environmental rights. It also establishes a right to a clean and healthy environment and makes provision for the carrying out of EIA. It also provides for public participation in environmental law and access to justice. The act establishes the National Environment Council (NEC), the NEMA, the Provincial and District Environment Committees, the National Environment Tribunal (NET), and the Public Complaints Committee (PCC). In all these administrative structures, provision is made for public participation, and there is concern to ensure that justice is accessible.

Significantly, EMCA provides for the right of every person to a clean and healthy environment. It also makes it the obligation of every person to protect and manage the environment. Any person may bring an action in the high court to enforce the right to a clean and healthy environment.

Redress may be sought if the right has been violated, is being violated, or is likely to be violated. In determining the dispute, the court will be guided by the principles of sustainable development such as public participation in the development policies plans and processes for the management of the environment.

According to Kameri-Mbote (2005), one great innovation of the EMCA is that it overcomes most of the limitations on standing to sue. It explicitly provides that an aggrieved person need not show special damage or peculiar injury beyond that which is suffered by other affected people. Effectively, this provision grants to every person the right to protect the environment.

EMCA is an Act of Parliament that provides for the establishment of an appropriate legal and institutional framework for the management of the environment and for matters connected therewith and incidental thereto. The act recognizes the fact that the environment constitutes the foundation of national economic, social, cultural, and spiritual advancement in Kenya.

Section 51 of the act provides for the conservation of biological resources *in situ* and mandates NEMA to issue guidelines that can be used to ensure that biological resources are protected. These include the development of land use guidelines that are compatible with the conservation of biological resources and selection and management of buffer zones including special arrangements for the protection of species, ecosystems, and habitats threatened with extinction. Sections 112, 113, 114, and 115 provide for the application, granting, enforcement, and compensation for environmental easements, which may be done if found necessary for purposes of conserving and enhancing the environment.

This act requires every development likely to have an impact on the environment to undertake an EIA. The Second Schedule of the act says any activity out of character with its surrounding, or any structure of a scale not in keeping with its surrounding, or an activity leading to major changes in land use must undergo an EIA.

Section 42 prohibits any person from depositing any substance in a lake, river, or wetland or in, on, or under its bed if that substance would adversely affect the river, lake, or wetland. Section 86 outlines the standards for waste and empowers NEMA to authorize any person to operate a disposal site or plant.

Section 87 deals with prohibition against dangerous handling and disposal of waste; subsection 1 states that "No person shall discharge or dispose any waste whether generated within or outside Kenya in such a manner as to cause pollution to the environment or ill health to any person". Section 88 provides for application of licence for managing wastes. It also prohibits transportation of solid wastes within the country without a licence.

5.4 Environmental (Impact Assessment and Audit) Regulations, 2003

These environmental regulations were developed and gazetted by NEMA on 13 June 2003. The regulations apply to all policies, plans, programmes, and projects and provide the guidelines on terminologies widely used in environmental management and assessment, preparation of a project report, the EIA study, environmental audit and monitoring, miscellaneous provisions like on environmental assessment and regional and international issues, appeal to the tribunal and fees paid to NEMA for submitted EIA reports, and five schedules. Thus, the regulations document is an excellent reference on the implementation of provisions in EMCA.

5.5 Regulatory Framework for Environmental Management in Kenya

For a long time, several ministries and parastatals as lead agencies have been responsible for environmental protection and natural resource management in Kenya. This fragmented responsibility led to conflicting ministerial objectives and created gaps in coverage and became one of the main constraints to effective environmental management. Despite the Kenyan Parliament having passed the Environmental Management and Coordination Act in 1999, EMCA did receive presidential assent on 6 January 2000 and came into force on 14 January 2000. Soon thereafter, it became evidently necessary to put in place an umbrella principal government institution responsible for the implementation of all policies relating to the environment. The Kenyan Government realized about this need and in 2002 created the NEMA as the supreme regulatory and advisory body on environmental management in Kenya. NEMA was required to coordinate and supervise the various environmental management activities being undertaken by statutory organs with a view to promoting their integration into development policies, programmes, plans, and projects that provide sustainable development and a safe and healthy environment to all Kenyans. The key functions of NEMA through the NEC include responsibility for policy formulation and direction for the purposes of the act; setting national goals and objectives and determining policies and priorities for the protection of the environment; promotion of cooperation among public departments, local authorities, private sector, non-governmental organizations, and such other organizations engaged in environmental protection programmes; and performing such other functions as are assigned by the act. EMCA provides for seven to eight regulatory and advisory institutions that have since been set up and are now operational. These regulatory and advisory institutions include

1. National Environmental Council (NEC),
2. National Environmental Management Authority (NEMA),
3. Provincial and District Environment Committees (PECs and DECs),
4. Public Complaints Committee (PCC),
5. National Environment Tribunal (NET),
6. National Environment Action Plan (NEAP) Committee,
7. Standards and Enforcement Review Committee (SERC).

The EMCA also empowers the Director General of NEMA to establish TACs to advise on environmental audit/impact assessment-related decisions to be made from time to time.

5.5.1 National Environment Council

The NEC is the top policymaking body under the act charged with the responsibility of formulating policy on matters relating to environment management in Kenya. The council is composed of two representatives of public universities in Kenya, two representatives of specialized research institutions in Kenya, three representatives of the business community, and two representatives of NGOs active in the environmental field. The council regulates its own procedure and may invite any person to attend and participate in its deliberations, but the invited person is not entitled to vote.

5.5.2 National Environment Management Authority

The NEMA was established under the Environmental Management and Coordination Act (EMCA) as a Government Parastatal and became operational on 1 January 2002. The NEMA is an advisory and regulatory body and therefore is the principal government institution responsible for the implementation of all policies relating to the environment. The authority is the one responsible for dealing with EIA. NEMA has a board of management and a director general who is appointed by the president. The board of the authority includes at least seven members who are not public servants. It is noteworthy that NEMA has over recent years developed its human resource capacity to enable it to discharge its duties in the environmental realm.

5.5.3 Provincial and District Environment Committees

The following civil society and business sector representatives sit on the Provincial Environment Committees: a representative of each local authority within the province, two representatives of farmers or pastoralists, two representatives of NGOs involved in environmental management programmes in the province, and a representative of every regional development authority in the province.

The District Environment Committees also include a representative of each local authority within the district; four representatives of farmers, women, youth, and pastoralists; two representatives of NGOs involved in environmental management programmes in the district; two representatives of community-based organizations involved in environmental management programmes in the district; and two representatives of the business community in the district. The function of the Provincial and District Environment Committees is the proper management of the environment within the province and district.

5.5.4 *Public Complaints Committee*

The PCC, established under Article 31 of EMCA, was formally constituted in August 2001 but only started serious operations in January 2003. The membership of the PCC comprises of a chairman appointed by the minister who must be a person duly qualified to be appointed as a judge of the High Court of Kenya, a representative of the attorney general, a representative of the Law Society of Kenya (LSK), a representative of nongovernmental organizations appointed by the NGO Council who acts as the secretary to the committee, a representative of the business community appointed by the minister, and two members appointed by the minister for their active role in environmental management. Their functions are to investigate any allegations or complaints against any person regarding the environment in Kenya and any allegations or complaints against NEMA regarding the environment in Kenya and to investigate on its own motion any suspected case of environmental degradation in Kenya.

The PCC is concerned with investigation of complaints relating to environmental damage and degradation generally, and it has powers to investigate NEMA. It can initiate investigations on its own without waiting for a complaint to be made. The PCC's findings are reported to the NEC. It is noteworthy that the NEC is not specifically requested to do anything with regard to the PCC's report and often will note and adopt its report with no follow-up action. Its members include representatives of the LSK, the NGO sector, and the business community. Some of the problems encountered by the PCC are nonappearance in response to summons, hostility between parties during hearings, hostility towards PCC investigators, lack of understanding of EMCA, and abdication of duty by lead agencies. Given that the duty of the PCC is only to file its report with the NEC, it lacks mandate to see the recommendations it makes carried through.

The PCC is expected to make a report of its findings to the NEC as well as to prepare periodic reports of its activities. Such reports form part of the annual report on the state of the environment. The role of the PCC in enabling the citizenry to access justice is immense given its standing *vis-à-vis* NEMA.

5.5.5 *National Environment Tribunal*

The National Environment Management Tribunal (NET) is established under Section 125 of EMCA. This is in addition to the recognition of the role of formal courts in environmental dispute settlement. The tribunal is set up to hear appeals from administrative decisions taken by organs responsible for enforcement of environmental standards. The appeals may be launched by the proponent of a project against the rejection of an environment impact assessment and against a denial of a licence. An appeal may also be launched by a local community against the grant of a licence by an administrative body such as NEMA to a project developer. The limited mandate of the NET has hampered its effectiveness because its work is predicated work that NEMA is doing. If NEMA and other administrative organs are inactive as has been the case until recently, NET has no work to do. Access to justice can be enhanced by giving the tribunal original jurisdiction in environmental matters. This would position it as the best forum for hearing environmental disputes before these are taken to the high court. The tribunal is supposed to be independent, but it continues to depend on the Ministry of Environment to supply it with the requisite personnel. It sits in Nairobi though it has powers to sit anywhere else in the country as long as it gives notice to the public to that effect. EMCA is silent on the enforcement of orders of the NET. It is necessary that NET is empowered to enforce its orders rather than being required to file these with formal courts.

5.5.6 National Environment Action Plan Committee

The EMCA also establishes the NEAP Committee. This cross-sectoral committee prepares the NEAP. It consists of, among others, representatives of the public universities and research institutions of Kenya, the NGO sector, and the business community.

5.5.7 Standards and Enforcement Review Committee

EMCA provides for the establishment of SERC as a committee of NEMA and consists of members set out in the third schedule of EMCA. The permanent secretary of the Ministry of Environment and Natural Resources is the chairman of SERC. The secretary of SERC is a director of NEMA appointed by the director general. SERC is allowed to regulate its procedure and may co-opt any person to attend its meetings and participate in the deliberations but has no vote. SERC shall meet once every 3 months for the transaction of its business and in consultation with the relevant lead agencies advise NEMA on how to establish criteria and procedures for the measurement and setting of environmental standards with specific reference to water quality and pollution control. SERC shall recommend minimum water quality standards for all the waters in Kenya and for different uses. SERC is also mandated to advise NEMA on the areas of research and investigations including collection of data. SERC shall also recommend to the director general measures to be taken regarding the monitoring and control of water pollution.

5.5.8 Technical Advisory Committee

EMCA authorizes NEMA to set up a TAC to advise it on EIA-related reports. In NEMA, it is the TAC that is responsible for the EIA review process. The TAC's mandate as provided for in EMCA is to advise the authority (NEMA) on whether to issue environmental management licences to developers/ proponents who have conducted EIA studies of their projects and submitted their EIA reports to NEMA. According to EMCA, the director general is the one to prescribe the terms of reference and rules of procedure for the TAC.

During the review of an EIA, the TAC should establish if the EIA team went to the field to collect the data and predict the impacts, establish if the team worked together, establish the standard or quality of primary data collected, establish if community and other stakeholders were involved in the EIA process, and examine the report for completeness.

5.6 Enactment Shortcomings and Institutional Weaknesses of EMCA

5.6.1 Shortcomings in the Enactment of EMCA

First, the EMCA as an act of parliament came into effect before an environmental policy was put in place. That is unprocedural and instead mandated the NEC to formulate policy in-house on matters relating to environmental management in Kenya. Thus, NEC lacks the consultation and participation of stakeholders required in policy formulation. Second, the EMCA lumps up projects (under the Second Schedule) that must undergo EIA without providing any guidelines on how to categorize such projects into categories A, B, and C as the World Bank and EC have done.

5.6.2 Institutional Weaknesses of EMCA

The institutional weaknesses of EMCA are evident in the PCC and the NET. The PCC has the opportunity to enhance people's access to justice in environmental matters. However, the limitation on the use of their reports denies the country an opportunity to deal with environmental issues that are very rampant and that impact significantly on people's enjoyment of the right to a healthy environment. In the PCC for instance, EMCA provides for the PCC findings to be reported to the NEC only for noting and adoption with no follow-up action. The NET established under Section 125 of EMCA seeks to offer expeditious and cheaper justice than the ordinary courts of law. By their nature, tribunals are designed to be more accessible, informal, and free from legal technicalities. As noted earlier, the NET lacks original jurisdiction and must await appeals from decisions of administrative organs under EMCA. This denies people a chance to go directly to the tribunal when they have environmental issues to be resolved. Likewise with the NET, EMCA provides for very limited mandate of the NET to the extent that its work is predicated on work that NEMA is doing. If NEMA and other administrative organs are inactive, then NET has no work to do.

Access to justice can be enhanced by giving the tribunal original jurisdiction in environmental matters. By their nature, tribunals are designed to be more accessible, informal, and free from legal technicalities. As noted earlier, the NET lacks original jurisdiction and must await appeals from decisions of administrative organs under EMCA.

It has also been observed that whereas the tribunal is supposed to be independent, it is not as it continues to depend on the Ministry of Environment to supply it with the requisite personnel. The lack of autonomy of the NET from the ministry and inadequate financial resources also hamper the effectiveness of NET as it can only carry out so many awareness-raising workshops on the budget it has. EMCA is also silent on the enforcement of orders of the NET. It is necessary that NET is empowered to enforce its orders rather than being required to file these with formal courts.

NET is based in Nairobi but can sit anywhere in Kenya as pointed out earlier. It is important to consider devolution of the tribunal's roles to district and provincial levels. This can, however, only make a difference if NET is given wider latitude to deal with environmental matters.

5.7 Barriers to Accessing Environmental Justice in Kenya

The processes of accessing environmental justice are yet to be articulated even though the procedural rights have been promulgated. Among the barriers to access to environmental justice is the absence of an Access to Information Act. There is also lack of awareness of environmental rights provided for under EMCA and other laws both on the part of the citizens and legal practitioners. Indeed, some lawyers do not know of the existence of the structures under EMCA and have at times appealed against NEMA in the high court instead of going to the NET, while judges have at times entertained appeals from decisions of NEMA before these have been taken to the NET. This raises the question as to whether judges and lawyers are adequately equipped to handle environmental matters. Most lawyers, magistrates, and judges went to law school before environmental law was taught as a discipline. Moreover, precedents on environmental law from the courts have been very reductionist in their interpretations of concepts such as *locus standi*. Lack of awareness of EMCA may lead to courts' reliance on such old and bad law and thus stifle developments in the law. In more concrete terms, access to court is limited, owing to the complexity of court processes and the inaccessibility thereto without legal representation, which is expensive. The geographical location of courts and transport and related costs are indeed structural barriers to access to justice (Ojwang, 1993; Ringla et al., 1999).

Further compounding the situation are the multiple sectoral laws that are yet to be synchronized with EMCA. The sectoral laws provide for access to formal courts as the dispute resolution mechanism. The establishment of both the PCC and NET under EMCA was geared to making justice more accessible in environmental matters. However, both institutions' effectiveness has been hampered by a variety of factors. Firstly, NET is still anchored to a ministry and lacks the requisite independence even to hire its own staff. It has to rely on staff from the Ministry of Environment who may not be skilled to discharge NET functions. Second and related to this is the limitation of NET's jurisdiction to appeals against NEMA decisions. If NEMA does not make any decisions or if no person appeals against such decisions, NET has no work.

Another major barrier to access to justice is the lack of awareness of and/or recognition of traditional governance systems like Councils of Elders that can foster sustainable environmental management. These institutions are very close to the people and need to be linked to formal institutions, for instance, among the pastoral communities of northeastern Kenyan, environmental security is closely linked with traditional governance structures owing to the fragility of the ecosystem, water scarcity, and historical conflicts over resources among the communities. Any interventions to enhance sustainable environmental management and access to justice must of essence include these structures if they are to be effective. Noninclusion of these structures creates parallel structures with token and contemptuous observance of national systems, while allegiance typically is to traditional norms and institutions.

The implication of not considering traditional governance institutions is that the deficiencies in these systems remain and can become further barriers to access. These barriers include *ad hoc* procedures, lack of awareness of and reverence for national and regional environmental management laws, no support information on status of resources such as water, and discrimination on the basis of gender and age. The recognition of these institutions would entail investment in institutional supply to deal with the barriers while establishing means of articulation between them and other access to justice.

5.8 Opportunities for Accessing Environmental Justice in Kenya

The promotion of access to justice needs to take a multifaceted approach so as to deal with supply and demand issues. The barriers identified earlier include both structural and mechanistic barriers. The latter can be dealt with in the short term, while the former require a longer time frame. The first action that needs to be taken is raising awareness on EMCA and the institutions established under it. EMCA was promulgated in 2000. While a lot of work has been done to put in place the institutional frameworks for operationalizing it, there is still not much awareness of its provisions. This might explain the dearth of environmental cases being dealt with in the courts. The situation is aggravated by the fact that environmental law has only been taught as part of the law curriculum in Kenyan Universities from the late *1990s* and even then as an optional course.

To fully realize the objectives of EMCA, it is critical that its users be well versed with its provisions and put it to use. Access to justice provisions in EMCA will remain inoperative if there is no critical mass of judicial functionaries utilizing the provisions. Use of the provisions can be hampered by lack of understanding of the provisions.

Raising the awareness of judges, magistrates, and lawyers on the provisions of EMCA and the institutional framework for its operationalization is therefore critical. Some work has been done by UNEP, and it would be good to add value to that work rather than replicate what has been done. Others could concentrate on training magistrates and lawyers since UNEP is training judges. In the longer term, awareness raising for the public is critical to access to justice since law does not go to people.

A related action is the training of staff at NEMA and related institutions to deal with issues that are raised under EMCA. The requirements in EMCA presuppose a cadre of trained staff to implement and enforce the provisions. Training of staff at NEMA to equip them with skills to handle various issues will promote access to justice. NEMA has been bringing on board new staff members. Short-term training courses based on a needs assessment will facilitate access to justice. This can, for instance, focus on EIA. The same kind of training would be beneficial to the members of the NET and the PCC.

Thirdly, environmental law reporting should be established within the National Council for Law Reporting. Indeed it would be useful to get at one place environmental decisions reached by the courts and also before the NET and the matters investigated by the PCC. This can be done through specialized reports of the cases within the realm of the National Council for Law Reporting.

Fourthly, specialized environmental courts manned with qualified personnel could be established to deal with environmental disputes. To ensure access to justice, it is necessary that the courts have personnel that can competently handle environmental matters. Specialized environmental courts can go a long way to help achieve this objective. Given the human resource problems currently plaguing the judiciary, a stop-gap measure would be to ensure that there are magistrates who have gone through the training recommended earlier at each district level. The approach taken with land tribunals can be adopted.

Fifthly, public interest litigation should be encouraged and promoted to deal with demand side of access to justice. Public interest litigation is an effective medium for promoting access to justice since most environmental matters take on a public character. It is also indispensable in a situation where the majority of the citizenry is illiterate and in most cases unaware of their rights. Moreover, most people live in abject poverty and do not have the resources to engage the services of lawyers.

Within most communities, there are institutions that the communities use to manage their environmental resources and for adjudication of disputes. Some of these traditional governance institutions impact significantly on access to environmental justice. Studies should be carried out in select representative areas to map out the nature and operations of these institutions with a view to identifying how they can be synergized with institutions established under EMCA. Following on such studies, there should be investment in institutional supply and capacity building of local communities to enable them to interact with and engage national and regional structures such as water boards, NEMA, KWS, and Forest Department. To ensure that the local community institutions foster access to justice, training, and capacity building is critical.

Some of the areas where interventions are needed are (a) development of clear operational procedures, (b) mechanisms for inclusion of all members of community in decision-making to ensure inclusiveness, (c) record keeping for justice delivery systems to ensure development of guiding rules for making decisions on specific matters and to link with appeals to formal courts, and (d) training on diverse issues relating to EMCA and its relationship with local structures.

In the long term, it is necessary to review EMCA and take stock of its operations and efficacy in 7–10 years. Some of the issues that could be considered are (a) jurisdiction for NET, (b) PCC reports, and (c) devolved PCC and NET gradually to align to devolved structures of government.

The provisions of the new constitution that was promulgated in August 2010 have far reaching implications for access to justice in terms of both substantive provisions as well as the mode of governance (ADB 1997; 2000). The constitutional provisions on environmental rights and access to justice can promote access to justice. Furthermore, the new devolved structures of governance would call for location of institutions promoting access to justice in environmental matters at the appropriate levels. These need to be captured in both the framework law as well as in the sectoral laws.

REFERENCES

ADB, 1993. Guidelines for Incorporation of Social Dimensions in Bank Operations. Social Dimensions Unit, ADB, Manila, Philippines, October 1993.

ADB, 1995. Involuntary Resettlement. ADB, Manila, Philippines, August 1995.

ADB, 1997. Governance: Sound Development Management (R151-95); OM 54, Governance, issued 13 January, 1997, Manila.

ADB, 1998. Policy on Gender and Development. ADB, Manila, Philippines, May 1998.

ADB, 1999a. Fighting Poverty in Asia and the Pacific: The Poverty Reduction Strategy of the Asian Development Bank. ADB, Manila, Philippines, October 1999.

ADB, 1999b. Reducing Poverty: Major Findings and Implications. A Report Based on Consultations in Selected DMCs of ADB. ADB, Manila, Philippines, September 1999.

ADB, 1999c. Poverty Reduction Strategy. ADB, Manila, Philippines, Working Paper 5–99, 25 August 1999.

ADB, 2000. Promoting Good Governance—ADB's Medium-Term Agenda and Action Plan. ADB, Manila.

Banerjee, S.P., Rathore, C.S., 1994. Assessing environmental impacts of mining projects: a quantitative perspective. In: Singh, T.N. (Ed.), Proceedings of the Fourth Asian Mining. An International Conference on Geology, Mining, Ecology, Mineral Processing, 24–28 November 1993, Calcutta, India. Oxford and IBH Publishing Co., Ltd., New Delhi, pp. 75–87, 821 pp.

Biswas, A.K., Geping, Q., 1987. Environmental Impact Assessment for Developing Countries. Tycooly, London.

Bonde, J., Cherp, A., 2000. Quality review package for strategic environmental assessments of land-use plans. Impact Assess. Project Appraisal 18 (2), 99–110.

Clerk, B.D., 1996. Lecture Notes for the 13th Intensive Training Course on Environmental Assessment and Management. University of Aberdeen, Scotland.

Environmental Management and Coordination Act, 1999. In: Wamukoya, G.M., Situma, F.D.P. (Eds.), Environmental Management in Kenya: A Guide to the Environmental Management and Coordination Act. Centre for Research and Education Environmental Law 2000.

Erickson, A.P., 1994. A Practical Guide for Environmental Impact Assessment. Academic Press Inc., New York, 266 pp.

Glasson, J., Therivel, R., Chadwick, A., 1994. Introduction and principles. In: Glasson, J., et al. (Ed.), Introduction to Environmental Impact Assessment. UCL Press, London, pp. 2–24.

Goodman, R., Tillman, R., 1996. Strategic Environmental Assessment: Strengthening the EA Process. In: Goodland, R., Mercier, J.R., Muntemba, S. (Eds.), Environmental Assessment (EA) in South Africa—A World Bank Commitment, Proceedings of the Durban World Bank Workshop, June 25, 1995. The World Bank, Washington, DC.

Heer, J.E., Hagerty, D.J., 1977. Environmental Assessments and Statements. Van Nostrand Reinhold, New York.

Jain, R.K., Urban, L.V., Stacey, G.S., 1977. Environmental Impact Analysis: A New Dimension in Decision Making. Van Nostrand Reinhold, New York.

Kameri-Mbote, P., 2005. Towards greater access to justice in environmental disputes in Kenya: opportunities for intervention. IELRC Working Paper 2005-1.

Lawrence, D.P., 1994. Designing and adapting the EIA planning process. Environ. Profession. 16, 2–21.

Leopold, L., Clarke, F.E., Hanshaw, B.B., Balslay, J.R., 1971. A Procedure for Evaluating Environmental Impact. USGS, Washington, DC, US Geological Survey (USGS) Circular 645.

NEMA, 2002. Environment Impact Assessment Guidelines and Administrative Procedures. EMCA (No 8 of 1999), Kenya.

NEMA, 2007. Customer Service Charter, 2012–2014. National Environment Management Authority, Kenya.

Ojwang, J.B., 1993. Environmental Law and the Constitutional Order. ACTS Press, Nairobi.

Peterson, G.L., Gemmell, R.S., Schofer, J.L., 1974. Assessments of environmental impacts: multidisciplinary judgements of large-scale projects. Ekistics 218, 23–30.

Rees, W.E., 1988. A role for environmental assessment in achieving sustainable development. Environ. Impact Asses. Rev. 8, 273–291.

Ringla, D., et al., 1999. Legal Hurdles to Accessing Environmental, Information 6(2) Innovation (October 16–18).

Sadler, B., 1996. Final Report of the International Study of the Effectiveness of Environmental Assessment. CEAA and IAIA, Canada.

Sadler, B., Verheem, R., 1996. Strategic Environmental Assessment: Status, Challenges and Future Directions. Ministry of Housing, Spatial Planning and the Environment, The Netherlands.

Sinha, S., 1998. Environmental Impact Assessment: an effective management tool. TERI Inform. Monit. Environ. Sci. 3 (1), 1–7, New Delhi, India.

Smith, B.D., Wansem, M., 1995. Strengthening EIA capacity in Asia: environmental impact assessment in The Philippines, Indonesia, and Sri Lanka. World Resources Institute, Washington, DC, 90 pp.

United Nations Environment Programme (UNEP), 1988. Environmental Impact Assessment: Basic Procedures for Developing Countries. UNEP Regional Office for Asia and the Pacific, Bangkok, 16 pp.

Wandesforde-Smith, G., Carpenter, R.A., Horberry, J., 1985. EIA in developing countries: an introduction. Environ. Impact Asses. Rev. 5 (3), 201–206.

Wathern, P., 1988. An introductory guide to EIA. In: Wathern, P. (Ed.), Environmental Impact Assessment: Theory and Practice. Unwin Hyman, London, pp. 3–30.

Weiner, K.S., 1997. Basic purposes and policies of the NEPA regulations. In: Clark, R., Canter, L. (Eds.), Environmental Policy and NEPA. St. Lucie Press, Boca Raton, FL, USA.

World Bank, 1999a. Pollution Prevention and Abatement Handbook: Toward Cleaner Production. World Bank Group, Washington, DC, USA.

World Bank, 1999b. World Bank Operational Manual—OP 4.01 on Environmental Assessment, January 1999.

Section II

Natural Hazards

Chapter 19

Seismic Hazard

Josphat K. Mulwa*, Fumiaki Kimata† and Nguyen Anh Duong‡
*Department of Geology, University of Nairobi, Nairobi, Kenya
†Research Center for Seismology, Volcanology and Disaster Mitigation, Nagoya University, Nagoya, Japan
‡Institute of Geophysics, Vietnamese Academy of Science and Technology, Hanoi, Vietnam

1 INTRODUCTION

Kenya is dissected longitudinally by the eastern branch of the East African Rift System (EARS), which has been suggested to represent an early stage in the evolution of a divergent lithospheric plate boundary (Baker et al., 1972). Elsewhere, the EARS is considered as an intra-continental ridge system comprising an axial rift, with the most characteristic features being narrow elongate zones of thinned continental lithosphere related to asthenospheric intrusions in the upper mantle (Chorowicz, 2005). This hidden part of the rift structure is expressed on the surface by thermal uplift of the rift shoulders. The graben valleys and basins are organized over a major failure in the lithospheric mantle and in the crust comprise a major border fault, linked in depth to a low angle detachment fault, inducing asymmetric rollover pattern, eventually accompanied by smaller normal faulting and tilted blocks (Chorowicz, 2005).

The EARS forms two main branches, the eastern and western branches, and a third, southeastern branch is in the Mozambique Channel. The eastern branch runs over a distance of 2200 km, from the Afar Triangle in the north through the main Ethiopian Rift, the Omo–Turkana lows, and the Kenya (Gregory) Rifts, and ends in the basins of Northern–Tanzanian divergence in the south (Chorowicz, 2005). In Kenya, the eastern branch of EARS is variously referred to as Eastern, Gregory (Gregory, 1921; Sikes, 1926; Saggerson, 1991) or the Kenya Rift Valley. The latter term is used in this chapter and is the basis upon which discussions in this chapter are based (Figure 1).

The EARS and by extension the Davie Ridge, which is considered as the seaward extension of eastern branch (Kenya Rift Valley) of the East African Rift Valley (Mougenot et al., 1986), are characterized by divergence whose maximum rate is estimated to be about 7 mm/year (Chase, 1978). This rate of divergence is somewhat much slower than that found at most active mid-ocean ridges or even the convergence of India–Burma plates or that between

Developments in Earth Surface Processes, Vol. 16. http://dx.doi.org/10.1016/B978-0-444-59559-1.00019-0

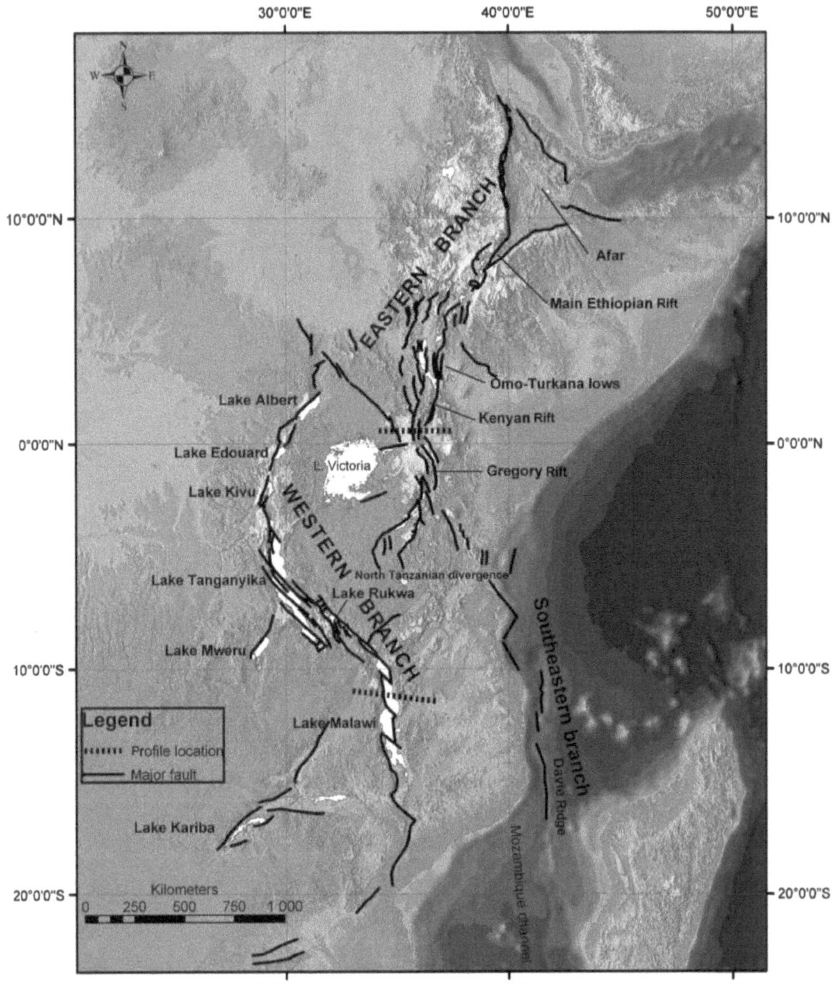

FIGURE 1 Map showing the extent of the EARS from Afar Triangle through the main Ethiopian Rift, the Omo–Turkana lows, the Kenyan (Gregory) Rifts, the basins of Northern–Tanzanian divergence and further south to Mozambique channel. *After Chorowicz (2005)*. (For colour version of this figure, the reader is referred to the online version of this chapter.)

the Australian and Sunda plates (Stein and Okal, 2006). Despite this slow rate of divergence, the Kenya Rift Valley and the Davie Ridge are characterized by frequent seismicity with large and shallow earthquakes occurring occasionally. The Davie Ridge/Fracture Zone is a 2200 km-long prominent relic fracture zone that cuts across the West Somali Basin (Coffin and Rabinowitz, 1987, 1988; Grimison and Chen, 1988). It ranges between 30 and 120 km wide, with a west-facing scarp along the lower half of its length, which rises as much as 2300 m above the sea floor (Scrutton, 1978; Mougenot et al., 1986).

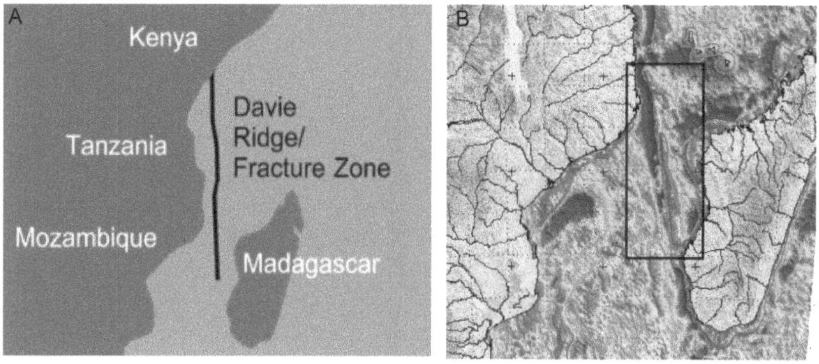

FIGURE 2 (a) Davie Fracture and the present-day position of Madagascar from Africa. (b) Satellite-derived free-air gravity anomaly of the Davie Ridge/Fracture Zone enclosed in box. *Panel (a): After Grimison and Chen (1988) and Chen and Grimison (1989); panel (b): After Bird (2001).* (For colour version of this figure, the reader is referred to the online version of this chapter.)

Seismic reflection, gravity, and magnetic data from offshore East Africa allow the Davie Fracture Zone to be traced from 11°S to its intersection with the Kenyan coast at 2°S, constraining the relative motion of Madagascar and Africa (Coffin and Rabinowitz, 1987) (Figure 2).

Further, numerous faults and fractures probably associated with the Davie Fracture have been mapped using recent gravity and magnetic data between latitudes 2°21′S and 3°03′S and longitudes 40°08′E and 40°45′E by Gippsland Offshore Petroleum Limited (2009). Seasat-derived free-air gravity anomalies and slope/rise positive magnetic anomalies observed in shipboard data help to locate the continent–ocean boundaries (COB) off the shore of East Africa and Madagascar.

Seismic reflection data further document a diapir province off Madagascar, presumably conjugate to that observed off the Kenya and Somalia coasts. The Dhow and Very Large Crude Carrier basement ridges are complex features and do not appear to be simple fracture zones owing their existence entirely to the separation of Madagascar and Africa. Coffin and Rabinowitz (1987) have determined from these data a predrift fit of Madagascar and Africa involving a 14.2° rotation of Madagascar to Africa about a pole at 10°N, 150°E. The geometry of the reconstruction adheres to seismic and potential field data, indicating the oceanic nature and extent of the Comoros Basin and of the Somali Basin between Kenya and the Seychelles, and it does not conflict with onshore or offshore stratigraphy. Timing of the opening of the Western Somali Basin is constrained by Mesozoic marine magnetic anomalies and extrapolation to the interpreted COB and occurred between approximately 165 and 130 Ma (Coffin and Rabinowitz, 1987).

According to Grimison and Chen (1988) and Chen and Grimison (1989), plate tectonics envisions Madagascar starting near Kenya. Madagascar, and the lithosphere it is sitting on, then drifts south as a ‘microplate’ with Davie Ridge as the west boundary of the microplate (Figure 3a). However, many years

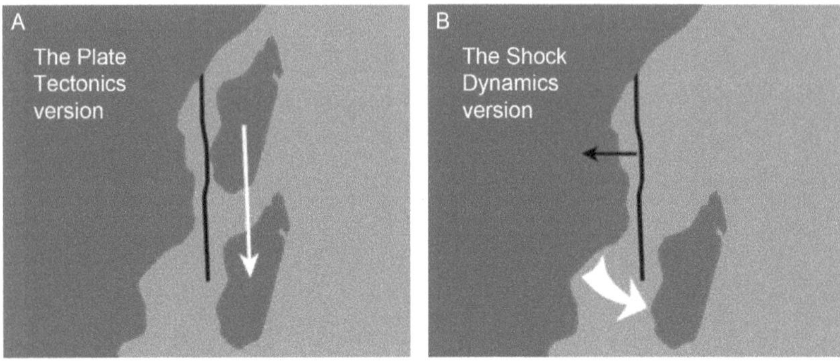

FIGURE 3 (a) Initial hypotheses to explain the drifting of Madagascar from the Kenyan coast by strike-slip movement along the Davie Fracture. (b) Shock dynamics hypotheses for the detachment of Madagascar from Africa (http://www.newgeology.us/presentation6.html). *Panel (a): After Grimison and Chen (1988) and Chen and Grimison (1989).* (For colour version of this figure, the reader is referred to the online version of this chapter.)

ago, it was determined that there is no evidence that Davie Ridge has ever been a transform fault (where plate sections slide past each other). Although the nodal [fault] planes are parallel to the ridge, 'all the large to moderate-sized earthquakes have focal mechanisms of pure normal [pull-apart] faulting and there is no significant component of strike-slip [slide by] motion'. This region is under east–west extension and seems to be within the same regional stress field as the EARS north of 5°S. However, according to Chen and Grimison (1989), no continuous spreading rift has developed north of Madagascar. Also, based on the geology alone, Madagascar moved to the east and only a short distance (Boast and Nairn, 1982). Based on this perspective and just looking at it in the starting position of the plate tectonics version, Madagascar seems a rather poor fit to the East African (Kenyan) coast.

A plausible hypothesis to explain the movement and present-day position of Madagascar is from shock dynamics (http://www.newgeology.us/presentation6.html). In shock dynamics, Madagascar began attached to Africa and swung slightly to the east as Australia pulled away. Then westward tension on Africa from the separation of South and North America, the same tension that opened the East African Rift Zone, fractured the lithosphere at its weakest point at that moment—the impact site (Figure 3b). This version fits the evidence better.

2 TECTONIC SETTING, SEISMICITY, AND SEISMIC HAZARDS IN KENYA

2.1 Tectonic Setting

Tectonics and volcanism since the Tertiary have concentrated on the region encompassing the present Kenya Rift Valley which is one of the better known parts of the EARS. The geologic evolution of the Kenya Rift Valley has been described by a number of authors (e.g. Baker and Wohlenberg, 1971; Baker et al., 1972, 1988; Logatchev et al., 1972; King and Chapman, 1972; Baker, 1987) and reviewed recently by Smith (1994) and Smith and Mosley (1993). Chorowicz (2005) describes in detail the overall morphology of the EARS and notes that it resembles that of mid-ocean ridges with the central rift valleys acting as depositional features mainly resulting from volcanic eruptions. The thicknesses of this volcanic material within the Kenya Rift floor vary between 2 and 5 km and are characterized by low density of 2.3–2.4 g/cm^3 (Simiyu and Keller, 2001; Mariita, 2003; Mulwa et al., 2009) (Figure 4).

The formation of the Kenya Rift started about early Miocene in the north around Lake Turkana and migrated southwards being active from about middle

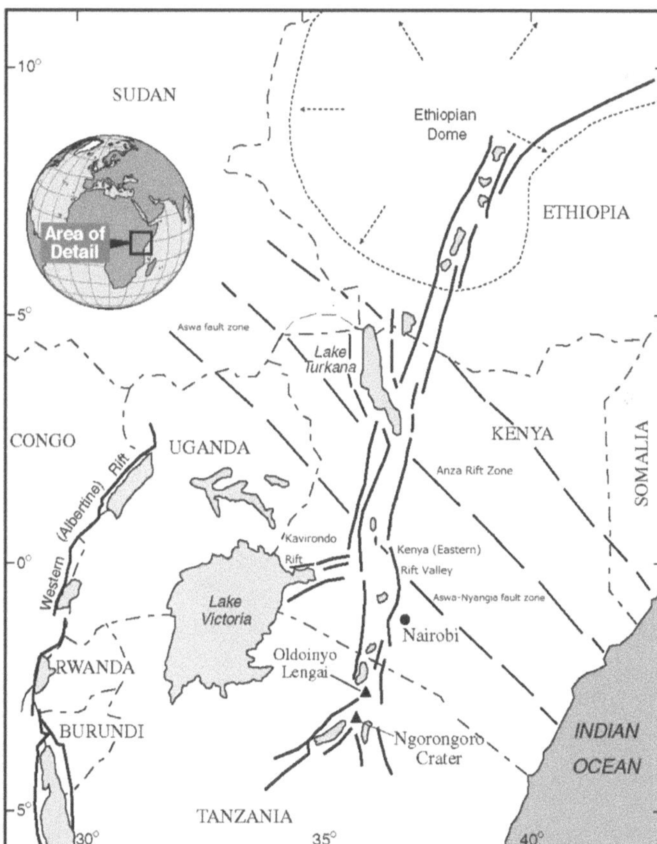

FIGURE 4 Location map of Kenya showing the NW–SE trending fault zones. *Modified after Ashley et al. (2004).* (For colour version of this figure, the reader is referred to the online version of this chapter.)

to late Miocene in the central segment (Baker et al., 1972, 1988; Baker, 1986; Smith and Mosley, 1993). The formation of the rift started by up-doming and volcanism on the crest of uplift and followed by faulting to form a half graben. The formation of a full graben occurred during the early Pleistocene and on the floor was erupted lava flows of basaltic and trachytic composition and intercalated with tuffs (Gregory, 1921; Sikes, 1926; Saggerson, 1991). Subsequently, sheet trachytes were grid faulted with dominant north–south closely spaced faults. Baker (1987) identified three major stages in the tectonic development of the Kenya Rift: (i) the pre-rift stage from 30 to 12 Mya with the formation of a depression and minor faulting, (ii) half-graben stage from 12 to 4 Mya with the formation of the main boundary faults, and (iii) the graben stage (<4 Mya) with an increase and inward migration of faulting. All stages were accompanied by intense Quaternary volcanism mostly of alkaline nature. Apart from Ol Doinyo Lengai in northeastern Tanzania, all the rift volcanoes are classified as extinct. However, some post volcanic hydrothermal activity is evident in a number of the rift volcanoes (Schlüter, 1997). Late Quaternary volcanic activity within the EARS was concentrated in the axial region of the Kenya Rift Valley and resulted in formation of caldera volcanoes and volcanic cones. Figure 5 shows the eastern rift volcanoes.

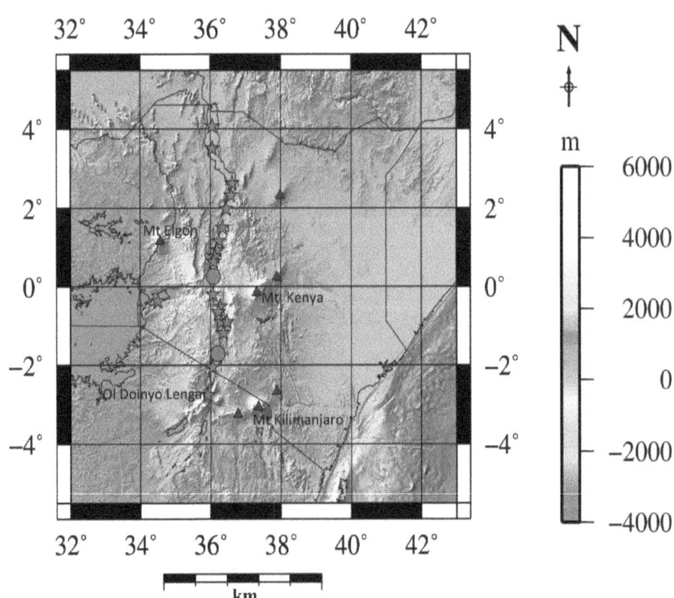

FIGURE 5 Volcanic centres aligned along the Kenya Rift Valley (red stars) with some off-rift volcanoes (blue triangles). The red circles show areas characterized by the occurrence of hot springs, geysers, and deposition of silica sinter but where there has not been any volcanic eruption in the past. *After Mulwa (2011).* (For interpretation of the references to colour in this figure legend, the reader is referred to the online version of this chapter.)

There is strong evidence to suggest that the development of the Cenozoic Kenya Rift and the present-day framework of fault lines were influenced by pre-existing basement structures (Braile et al., 1995). Particularly, the central and southern parts of the generally NNE trending Kenya Rift developed inside a broad, NW trending zone that is related to the suture between the Archean Tanzania Craton in the south–west and the Proterozoic Mozambique belt in the north–east. The craton margin zone is accompanied by several first-order NW trending and second-order NNW–SSE trending lineaments, characterized by ductile and brittle shears and well documented in the Proterozoic rocks east and west of the rift (Figure 4). Smith and Mosley (1993) have interpreted the shear zones as major zones of weakness that were controlling rift propagation and accommodation of tectonic stress within several rift sectors. The course and faulting style of the rift often changes at the intersections with the shear zones. On a regional scale, the rift sections influenced by the reworked craton margin are located roughly between Lake Baringo in the north and Lake Magadi in the south. However, on a local scale, NW–SE and NNW–SSE trending fractures presumed to be under basement control are evident. Such zones have been described and referred to by Smith and Mosley (1993) as 'accommodation zones' and include the Kerio–Bogoria–Marmanet zone in the Lake Bogoria/Lake Baringo area and the Engorika–Magadi–Lembolos zone in Lake Magadi area. Baker et al. (1972) and Atmaoui and Hollnack (2003) have further discussed oblique shear zones crossing the Kenya Rift.

3 SEISMICITY AND SEISMIC HAZARDS

The East African region is characterized by a moderate level of seismicity that is mainly controlled by the structural trend of the EARS (Figure 6). Usually small-magnitude earthquakes are common on the eastern branch (Kenya Rift) of EARS and these have been described as earthquake swarms by Ibs-von Seht et al. (2008), who have also further given an overview of the occurrence of earthquake swarms in the Kenya Rift. In comparison to the whole of the EARS, the Kenya Rift shows a relatively low seismic activity. The most significant earthquakes in Kenya up to now have been the M_S 6.9 earthquake at Subukia area in the central part of the Kenya Rift in 1928 and an aftershock M_S 6.0 which occurred 4 days later and the 1913 Turkana region earthquake having a surface wave magnitude M_S 6.0. According to Ambraseys (1991), the 1928 Subukia earthquake is a little-known earthquake that occurred in an area of relatively sparse seismicity but high tectonic activity. It was associated with a 38-km-long surface break that showed normal faulting with a small component of left lateral motion. The earthquake was characterized by a maximum throw of about 240 cm with an average along the rupture of less than 100 cm. It triggered rock falls and minor landslides and exhibited

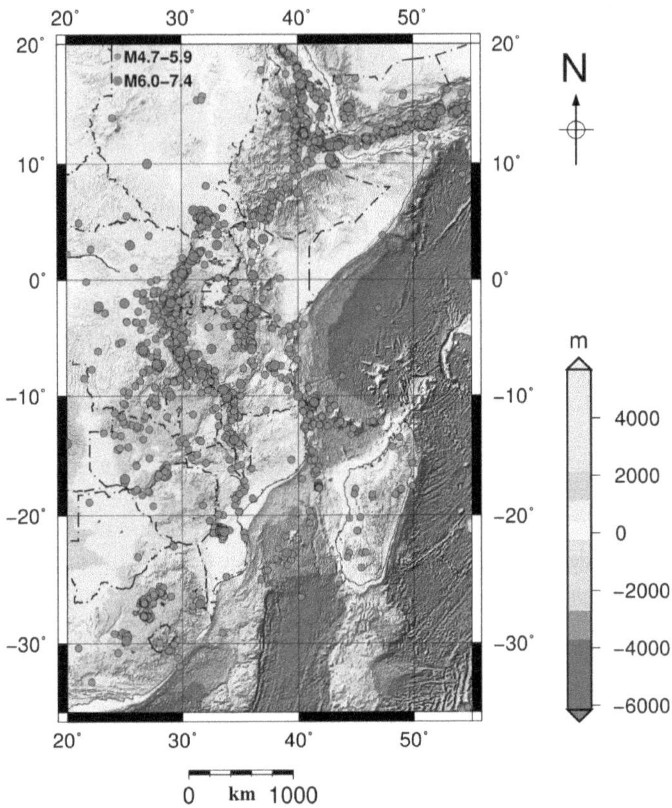

FIGURE 6 Seismicity in the eastern part of Africa for body-wave magnitudes mb ≥ 4.7 from 1900 to 2010. All focal depths are within the crust (≤30 km). *After Mulwa (2011).* (For colour version of this figure, the reader is referred to the online version of this chapter.)

long-period effects at large distances within a radius of 460 km. Considering the magnitude of the Subukia earthquake, the amount of damage it caused was not as great as it could have been, and this is directly attributed to the sparsity of dwellings and the inherent resistance to earthquake shaking of the local type of huts (Ambraseys, 1991). Except these three significant earthquakes, therefore, no $M > 5$ earthquakes in the Kenya Rift have been reported since 1928 and this explains why Kenya has been relatively free from seismicity (earthquake) hazards over the years.

Earthquake records, shown in Figure 6, for the period 1900–1930 have been obtained from published macroseismic information while those for the period 1973–2010 have been obtained from the USGS/National Earthquake Information Center (NEIC) catalogue. Consequently, a significant lack of information on earthquakes exists for the period 1931–1972.

Numerous local earthquake studies by Wohlenberg (1968), Rykounov et al. (1972), Pointing et al. (1985), Maguire et al. (1986, 1988), Young et al. (1991),

Tongue et al. (1992, 1994) and Tongue (1992) have, however, established a high microearthquake activity in the Kenya Rift. Wohlenberg (1968) particularly noted that all tectonic structures are seismically active, and in addition, seismically active zones are found, which are not connected with any known tectonic structures. Ibs-von Seht et al. (2001) reported a constant rate of $\sim 10\, M < 3$ earthquakes per day for the southernmost part of the Kenya Rift. The background seismicity of the Kenya Rift in this southern part was restricted to depths of 10 km whereas the earthquake swarms were generally shallow with most of the hypocentres between 0 and 6 km deep. For the central Kenya Rift, Young et al. (1991) reported more than $500\, M < 2.7$ earthquakes during a 3-month period in Lake Bogoria region. Tongue et al. (1994) observed a short-lasting (less than 1 day) $M < 2$ earthquake swarm in Lake Baringo region. The swarm earthquakes were found to form a narrow, elongated cluster in the centre of the rift at ~5 km depth. Young et al. (1991) and Tongue et al. (1994) attributed the earthquake swarm activity to emplacement of dikes.

A review of the seismicity in Kenya for the period 1906–2010 (Figure 7) has been undertaken by the principle author for the purpose of this write-up.

The seismicity in Kenya for the period 1906–1963 is mainly based on macroseismic data and the magnitudes are expressed as surface wave magnitudes, M_s, for example, Shah (1986) and Ambraseys (1991). The seismicity

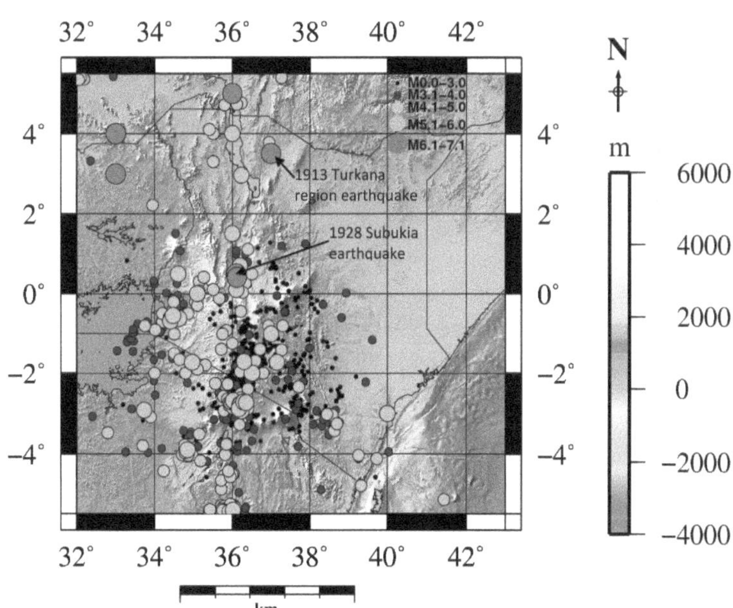

FIGURE 7 Seismicity in Kenya for the period 1906–2010 for local magnitudes $0.0 < M_l \leq 7.1$. All focal depths are within the crust (≤ 30 km). *After Mulwa (2011)*. (For colour version of this figure, the reader is referred to the online version of this chapter.)

from 1963 to 2010 is based on data from instrumental recordings and the magnitudes are expressed as local magnitudes, M_l. We wish therefore to point out that, homogenization of the magnitude scale to local magnitude in Figure 7 has been undertaken using M_s to M_l conversion formula of Scordilis (2006), which he obtained by making use of a very large sample of collected and processed data from international seismological sources such as ISC, NEIC, and HRVD.

Even though the occurrence of large magnitudes earthquakes in the Eastern African region and Kenya, in particular, is rather low, historical records based on macroseismic information, supplemented by re-examination of instrumental reports to re-evaluate the position and size of major earthquakes and recent deployments of seismic monitoring networks by United States Geological Survey (USGS) and International Monitoring System show an increase in the seismic activity in the region. Figure 6 shows that the Eastern African region, where Kenya is located, has in the past experienced strong ($mb \geq 6.0$) earthquakes between the period 1900 and 2010. Furthermore, along the East African coast in the Indian Ocean, earthquakes as deep as 40 km have been recorded below the Davie Ridge (Grimison and Chen, 1988). However, evaluation of historical and recent earthquake data shows that magnitude 6.0–7.2 earthquakes at relatively shallow depths of 10–30 km are a common occurrence along the Kenya Rift Valley and the Davie Ridge in the Mozambique Channel as shown in Figure 6 (Mulwa, 2011). The focal mechanism of these earthquakes (Figure 8) supports what has previously been proposed by Mougenot et al. (1986) and Chase (1978) that the Davie Ridge is a southward extension of the eastern arm (Kenya Rift) of the EARS. The earthquake focal mechanism indicates that the Davie Ridge is characterized by predominantly normal faulting with occasional oblique faulting. The seismic activity along the Kenyan–Tanzanian coast along the Davie Ridge is, however, relatively 'quiet' with respect to earthquakes of body-wave magnitude $mb \geq 6.0$ as shown in Figure 7.

However, if we consider two recent Indonesian scenarios, the 26 December 2004 Sumatra earthquake ($M_w = 9.3$) responsible for generating a mega tsunami with devastating consequences along the coasts of Indian Ocean occurred in the northern part of Sunda megathrust (Figure 9). Along the Sunda megathrust, the Indian Plate is being subducted beneath the Burma Plate, a small sliver or microplate, between the Indian Plate and the Sunda Plate that contains much of Southeast Asia (Stein and Okal, 2006). The Sunda megathrust, which had initially been assumed dormant, ruptured along a length of ~1300 km, and this was responsible for the $M_w = 9.2$ earthquake that generated a 30 m-high tsunami. Along this thrust belt, about 20 mm of convergence occurs every year between the Indian and Burma plates. However, the megathrust fault is locked, so strain builds up on it. Eventually the accumulated strain exceeds the frictional strength of the fault, and it slips in a great earthquake, such as that which occurred in 26 December 2004 (Stein and Okal, 2006). The Mentawai region (Pulau Pagai Selatan) earthquake

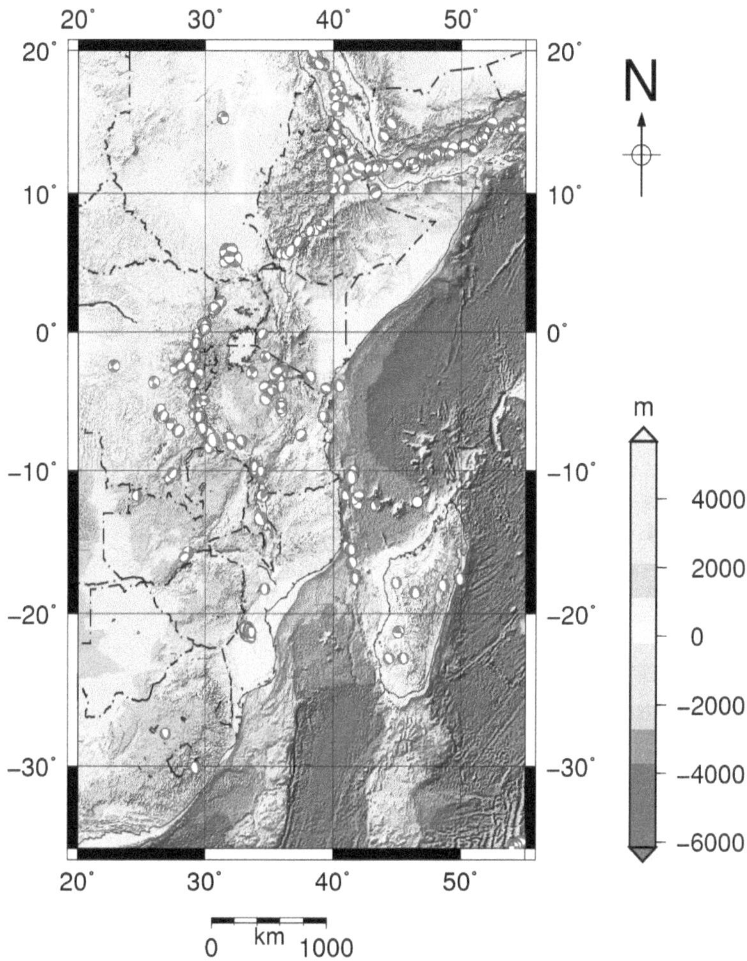

FIGURE 8 Focal mechanism (fault plane) solutions for some large earthquakes in Eastern African region. The focal mechanism shows predominantly normal faulting especially along the coastal region in the Indian Ocean. The red and green 'beach balls' represent focal depths of 10–30 km and $30 \le d \le 40$ km, respectively. *Adapted from the CMT Catalogue (1973–2010).* (For interpretation of the references to colour in this figure legend, the reader is referred to the online version of this chapter.)

$M_w = 7.2$ (BMKG) and $M_w = 7.8$ (USGS) of 25 October 2010 occurred as a result of thrust faulting on or near the subduction interface plate boundary between the Australia and Sunda Plate, where the Australia Plate is moving north–northeast with respect to the Sunda Plate at a velocity of approximately 57–69 mm/year. The focal depth of this earthquake was 20.6 km and is the latest in a sequence of large ruptures along the Sunda megathrust, including the M_w 9.3 Sumatra earthquake of 2004 discussed earlier (S. Ajat, personal communication).

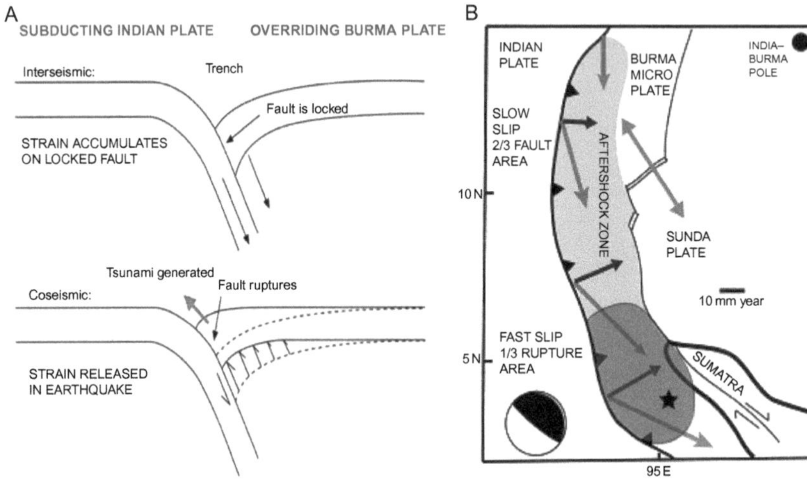

FIGURE 9 (a) Earthquake cycle at a subduction zone and (b) India–Burma convergence mechanism; the red and blue arrows show total and orthogonal convergence, respectively; the focal mechanism is purely thrust faulting. *After Stein and Okal (2006).* (For colour version of this figure, the reader is referred to the online version of this chapter.)

Both the Sumatra and Mentawai earthquakes generated tsunamis and the casualties from the two earthquakes were 280,000 and 340 people dead, respectively, with a significant number of people left homeless in addition to severe destruction of property.

The eastern branch (Kenya Rift Valley) of the EARS and, by extension, the Davie Ridge, are characterized by divergence whose maximum rate is estimated to be about 7 mm/year (Chase, 1978). This rate of divergence is somewhat much slower than that found at most active mid-ocean ridges or even the convergence of India–Burma plates or that between the Australian and Sunda plates (Stein and Okal, 2006). Despite this slow rate of divergence, the East African Rift Valley and the Davie Ridge are characterized by frequent seismicity with large (mb $\geq$ 6.0) and shallow (10–30 km) earthquakes occurring occasionally as shown in Figures 7 and 8.

Furthermore, the EARS, and precisely the Kenya Rift Valley, is characterized by $\sim$3-km-thick sediments and normal-faulting mechanism. Deformation has been active along the Kenya Rift Valley as evidenced by high seismic activity. Surface deformation studies from SAR Interferometry show that the southern Kenya Rift Valley, in Magadi area, is characterized by 14 cm of deformation over a 10-km-long stretch (Kuria et al., in press). If the Davie Ridge is an extension of eastern branch (Kenya Rift Valley) of the East African Rift Valley, we cannot rule out the occurrence of tsunami-generating earthquakes, which are bound to have devastating consequences on the eastern coast of Africa, especially in Kenya and Tanzania.

4 TSUNAMI POTENTIAL ALONG THE EAST AFRICAN COAST

To assess the tsunami hazard potential along the East African coast, we considered near-field tsunami source models that may be occasioned by rupture along the Davie Ridge. We used TSUNAMI-2 simulation program written by M. Nakamura (personal communication) and modified by N. Duong (personal communication). Tsunami propagation was calculated using ETOPO2v2 two arc-minute gridded global bathymetry data (NOAA, 2006). From Figure 8, the focal mechanism of earthquakes is purely due to normal faulting. We therefore considered normal-faulting type of events in the Davie Ridge at exactly the same epicentral locations as the mb 6.0–7.4 events shown in Figure 6. Further, we considered three earthquake case scenarios of M_w 7.2, 8.0, and 9.0 and adopted fault parameters such as rupture length, width, and surface displacement (fault dislocation) for normal faulting from Wells and Coppersmith (1994) (Table 1). These fault parameters are tabulated in Table 1.

The results of tsunami simulation for the three earthquake case scenarios are discussed later.

Figure 10a and b shows the maximum sea-floor displacement along the Davie Ridge and the maximum tsunami height due to a source model of M_w 7.2.

Figure 10a shows that the maximum expected sea-floor deformation due to a magnitude M_w 7.2 earthquake along the Davie Ridge is not likely to exceed 0.5 m. Such sea-floor deformation is likely to generate a tsunami of maximum height of about 1 m. The highest tsunami will hit the coast of Tanzania near its border with Mozambique. In our view, such a tsunami is not likely to be devastating and, consequently, the Tanzania/Mozambique coast would not be at high risk if events of this size happened.

Figure 11a and b shows the maximum sea-floor deformation along the Davie Ridge and the maximum tsunami height in metres due to a source model of moment magnitude M_w 8.0. The maximum expected sea-floor deformation due to a magnitude M_w 8.0 earthquake along the Davie Ridge is not

TABLE 1 Fault Parameters Used for Tsunami Simulation Along the Davie Ridge (Wells and Coppersmith, 1994)

Magnitude (M_w)	Strike[a]	Dip[a] (west)	Slip (m)	Length (km)	Width (km)
7.2	0° (360°)	70°	3.5	60	35
8.0	0° (360°)	70°	7.6	200	50
9.0	0° (360°)	70°	20	1200	200

[a] *Values of strike and dip are adapted from continental studies of the Kenya Rift Valley, for example, Omenda (1994) and Nyambok (2001).*

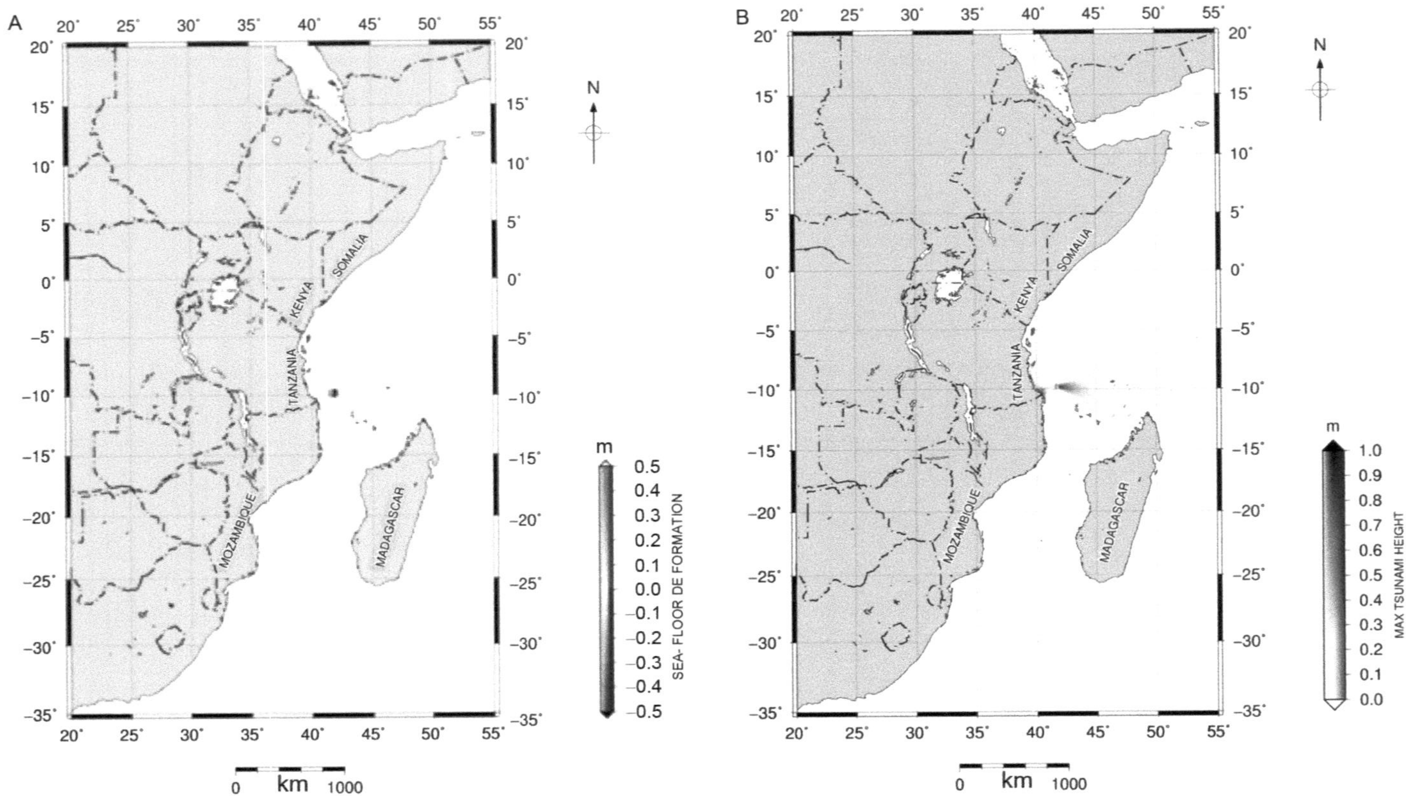

FIGURE 10 (a) Sea-floor deformation and (b) maximum tsunami height due to an earthquake of moment magnitude M_w 7.2 along the Davie Ridge. (For colour version of this figure, the reader is referred to the online version of this chapter.)

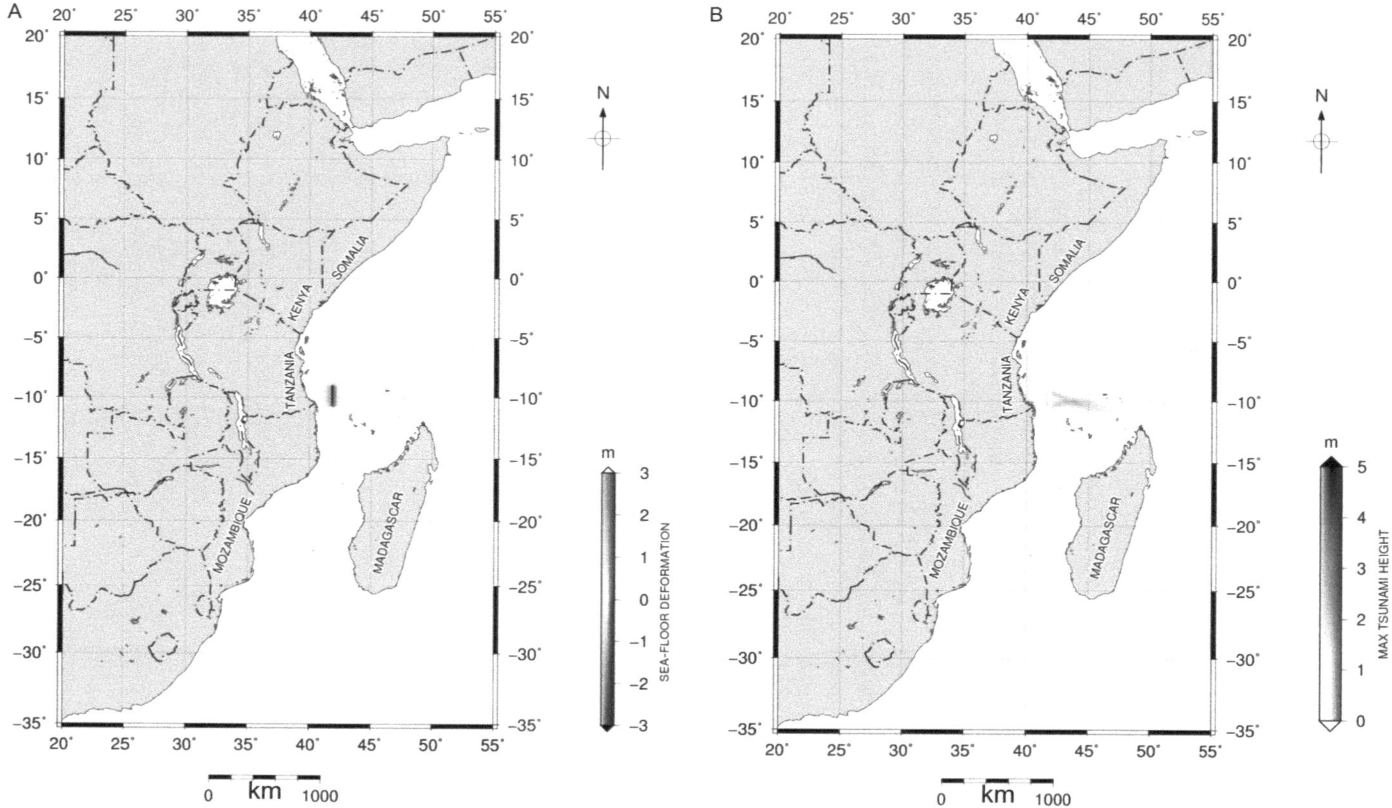

FIGURE 11 (a) Sea-floor deformation and (b) maximum tsunami height due to an earthquake of moment magnitude M_w 8.0 along the Davie Ridge. (For colour version of this figure, the reader is referred to the online version of this chapter.)

likely to exceed 3.0 m, and such sea-floor deformation is likely to generate a tsunami of maximum height of about 5 m (Figure 11b). Like in the previous scenario, the highest tsunami will hit the coast of Tanzania near its border with Mozambique. Figure 12a and b shows the tsunami time map and a pictorial view of maximum tsunami height due to a magnitude M_w 8.0 earthquake. From Figure 12a, the tsunami is expected to reach the Tanzanian as well as the Mozambique coastlines in about 30 min. The maximum tsunami wave height of about 5 m will be expected to hit the southern Tanzanian coastline. Tsunami wave height along the Mozambique coastline will be about 1 m. We would like to mention that the simulation time for M_w 8.0 scenario was 5400 s or 1 h and 30 min. Consequently, using this simulation time, the tsunami is expected to reach the Kenyan coast more than 30 min after the initial rupture, and Figure 12b shows that the tsunami height is likely to be less than 1 m.

Figure 13a and b shows the tsunami wave height in metres propagating after rupture on the Davie Ridge and the ocean floor bathymetry in metres both plotted against the horizontal distance between the Kenyan and Madagascar coastlines at $\cong 345°$ from the Madagascar coastline. Figure 13a shows that the initial tsunami wave height is only about 2.1 m but this wave height increases as the tsunami wave approaches the East African coastline reaching a maximum of 5 m along the southern Tanzanian coastline and about 1 m along the Mozambique coastline (Figure 12b).

In our view, such a tsunami is likely to have devastating consequences, and therefore, the Tanzanian and Mozambique coasts would be at high risk if earthquakes of M_w 8.0 ever happened along the Davie Ridge. On the contrary, the Kenyan coastline is not likely to be at high risk apart from an abnormal but short-lived ocean transgression and regression as ocean water surges onto the mainland.

Figure 14a and b shows the maximum sea-floor deformation along the Davie Ridge as well as the maximum tsunami height due to an earthquake of moment magnitude M_w 9.0. For such a magnitude M_w 9.0 earthquake, the maximum expected sea-floor deformation along the Davie Ridge is not likely to exceed 10.0 m, and such sea-floor deformation is likely to generate a tsunami of a height of about 10 m (Figure 14b). The highest tsunami wave will hit the coasts of Kenya, Tanzania, Mozambique, and Madagascar. Figure 15a and b shows the tsunami time map and a pictorial view of maximum tsunami height due to an earthquake of magnitude M_w 9.0 along the Davie Ridge. From Figure 15a, the tsunami is expected to reach the coastlines of Kenyan, Tanzanian, Mozambique, Madagascar, and southern Somalia in about 5 min. Digitization of the resulting map to display a pictorial view of the maximum tsunami wave height (Figure 15b) was done for the coastlines of Kenya, Tanzania, and southern part of Somalia. Consequently, from Figure 15b, the maximum tsunami wave height of about 10 m will hit the coastlines of Kenya, Tanzania, and Somalia. However, from Figure 14b, the

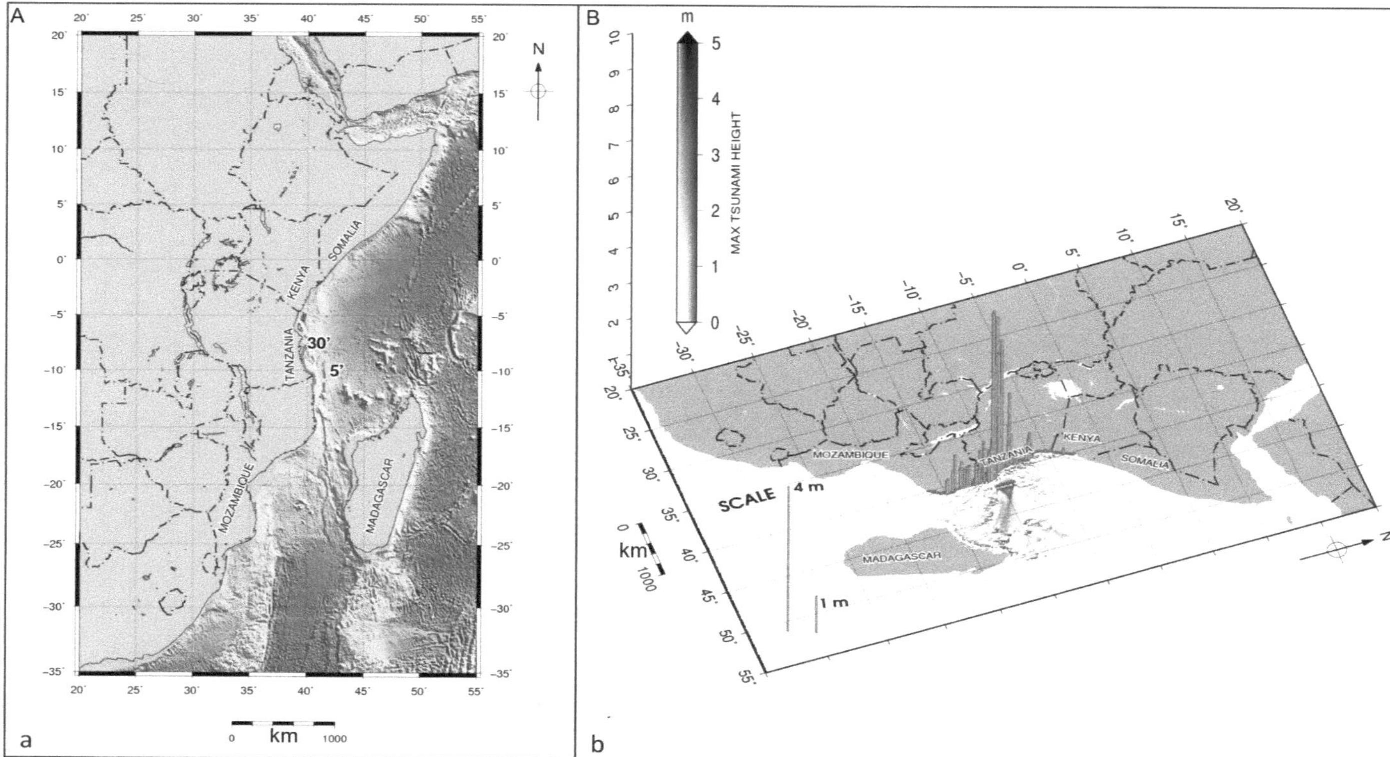

FIGURE 12 (a) Tsunami time map and (b) pictorial view of maximum tsunami height due to an earthquake of moment magnitude M_w 8.0 along the Davie Ridge. (For colour version of this figure, the reader is referred to the online version of this chapter.)

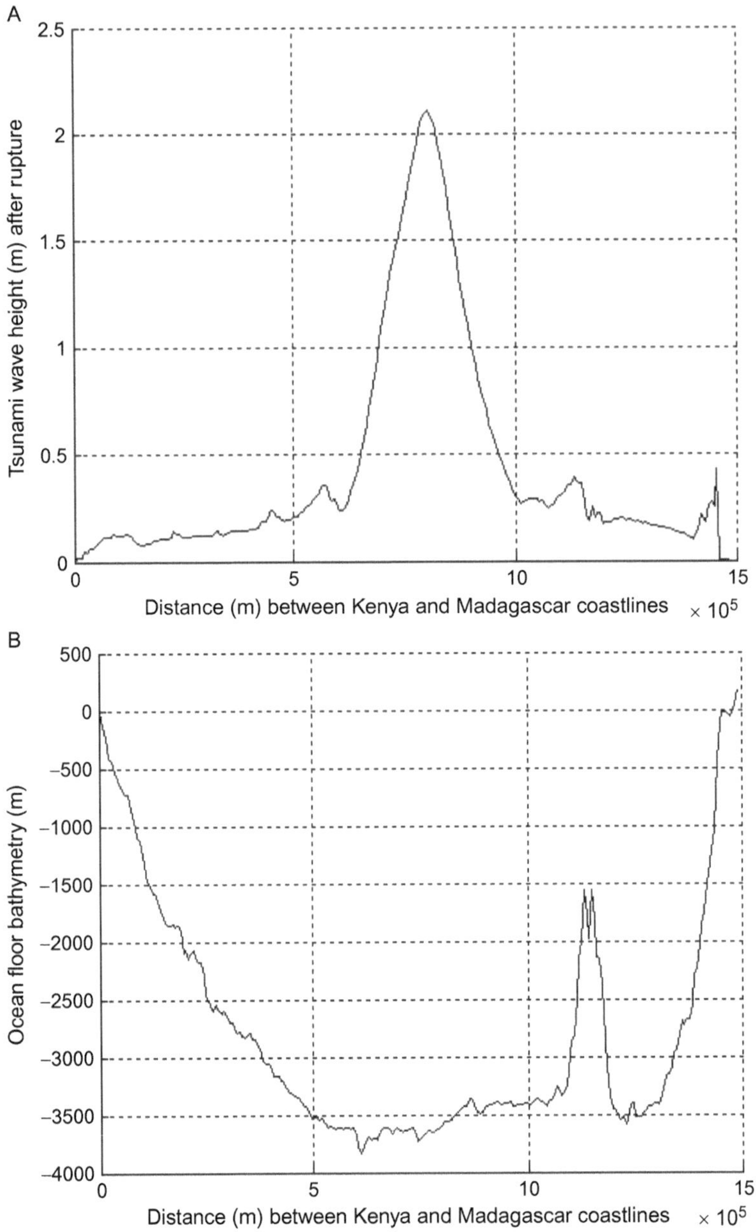

FIGURE 13 (a) Tsunami wave height (m) propagating after rupture on the Davie Ridge due to M_w 8.0 earthquake. Distance (m) between Kenya and Madagascar coastlines. (b) Ocean floor bathymetry (m) between the Kenyan and Madagascar coastlines.

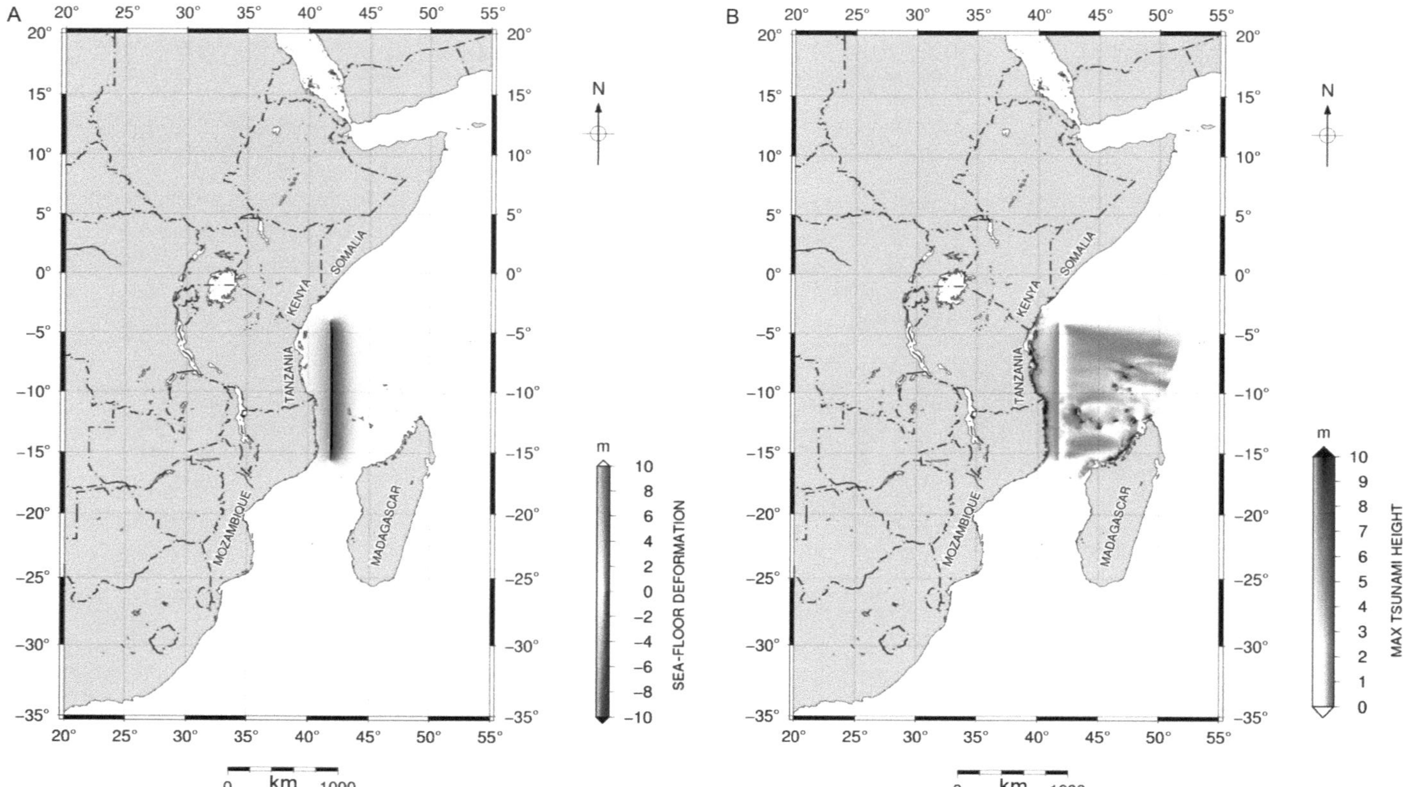

FIGURE 14 (a) Sea-floor deformation and (b) maximum tsunami height due to an earthquake of moment magnitude M_w 9.0 along the Davie Ridge. (For colour version of this figure, the reader is referred to the online version of this chapter.)

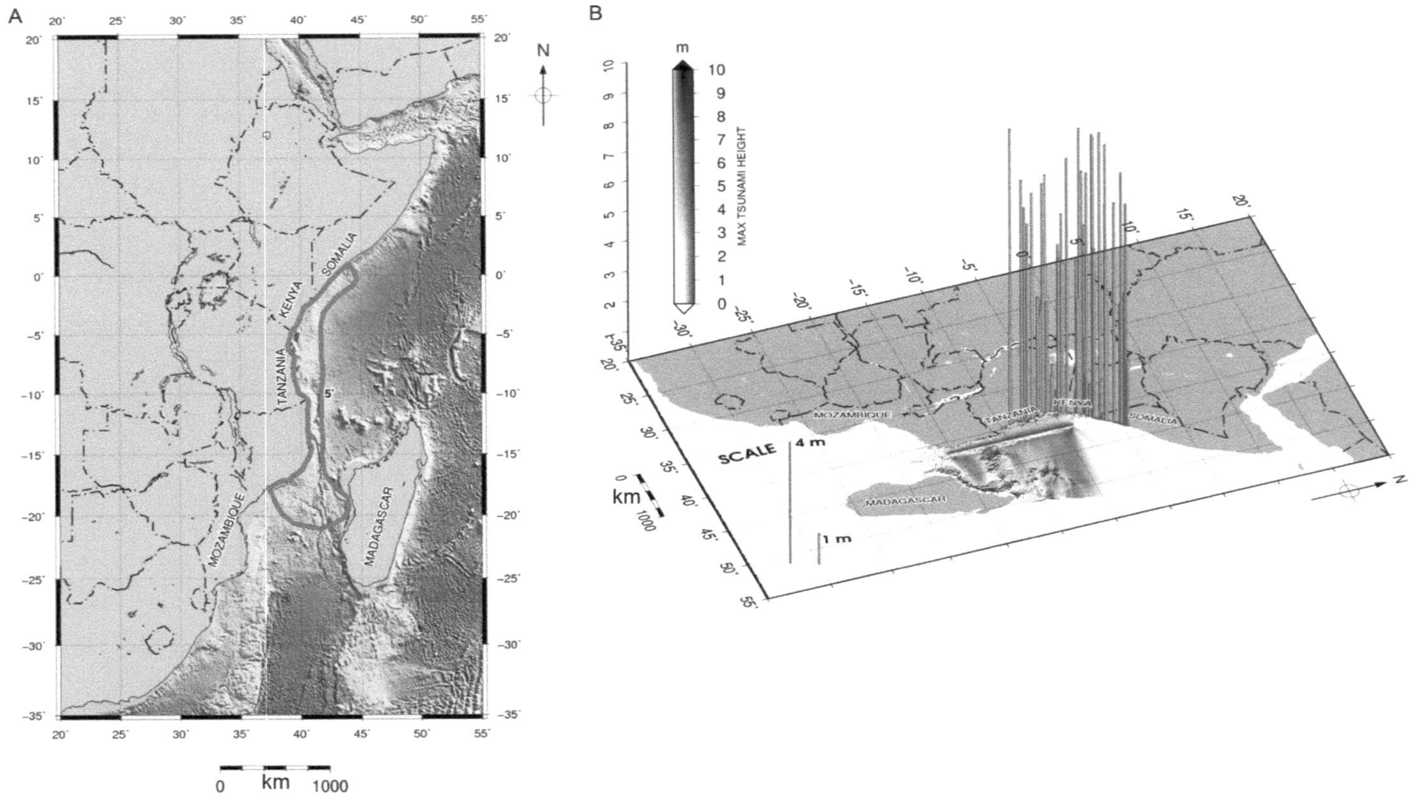

FIGURE 15 (a) Tsunami time map and (b) pictorial view of maximum tsunami height due to an earthquake of moment magnitude M_w 9.0 along the Davie Ridge. (For colour version of this figure, the reader is referred to the online version of this chapter.)

coastlines of Mozambique and Madagascar will not be exceptional and are therefore likely to be hit by the 10-m-high tsunami wave. Like in the previous scenario, we would like to mention that the simulation time for M_w 9.0 was also 5400 s or 1 h and 30 min. For this simulation time and considering a magnitude M_w 9.0 earthquake on the Davie Ridge, the tsunami is expected to reach the coasts of Kenya, Tanzania, Somalia, Mozambique, and Madagascar in just about 5 min after the initial rupture as shown in Figure 15a.

Figure 16a and b shows the tsunami wave height in metres propagating after rupture on the Davie Ridge and the ocean floor bathymetry in metres both plotted against the horizontal distance between the Kenyan and Madagascar coastlines at $\cong 345°$ from the Madagascar coastline. Figure 16a and b shows that in the deep ocean (depth 2.5–3.5 km), the initial tsunami wave height is between 4.0 and 6.0 m but this wave height increases as the tsunami wave approaches the East African coastline, reaching a maximum of about 10.5 m on the Kenyan coastline and about 12.0 m on the north and northwestern Madagascar coastline.

In our view, such a tsunami is likely to have devastating consequences and result to a significant number of fatalities. Therefore, the Kenyan, Tanzanian, Somalia, Mozambique, and Madagascar coastlines would be at high risk if earthquakes of M_w 9.0 ever happened along the Davie Ridge.

5 CONCLUSION

The East African region is characterized by a moderate level of seismicity that is mainly controlled by the structural trend of the EARS and its southward extension into the Indian Ocean forming the Davie Ridge. Most of the significant earthquakes in Kenya up to now have been the 1928 M_s 6.9 earthquake at Subukia area in the central part of the Kenya Rift and the M_s 6.0 aftershock that occurred 4 days later, and the 1913 Turkana region earthquake having a surface wave magnitude M_s 6.0. Apart from these three earthquakes, no $M>5$ earthquakes have been reported up to date. According to Ambraseys (1991), the amount of damage caused by the 1928 M_s 6.9 Subukia earthquake was not as great as it could have been due to the sparsity of dwellings and the inherent resistance to earthquake shaking of the local type of huts. In Kenya, for example, there has been mushrooming of skyscrapers since independence and concerns have been raised over the possible damage to buildings and loss of life should a major earthquake hit Kenya. According to Opiyo (c.f. Aluanga, 2010), 50% of buildings in Kenya, especially Nairobi City, may not withstand an earthquake of magnitude 6.0 and this could easily flatten buildings in several residential areas within the city and its environs. Further, according to Kianji (c.f. Aluanga, 2010), should an earthquake similar to the 1928 Subukia earthquake occur today, with the kind of buildings and population densities in our Kenyan cities, the expected damage is likely to be greater than that which occurred in 1928 and reported by Ambraseys (1991). The lack

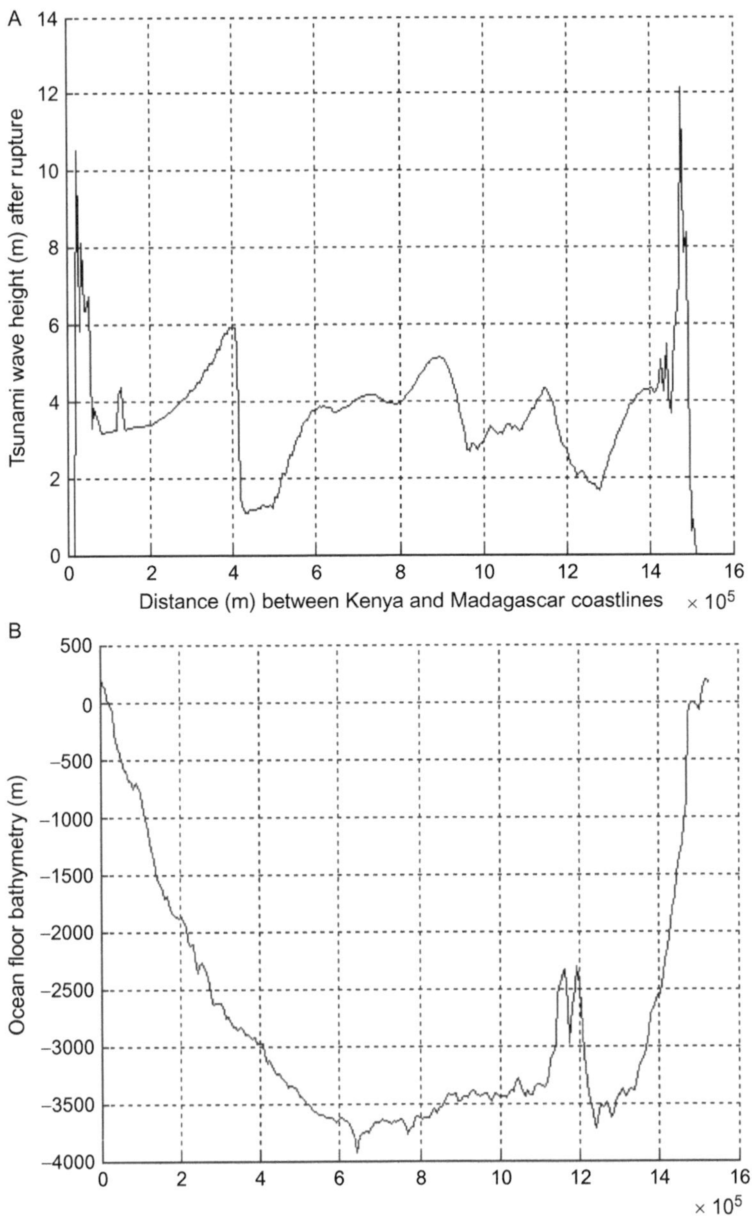

FIGURE 16 (a) Tsunami wave height (m) propagating after rupture on the Davie Ridge due M_w 9.0 earthquake. (b) Ocean floor bathymetry (m) between the Kenyan and Madagascar coastlines.

of earthquake-related hazards does not, therefore, necessarily imply that the East African region, and Kenya in particular, is safe from earthquake hazards.

Tsunami simulation has been undertaken for near-field earthquake source models of M_w 7.2, 8.0, and 9.0. For M_w 7.2 source model, the expected sea-floor deformation is about 0.5 m and the calculated tsunami heights do not exceed 1.0 m. The East African coastline would therefore not be at high risk if events of this size happened on the Davie Ridge. With regard to a source model of M_w 8.0, the sea-floor deformation would be 3.0 m and the maximum tsunami heights would be about 5.0 m along the Tanzanian and Mozambique coastlines. The tsunami wave will reach these coastlines in about 30 min after rupture along the Davie Ridge. The tsunami wave height would be much less than 1.0 along the Kenyan coastline and this will arrive after more than 30 min. Consequently, the coasts of Tanzania and Mozambique would be at risk of damages if earthquakes of M_w 8.0 happened along the Davie Ridge. For M_w 9.0 source model, the expected sea-floor deformation would be about 10.0 m and the calculated tsunami wave heights of about 10.0 m will hit the coasts of Kenya, Tanzania, southern Somalia, Mozambique, and Madagascar. The tsunami wave heights on the Kenyan and Madagascar coastlines would be 10.5 and 12 m, respectively. The travel time of the tsunami will be about 5 min after rupture along the Davie Ridge. Owing to the this short travel time of the tsunami, the effects on the coastlines of Kenya, Tanzania, Somalia, Mozambique, and Madagascar would be devastating with a significant number of casualties due to the fact that there will not be enough time for people to evacuate.

Tsunami simulations studies for far-field earthquakes such as in Indonesia, Japan, and elsewhere in the circum-pacific belt have shown that such tsunamis arrive at the East African coasts in about 9 h and the tsunami height is only about 1–3 m (Ngunjiri, 2007; Odhiambo, 2010; Oloo, 2009). The long travel time for the far-field tsunamis would allow people to evacuate from the coastal areas of the East African coast. Therefore, the greatest tsunami threat is likely to be posed by the near-field earthquakes along the Davie Ridge.

To sum up, we have got to ask ourselves whether earthquakes pose a serious enough threat in Kenya and generally in the East African region such that a share of the meagre resources available should be used to make structures earthquake-resistant? According to Asfaw (1992), such a question is nowhere more important than in East Africa where we have a number of competing natural and man-made disasters.

REFERENCES

Aluanga, L., 2010. What would happen if an earthquake hit Kenya? East African Standard Newspaper, 14.03.2010.

Ambraseys, N.N., 1991. Earthquake hazard in the Kenya Rift: the Subukia earthquake 1928. Geophys. J. Int. 105, 253–269.

Asfaw, L.M., 1992. Seismic risk at a site in the East African rift system. Tectonophysics 209, 301–309.

Ashley, G.M., Mworia, J.M., Muasya, A.M., Owen, R.B., Driese, S.G., Hover, V.C., Renaut, R.W., Goman, M.F., Mathai, S., Blatt, S.H., 2004. Sedimentation and recent history of a freshwater wetland in a semi-arid environment: Loboi Swamp, Kenya, East Africa. Sedimentology 51, 1–21.

Atmaoui, N., Hollnack, D., 2003. Neotectonics and extension direction of the Southern Kenya rift, Lake Magadi. Tectonophysics 364, 71–83.

Baker, B.H., 1986. Tectonics and volcanism of the southern Kenya rift valley and its influence on sedimentation in the African rifts. In: Frostick, L.E., Renaut, R.W., Reid, I., Tiercelin, J.J. (Eds.), Geol. Soc. Lond. Spec. Publ., 25, pp. 45–57.

Baker, B.H., 1987. Outline of the petrology of the Kenya rift alkaline province. Geol. Soc. Lond. Spec. Publ., 30, pp. 293–311.

Baker, B.H., Wohlenberg, J., 1971. Structure and evolution of the Kenya rift valley. Nature 229, 538–542.

Baker, B.H., Mohr, P.A., Williams, L.A.J., 1972. Geology of the eastern rift system of Africa. Geol. Soc. Am. Spec. Pap. 136, 1–67.

Baker, B.H., Mitchell, J.G., Williams, L.A.J., 1988. Stratigraphy, geochronology and volcano-tectonic evolution of the Kedong-Naivasha-Kinangop region, Gregory rift valley Kenya. J. Geol. Soc. Lond. 145, 107–116.

Bird, D., 2001. Shear margins: continent-ocean transform and fracture boundaries. The Leading Edge, 150–159, February 2001.

Boast, J., Nairn, A.E.M., 1982. An outline of the geology of Madagascar. In: Nairn, A.E.M., Stehli, F. (Eds.), The Ocean Basins and Margins. The Indian Ocean, vol. 6. Plenum Press, New York, pp. 649–696, (Chapter 14).

Braile, L.W., Keller, G.R., Wendland, R.F., Morgan, P., Khan, M.A., 1995. The East African rift system. In: Olsen, K.H. (Ed.), Developments in Geotectonics, Continental Rifts: Evolution, Structure, and Tectonics, vol. 25, pp. 213–231.

Chase, C.G., 1978. Plate kinematics: the Americas, East Africa and the rest of the world. Earth Planet. Sci. Lett. 37, 355–368.

Chen, W.-P., Grimison, N.L., 1989. Earthquakes associated with diffuse zones of deformation in the oceanic lithosphere: some examples. Tectonophysics 166, 133–150.

Chorowicz, J., 2005. The East African rift system. J. Afr. Earth Sci. 43, 379–410.

Coffin, M.F., Rabinowitz, P.D., 1987. Reconstruction of Madagascar and Africa: evidence from the Davie Fracture Zone and Western Somali Basin. J. Geophys. Res. 92 (B9), 9385–9406.

Coffin, M.F., Rabinowitz, P.D., 1988. Evolution of the Conjugate East African-Madagascan Margins and the Western Somali Basin. The Geological Society of America, Boulder, pp. 1–78, Special Paper 226.

Gippsland Offshore Petroleum Ltd., 2009. Gravity and magnetic surveys in the Lamu embayment (Unpublished).

Gregory, J.W., 1921. The Rift Valleys and Geology of East Africa. Seeley Service, London.

Grimison, N.L., Chen, W.-P., 1988. Earthquakes in the Davie Ridge-Madagascar Region and the Southern Nubian-Somalian Plate Boundary. J. Geophys. Res. 93 (B9), 10439–10450.

Ibs-von Seht, M., Blumenstein, S., Wagner, R., Hollnack, D., Wohlenberg, J., 2001. Seismicity, seismotectonics and crustal structure of the southern Kenya rift—new data from the Lake Magadi area. Geophys. J. Int. 146, 439–453.

Ibs-von Seht, M., Plenefisch, T., Klinge, K., 2008. Earthquake swarms in continental rifts—a comparison of selected cases in America, Africa and Europe. Tectonophysics 452, 66–77.

King, B.C., Chapman, G.R., 1972. Volcanism of the Kenya rift valley. Phil. Trans. Roy. Soc. A271, 185–208.

Kuria, Z.N., Woldai, T., Van der Meer, F.D., Barongo, J.O., Mathu, E.M. Characterisation of earth surface deformation from SAR Interferometry and Field Studies, Lake Magadi Area, Southern Kenya (in press).

Logatchev, N.A., Beloussov, V.V., Milanovsky, E.E., 1972. East African Rift development. Tectonophyiscs 15, 71–81.

Maguire, P.K., Cooke, P.A.V., Laffoley, N.d'A., Evans, J.R., 1986. Earthquake recording in the central part of the Kenya Rift Valley. EOS 67, 1103.

Maguire, P.K., Shah, E.R., Pointing, A.J., Cooke, P.A., Khan, M.A., Swain, C.J., 1988. The seismicity of Kenya. J. Afr. Earth Sci. 7, 915–923.

Mariita, N.O., 2003. An Integrated geophysical study of the northern Kenya rift crustal structure: Implications for geothermal energy prospecting for Menengai area. PhD dissertation, University of Texas at El Paso, USA, pp. 1–188.

Mougenot, D., Recq, M., Virlogeux, P., Lepvrier, C., 1986. Seaward extension of the East African Rift. Nature 321, 599–603.

Mulwa, J.K., Fairhead, D., Barongo, J.O., Mariita, N.O., 2009. Heat source mapping and evaluation of geothermal resource potential in Lake Bogoria basin, Kenya. SEG Expanded Abstract 28, 1294–1299.

Mulwa, J.K., 2011. Determination of source parameters for the May 20, 1990 Southern Sudan earthquake by inversion of teleseismic body-waves. In: JICA Training Programme, Operating Management of Earthquake, Tsunami and Volcano Eruption Observation System, Fiscal Year 2010. Research Center for Seismology, Volcanology and Disaster Mitigation, Graduate School of Environment Studies, Nagoya University/Chubu International Center, Japan International Cooperation Agency, Japan, pp. 15–31.

National Oceanic and Atmospheric Administration (NOAA), 2006. 2-Minute gridded global relief data (ETOPO2v2). http://www.ngdc.noaa.gov/mgg/global/etopo2.html, Retrieved on November 10, 2010.

Ngunjiri, C.M., 2007. Tsunami and seismic activities in Kenya. In: JICA Training Programme, Operating Management of Earthquake, Tsunami and Volcano Eruption Observation System, Fiscal Year 2006. Research Center for Seismology, Volcanology and Disaster Mitigation, Graduate School of Environment Studies, Nagoya University/Chubu International Center, Japan International Cooperation Agency, Japan, pp. 13–23.

Nyambok, I.O., 2001. Hydrogeology and geochemistry of groundwater resources of the Central Rift of Kenya. In: Proc. Regional Workshop on Research for Sustainable Water Resources in East African Rift Valley, Addis Ababa, Ethiopia, 27–29 November, pp. 1–16.

Odhiambo, J.A., 2010. East African rift system, seismic activity, ground deformation and tsunami hazard assessment in Kenya coast. In: JICA Training Programme, Operating Management of Earthquake, Tsunami and Volcano Eruption Observation System, Fiscal Year 2009. Research Center for Seismology, Volcanology and Disaster Mitigation, Graduate School of Environment Studies, Nagoya University/Chubu International Center, Japan International Cooperation Agency, Japan, pp. 1–21.

Oloo, P.N., 2009. Seismic and tsunami activities in Kenya. In: JICA Training Programme, Operating Management of Earthquake, Tsunami and Volcano Eruption Observation System, Fiscal Year 2008. Research Center for Seismology, Volcanology and Disaster Mitigation, Graduate School of Environment Studies, Nagoya University/Chubu International Center, Japan International Cooperation Agency, Japan, pp. 85–111.

Omenda, P.A., 1994. The geological structure and reservoir characteristics of the Olkaria West geothermal field. Proc. Stanford Geothermal Reservoir Eng. Workshop, 19, pp. 125–130.

Pointing, A.J., Maguire, P.K.H., Khan, M.A., Francis, D.J., Swain, C.J., Shah, E.R., Griffiths, D.H., 1985. Seismicity of the northern part of the Kenya rift valley. J. Geodyn. 3, 23–37.

Rykounov, L.N., Sedov, V.V., Savrina, L.A., Bourmin, V.J., 1972. Study of microearthquakes in the rift zones of East Africa. Tectonophysics 25, 123–130.

Saggerson, E.P., 1991. Geology of the Nairobi area. Geol. Surv. Kenya, Rep. No. 98.

Schlüter, T., 1997. Geology of East Africa. Beiträge zur Regionalen Geologie der Erde, 27, Gebrüder Bornträger, Berlin-Stuttgart, 484 pp.

Scordilis, E.M., 2006. Empirical global relations converting *M*S and *m*b to moment magnitude. J. Seismol. 10, 225–236.

Scrutton, R.A., 1978. Davie Fracture Zone and the movement of Madagascar. Earth Planet. Sci. Lett. 39, 84–88.

Shah, E., 1986. Seismicity of Kenya. PhD Thesis. University of Nairobi, Kenya.

Sikes, H.L., 1926. The structure of the eastern flank of the Rift valley near Nairobi. Geogr. J. 68, 385–402.

Simiyu, S.M., Keller, G.R., 2001. An integrated geophysical analysis of the upper crust of the southern Kenya rift. Geophys. J. Int. 147, 543–561.

Smith, M., 1994. Stratigraphic and structural constraints on mechanisms of active rifting in the Gregory Rift, Kenya. In: Prodehl, C., Keller, G.R., Khan, M.A. (Eds.), Crustal and upper mantle structure of the Kenya rift, Tectonophysics, vol. 236, pp. 3–22.

Smith, M., Mosley, P.N., 1993. Crustal heterogeneity and basement influence on the development of the Kenya Rift, East Africa. Tectonics 12, 591–606.

Stein, S., Okal, E.A., 2006. The 2004 Sumatra earthquake and Indian Ocean tsunami: what happened and why? Vis. Geosci. 10, 21–26.

Tongue, J.A., 1992. Tomographic study of local earthquake data from the Lake Bogoria region of the Kenya Rift Valley. Geophys. J. Int. 109, 249–258.

Tongue, J.A., Maguire, P.K.H., Young, P.A.V., 1992. Seismicity distribution from temporary earthquake recording networks in Kenya. Tectonophysics 204, 71–79.

Tongue, J., Maguire, P., Burton, P., 1994. An earthquake study in the Lake Baringo basin of the central Kenya Rift. Tectonophysics 236, 151–164.

Wells, D.L., Coppersmith, K.J., 1994. New empirical relationships among magnitude, rupture length, rupture width, rupture area, and surface displacement. Bull. Seismol. Soc. Am. 84 (4), 974–1002.

Wohlenberg, J., 1968. Remarks on the seismicity of East Africa between 4°N-12°S and 23-40°E. Tectonophysics 8, 567–577.

Young, P.A.V., Maguire, P.K.H., Laffoley, N.d'A., Evans, J.R., 1991. Implications of the distribution of seismicity near Lake Bogoria in the Kenya Rift. Geophys. J. Int. 105, 665–674.

Chapter 20

Overview of Landslide Occurrences in Kenya

Causes, Mitigation, and Challenges

Charles Maina-Gichaba*, Enoch K. Kipseba[†] and Moses Masibo[‡]
*Department of Geology, University of Nairobi, Nairobi, Kenya
[†]Mines and Geology Department, Ministry of Environment and Mineral Resources, Nairobi, Kenya
[‡]Commissioner of Mines and Geology, Ministry of Mining, PO Box 30009-00100, Nairobi, Kenya

1 INTRODUCTION

Landslides and other forms of ground failure affect communities all across the nation. Despite advances in science and technology, these events continue to result in human suffering, billions of dollars in property losses, and environmental degradation. As the population increases and the society becomes ever more complex, the economic and societal costs of landslides and other mass-wasting processes will continue to rise.

The term 'landslide' describes many types of downhill earth movements, ranging from rapidly moving catastrophic rock avalanches and debris flows in mountainous regions to more slowly moving earth slides and other ground failures. In addition to the different types of landslides, the broader scope of ground failure includes subsidence, permafrost, and shrinking soils. This chapter focuses on landslides, the most critical ground-failure problem facing most regions of the country.

Landslides are a significant component of many major natural disasters and are responsible for greater losses than is generally recognized. Landslide damage is often reported as a result of triggering events such as heavy torrential rains, floods, earthquakes, or volcanic eruptions even though the losses from landsliding may exceed all other losses from the overall disaster.

Landslide losses are increasing in Kenya and worldwide as development expands under pressures of increasing populations (Kenvironews, 2007). The resulting encroachment of developments into hazardous areas, expansion of transportation infrastructure, deforestation of landslide-prone areas, and changing climate patterns may lead to continually increasing landslide losses. However, an increase in the cost of landslide hazards can be curbed through

Developments in Earth Surface Processes, Vol. 16. http://dx.doi.org/10.1016/B978-0-444-59559-1.00020-7

better understanding and mapping of the hazards and improved capabilities to mitigate and respond to the hazards. Being among the most widespread geologic hazards on Earth, landslide losses and damages can amount to hundreds of millions of shillings and hundreds of deaths.

We have the capability as a nation to understand and identify these hazards and to implement mitigation measures. For many years, the Department of Mines and Geology under the Ministry of Environmental and Mineral Resources (recently renamed Ministry of Mining), the Kenya Red Cross Society, numerous universities, and the private sector have been grappling with understanding and reducing landslide hazards, and they have developed an extensive body of knowledge. However, to achieve the goal of significantly reducing losses from landslide hazards, we need a much more comprehensive scientific understanding of landslide processes and occurrence, a robust monitoring programme to warn of impending danger from active landslides, a much greater public awareness and understanding of the threat and the options for reducing the risk, and action at the local level.

A significant, sustained, long-term effort to reduce losses from landslides and other ground failures in Kenya will require a national commitment among all levels of government and the private sector. The central government, in partnership with county governments, must provide leadership, coordination, research support, incentives, and resources to encourage communities, businesses, and individuals to undertake mitigation to minimize potential losses and to employ mitigation in the recovery following landslides and other natural hazard events.

2 TYPES AND CAUSES OF LANDSLIDES

The term 'landslide', by definition, describes a wide variety of processes that result in the downward and outward movement of slope-forming materials including rock, soil/earth, artificial fill, or a combination of these. The materials may move by falling, toppling, sliding, spreading, or flowing due to gravity. Figure 1 shows a graphic illustration of a landslide, with the commonly accepted terminology describing its features.

The various types of landslides can be differentiated by the kinds of material involved and the mode of movement. A classification system based on these parameters is shown in Table 1. Other classification systems incorporate additional variables, such as the rate of movement and the water, air, or ice content of the landslide material.

Although landslides are primarily associated with mountainous regions, they can also occur in areas of generally low relief where such landslides occur as cut-and-fill failures (roadway and building excavations), river bluff failures, lateral spreading landslides, collapse of mine-waste piles (especially coal), and a wide variety of slope failures associated with quarries and

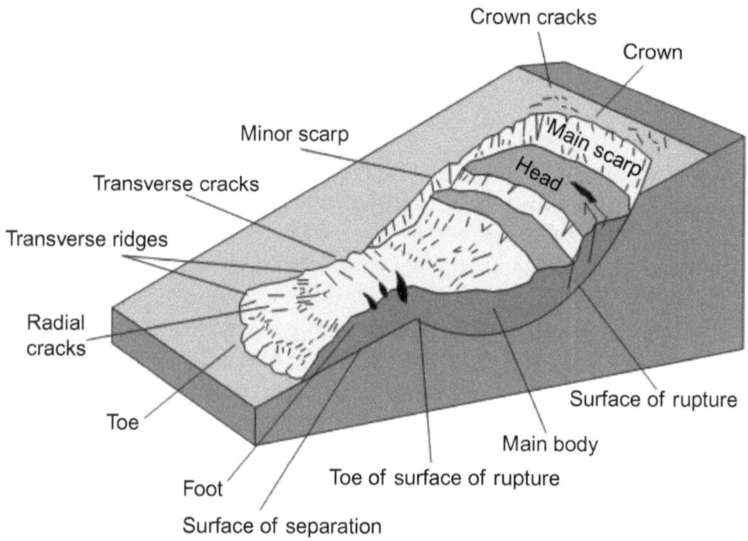

FIGURE 1 An idealized slump-earth flow showing commonly used nomenclature for labelling parts of a landslide. *Source: http://pubs.usgs.gov/fs/2004/3072.* (For colour version of this figure, the reader is referred to the online version of this chapter.)

TABLE 1 Types of Landslides

Type of Movement	Type of Material		
	Bedrock	Engineering Soils	
		Predominantely Coarse	*Predominantely Fine*
Falls	Rock fall	Debris fall	Earth fall
Topples	Rock topple	Debris topple	Earth topple
Slides			
Rotational	Rock slide	Debris slide	Earth slide
Translational			
Lateral spreads	Rock spread	Debris spread	Earth spread
Flows	Rock flow (deep creep)	Debris flow (soil creep)	Earth flow (soil creep)
Complex	Combinational of two or more principal types of movement		

Source: http://pubs.usgs.gov/fs/2004/3072.

open-pit mines. The most common types of landslides are described as follows and are illustrated in Figure 2.

2.1 Slides

Although many types of mass movements are included in the general term 'landslide', the more restrictive use of the term refers only to mass movements, where there is a distinct zone of weakness that separates the slide material from more stable underlying material. The two major types of slides are rotational slides and translational slides. Rotational slide: This is a slide in which the surface of rupture is curved concavely upward and the slide movement is roughly rotational about an axis that is parallel to the ground surface and transverse across the slide (Figure 2a). Translational slide: In this type of slide, the landslide mass moves along a roughly planar surface with little rotation or backward tilting (Figure 2b). A block slide is a translational slide in which the moving mass consists of a single unit or a few closely related units that move downslope as a relatively coherent mass (Figure 2c).

2.2 Falls

Falls are abrupt movements of masses of geologic materials, such as rocks and boulders, that become detached from steep slopes or cliffs (Figure 2d). Separation occurs along discontinuities such as fractures, joints, and bedding planes, and movement occurs by free fall, bouncing, and rolling. Falls are strongly influenced by gravity, mechanical weathering, and the presence of interstitial water.

2.3 Topples

Toppling failures are distinguished by the forward rotation of a unit or units about some pivotal point, below or low in the unit, under the actions of gravity and forces exerted by adjacent units or by fluids in cracks (Figure 2e).

2.4 Flows

There are five basic categories of flows that differ from one another in fundamental ways.

a. *Debris flow*: A debris flow is a form of rapid mass movement in which a combination of loose soil, rock, organic matter, air, and water mobilizes as a slurry that flows downslope (Figure 2f). Debris flows include $<50\%$ fines. Debris flows are commonly caused by intense surface-water flow, due to heavy precipitation that erodes and mobilizes loose soil or rock on steep slopes. Debris flows also commonly mobilize from other types of landslides that occur on steep slopes, are nearly saturated, and consist

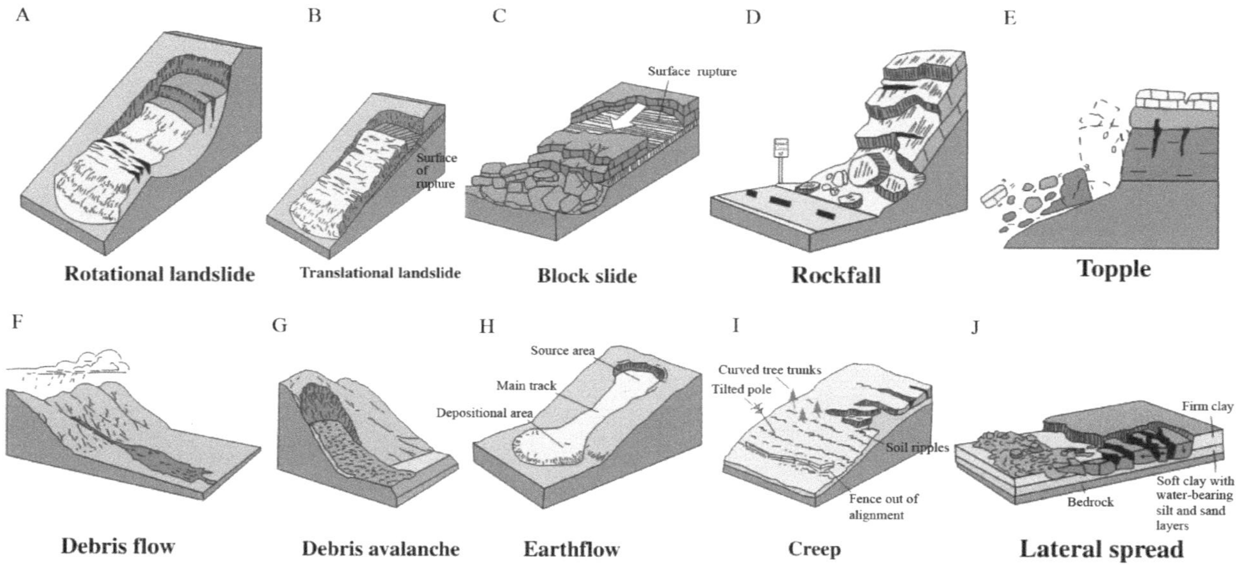

FIGURE 2 Schematic illustrations of the major types of landslide movements. *Source: http://pubs.usgs.gov/fs/2004/3072*. (For colour version of this figure, the reader is referred to the online version of this chapter.)

of a large proportion of silt- and sand-sized material. Debris-flow source areas are often associated with steep gullies, and debris-flow deposits are usually indicated by the presence of debris fans at the mouths of gullies. Fires that denude slopes of vegetation intensify the susceptibility of slopes to debris flows.

b. *Debris avalanche*: This is a variety of very rapid to extremely rapid debris flow (Figure 2g).

c. *Earth flow*: Earth flows have a characteristic 'hourglass' shape (Figure 2h). The slope material liquefies and runs out, forming a bowl or depression at the head. The flow itself is elongate and usually occurs in fine-grained materials or clay-bearing rocks on moderate slopes and under saturated conditions. However, dry flows of granular material are also possible.

d. *Mudflow*: A mudflow is an earth flow consisting of material that is wet enough to flow rapidly and that contains at least 50% sand-, silt-, and clay-sized particles. In some instances, for example, in many newspaper reports, mudflows and debris flows are commonly referred to as 'mudslides'.

e. *Creep*: Creep is the imperceptibly slow, steady, downward movement of slope-forming soil or rock. Movement is caused by shear stress sufficient to produce permanent deformation but too small to produce shear failure. There are generally three types of creep: (1) seasonal, where movement is within the depth of soil affected by seasonal changes in soil moisture and soil temperature; (2) continuous, where shear stress continuously exceeds the strength of the material; and (3) progressive, where slopes are reaching the point of failure as other types of mass movements. Creep is indicated by curved tree trunks, bent fences or retaining walls, tilted poles or fences, and small soil ripples or ridges (Figure 2i).

2.5 Lateral Spreads

Lateral spreads are distinctive because they usually occur on very gentle slopes or flat terrain (Figure 2j). The dominant mode of movement is lateral extension accompanied by shear or tensile fractures. The failure is caused by liquefaction, the process whereby saturated, loose, cohesionless sediments (usually sands and silts) are transformed from a solid into a liquefied state. Failure is usually triggered by rapid ground motion, such as that experienced during an earthquake, but can also be artificially induced. When coherent material, either bedrock or soil, rests on materials that liquefy, the upper units may undergo fracturing and extension and may then subside, translate, rotate, disintegrate, or liquefy and flow. Lateral spreading in fine-grained materials on shallow slopes is usually progressive. The failure starts suddenly in a small area and spreads rapidly. Often, the initial failure is a slump, but in some materials, movement occurs for no apparent reason. Combination of two or more of the types mentioned earlier is known as a complex landslide.

In Kenya, the characteristics of landslides may be presented in many forms shown in Figure 2 earlier as they result as effects of heavy storms, earthquakes, and volcanic activities.

3 CAUSATIVE FACTORS OF LANDSLIDES IN KENYA

While much research has been done on landslides around the world, the landslides in Kenya have received relatively little attention (Davies, 1996; Westerberg and Christianson, 1998; Ngecu and Ichang'i, 1999; Ngecu and Mathu, 1999; Westerberg, 1999; Inganga et al., 2001; Knapen et al., 2006; Claessens et al., 2007). Landslides as the most visible mass movements are recognized and well-documented global geomorphic hazards, owing to their major role in slope evolution in mountainous areas and the considerable economic, social, and geomorphological impacts (Mugagga et al., 2012).

There are multiple types of causes of landslides in Kenya, which can be classified in two categories, namely, inherent and triggering factors (Kipseba et al., 2013).

The inherent (natural/human-induced) ones comprise the following:

3.1 Geologic Causes

a. Weak or sensitive materials
b. Weathered materials
c. Sheared, jointed, or fissured materials
d. Adversely oriented discontinuity (bedding, schistosity, fault, unconformity, and contact)
e. Contrast in permeability and/or stiffness of materials

3.2 Morphological Causes

a. Tectonic or volcanic uplift
b. Glacial rebound
c. Fluvial, wave, or glacial erosion of slope toe or lateral margins
d. Subterranean erosion (solution and piping)
e. Deposition loading slope or its crest
f. Vegetation removal (by fire and drought)
g. Thawing
h. Freeze-and-thaw weathering
i. Shrink-and-swell weathering

3.3 Human Causes

a. Excavation of slope or its toe
b. Loading of slope or its crest
c. Drawdown (of reservoirs)
d. Deforestation
e. Irrigation
f. Mining
g. Artificial vibration
h. Water leakage from utilities

The main triggering factors include

1. rainfall
2. earth movements
   - earthquakes, volcanism, faulting, and thunder
3. human activities
   - clearing of vegetation, farming, undercutting, building, quarrying, blasting, construction, and other engineering works

In Kenya, majority of, if not all, the landslides are caused/triggered by water and/or human activities. Slope saturation by water is a primary cause of landslides. This effect can occur in the form of intense rainfall, changes in groundwater levels, and water-level changes along coastlines, earth dams, and the banks of lakes, reservoirs, canals, and rivers. Landsliding and flooding are also closely related because both are related to precipitation, runoff, and the saturation of ground by water (Ngecu et al., 2004). In addition, debris flows and mudflows usually occur in small, steep stream channels and often are mistaken for floods; in fact, these two events often occur simultaneously in the same area.

Though there are no records of landslides that have been caused by volcanic activity, it is suggested that there are landslides whose triggering may be caused by seismic activity. This could be expected in the mountainous areas that are vulnerable to landslides. The occurrence of earthquakes and/or earth tremors in steep landslide-prone areas greatly increases the likelihood that landslides will occur, due to ground shaking alone or shaking-caused dilation of soil materials, which allows rapid infiltration of water. This is expected in most parts of the Rift Valley where there have been increased reports of ground failure with devastating effects. Widespread rock falls also are caused by loosening of rocks as a result of ground shaking.

The distribution of the occurrence of landslides in Kenya as recorded by the Department of Mines and Geology Department, 2012 is shown in Figure 3. It is clear that most landslides recorded are in the central highlands of Kenya, the Rift Valley, and the western region of Kenya. The general elevation of the landscape in Kenya is shown in Figure 4.

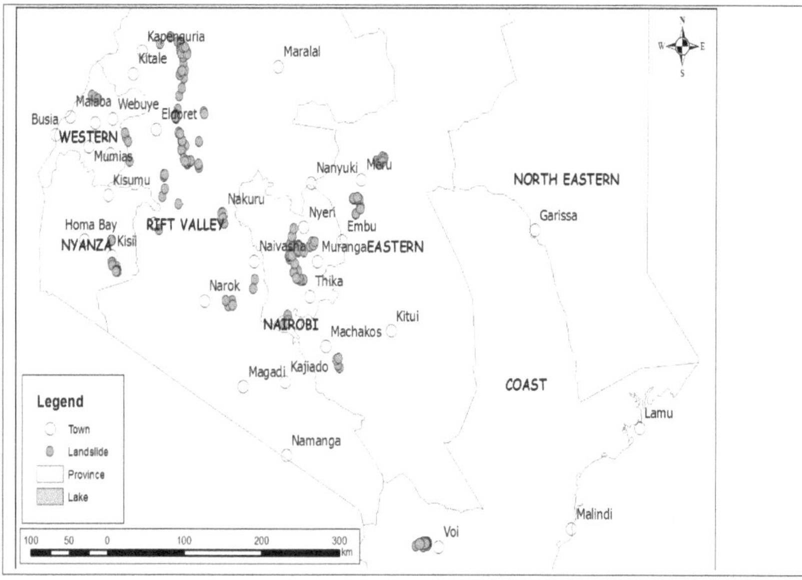

FIGURE 3 Landslide-prone areas in Kenya. *Source: Mines and Geology Department (2012).* (For colour version of this figure, the reader is referred to the online version of this chapter.)

These areas are characterized by high-precipitation regimes of over 1200 mm annually, high population density, and steep slopes of up to 80° as indicated in Figure 5, which shows the rainfall distribution in Kenya.

These areas are also found to have deep volcanic soils that become saturated with water during heavy downpour. In some places, the rocks are highly fragmented due to the effects of tectonic forces that have affected them and also weathered due to the constant presence of water in the area. The escarpment is characterized by stream valleys and ridges, as a result of erosion by small streams flowing downslope. The general drainage of the area is towards the west into the Rift Valley.

Landslides have also been observed in areas covered by rocks of the basement system such as the Yatta Area, Machakos County (Ogora and Kotut, 2013). In general, the rocks in such areas can be subdivided into four main age groups:

1. Archean—Mozambiquian rocks
2. Tertiary—volcanics and sediments
3. Pleistocene—sediments
4. Recent—soils and alluvial deposits

The metamorphic rocks vary in grain size from fine-grained schists to coarse gneisses and in composition from pure quartzites and marbles to varieties rich in biotite, muscovite, and hornblende, the intrusive being converted to

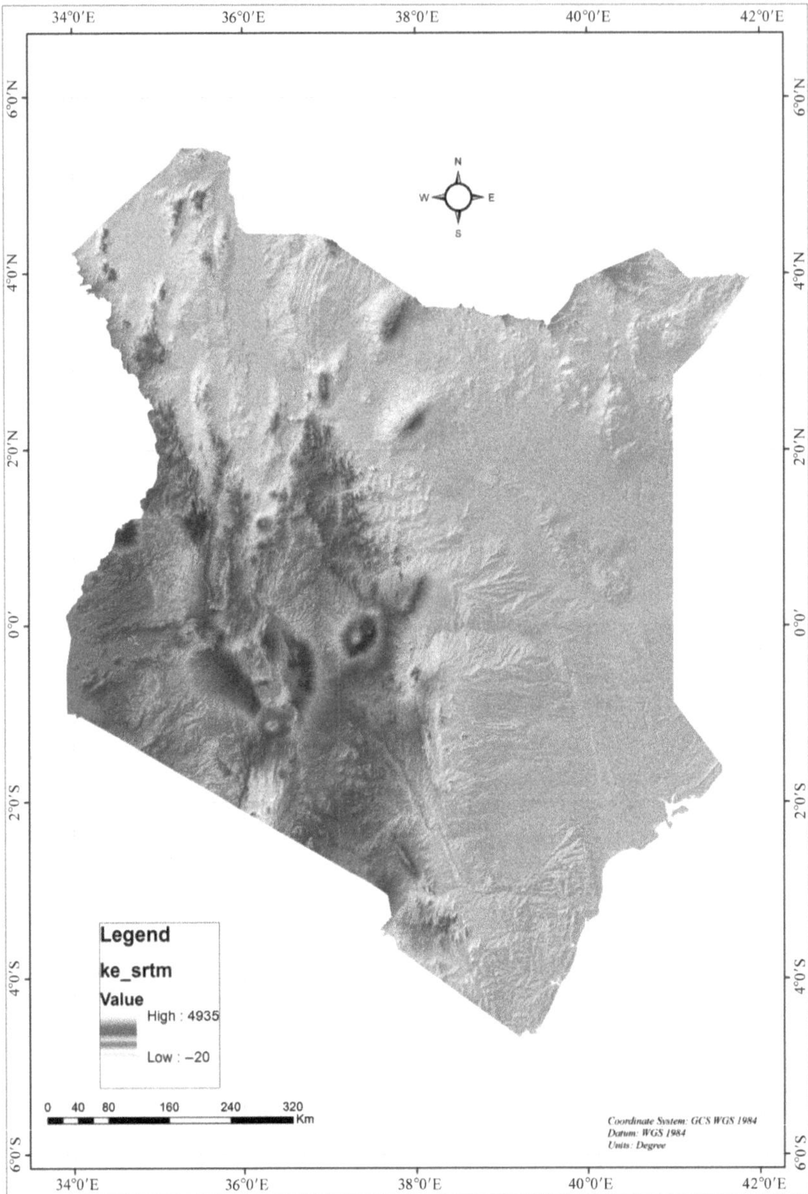

FIGURE 4 Digital elevation model of the landscape in Kenya (Compiled at the GIS lab, Department of Geology, University of Nairobi). (For colour version of this figure, the reader is referred to the online version of this chapter.)

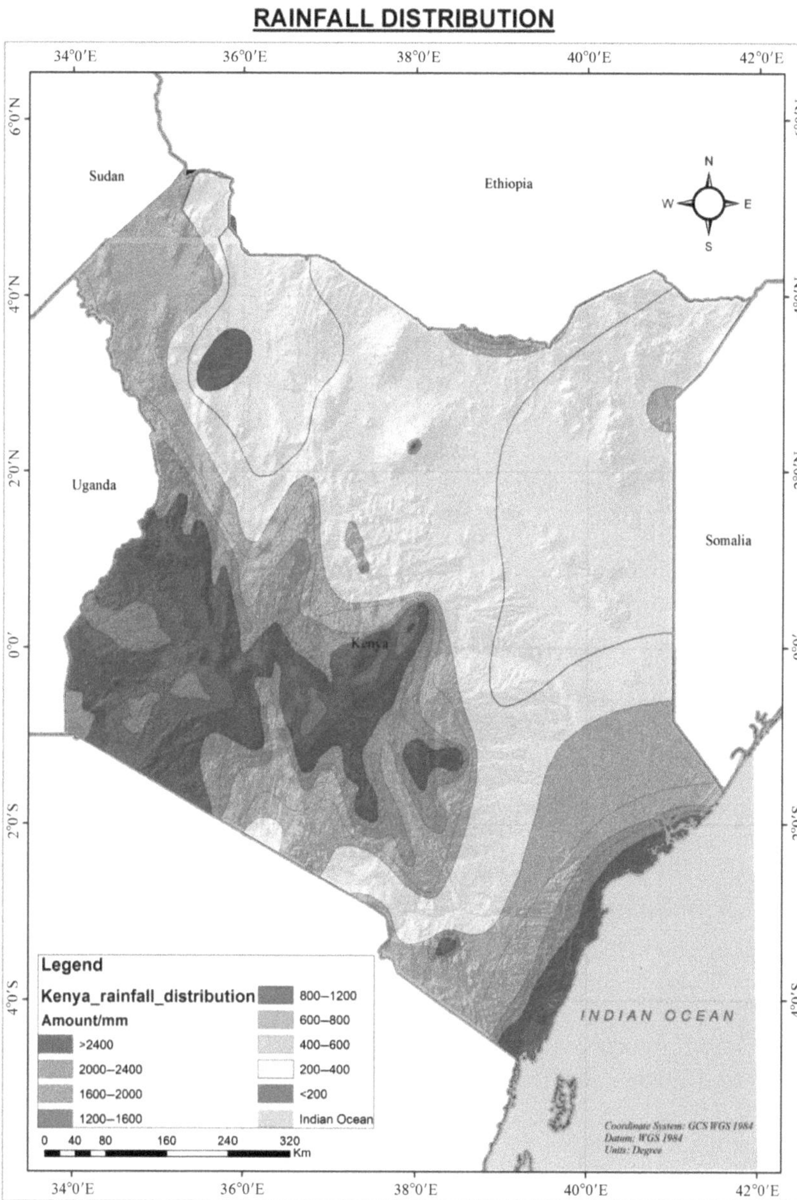

FIGURE 5 Annual rainfall distribution in Kenya (Compiled at the GIS Lab, Department of Geology, University of Nairobi). (For colour version of this figure, the reader is referred to the online version of this chapter.)

TABLE 2 Main Landslide-Prone Areas in Kenya

- Mt. Elgon
- Nandi Escarpment
- Gucha
- Nandi
- Kericho
- Rift Valley Escarpment
- Cherangani Hills
- Tugen Hills
- Nyambene Hills
- Muranga
- Taita Hills
- Nakuru Town

plagioclases. The water that triggers the slides infiltrates through the loose soils and the volcanic sediments before being intercepted by the often impervious Archean Mozambiquan/basement rocks (Table 2).

3.3.1 Anthropogenic Factors

Land tenure has been identified as the most important anthropogenic factor leading to landslides in the Kenya and the region. A driving force behind unsustainable land-use practices is the population growth and density in the areas bordering the mountainous regions.

3.3.2 Population Growth

The distribution of Kenya's population closely follows rainfall distribution patterns. Thus, areas of high rainfall tend to have high population density reflective of high population growth. Such high populations exert pressure to the land to produce food and other environmental services. This leads to land degradation, which, in turn, renders the land prone to landslides (Rop, 2011).

3.3.3 Land Tenure and Land Use

High population growth is the main cause of pressure on land, forest degradation, and land fragmentation. For example, high population makes families in most areas overcultivate their plots to maximize crop yield. It is generally noted that farming practices that do not allow the land to lie fallow from time to time are among the main causes of landslides. Furthermore, some crops have a better soil-holding capacity than others due to the size of their roots and also whether they are intercropped with other plants or not. The high population density and number of children per household also result in fragmentation of plot size for many families (see photo below). Once the children grow up and establish families of their own, it is expected that part of the plots of the parents is passed on to them. This means more pressure on the land and inability to lay parts of the land fallow, which is very important for regeneration and improving soil fertility.

Photo showing the patchwork agricultural plots in the hilly terrain of Kisii district, Western Kenya. The plots are further subdivided as the children in a household mature into adulthood.

3.3.4 Deforestation

Deforestation is a common problem in the districts bordering mountainous regions in Kenya and is also closely linked to population pressure. The only type of fuelwood used in the districts is charcoal that is collected from the local forests and produced locally. Historically, there has been a long-standing conflict with the authorities on local population's illegal resource harvest in the forest.

4 SOCIOECONOMIC AND ENVIRONMENTAL IMPACTS OF LANDSLIDES IN KENYA

Landslides in Kenya have been on the increase in the recent past. This has social, economic, and environmental impacts, mainly loss of life, agricultural land, and crops and destruction of infrastructure. Landslides tend to bury all that is their way, resulting in destruction of life and property. They may bury or sink buildings, rubble and boulders moved to block roads, railways, and lines of communication or waterways. They may destroy all property along

FIGURE 6 A bridge destroyed by the mudslide along Thingitungi stream. (For colour version of this figure, the reader is referred to the online version of this chapter.)

FIGURE 7 Household materials destroyed by the landslide. (For colour version of this figure, the reader is referred to the online version of this chapter.)

their way and render agricultural land unproductive. The casualties of mudflows, massive boulders, rocks, and all that can be moved by the landslides may be thousands of people and animals (Figures 6 and 7).

4.1 Socioeconomic Impacts of Landslides

There has not been any recorded effort to estimate the economic loss of landslides in Kenya despite the high frequency of the occurrence of the landslides. However, economic losses due to landslides can be classified as direct and indirect (Ngecu and Ichang'i, 1999; Ngecu and Mathu, 1999; Schuster and Highland, 2001). Direct costs can be defined as the costs of replacement, rebuilding, repair, or maintenance resulting from direct landslide-caused damage or destruction to property or installations. All other costs of landslides are indirect; examples are the following:

- Reduced real estate values in areas threatened by landslides.
- Loss of tax revenues on properties devalued as a result of landslides.
- Loss of industrial, agricultural, and forest productivity and of tourist revenues, as a result of damage to land or facilities or interruption of transportation systems (Knapen et al., 2006).
- Loss of human or domestic animal productivity because of death, injury, or psychological trauma (Kitutu et al., 2009).

Table 3 illustrates some of the common impacts of landslides in different parts of the country. Some of the impacts previously mentioned are also shown in Figures 8–12.

TABLE 3 Socioeconomic Impacts of Landslides in Kenya

Date	Landslide Location	Impact
April 2013	Kijabe Landslide	Blocked railway line, disrupted services
30.04.2002	Maua Landslide	11 People died
22.04.2001	Kitheu Landslide	4 People died, destroyed water pipes
30.04.1997 November 2008	Muranga Landslides	11 People died, 7 houses destroyed, 3 cows died 1 Child died
2002–2008	Timboiwo	Houses, crops, pasture destroyed Yabsoi's Farm, Kericho whole farm destroyed
2003	Kokwet, Kipkelion	4 People died, pasture lost, arable land rendered infertile
November 2008	Chesegon Landslide, Pokot Central	11 People died, pasture lost, arable land rendered infertile
2002	Kocholwa	Livestock killed
2010	Mt. Elgon, Pokot, Kessup, Subukia, and Timobo (continuous rock falls)	Lives lost Houses destroyed Roads blocked Crops destroyed
April 2013	Yatta Landslide	3 people died, houses destroyed

FIGURE 8 Uprooted plants and trees by a landslide. (For colour version of this figure, the reader is referred to the online version of this chapter.)

FIGURE 9 House destroyed by huge rock boulders down a slope due to a landslide. (For colour version of this figure, the reader is referred to the online version of this chapter.)

4.2 Environmental Impacts of Landslides in Kenya

4.2.1 Impacts of Landslides on Stream Environments

The main types of landslides that affect streams are debris flows, which may follow the stream channel for great distances (occasionally 100 km or more). Debris flows provide important sediment transport links between hillslopes and alluvial channels and thus are an important factor in drainage-basin sediment budgets (Schuster and Highland, 2001). In addition, debris flows influence the spatial and temporal distributions of sediment in stream channels,

FIGURE 10 Landslide debris flow in West Pokot County, Kenya. (For colour version of this figure, the reader is referred to the online version of this chapter.)

FIGURE 11 Yatta landslide of April 2013. (For colour version of this figure, the reader is referred to the online version of this chapter.)

either because they deposit sediment in the channels or because the deposits provide a source for accelerated transport of sediment farther downstream. The pollution of rivers and other water bodies adds on to the environmental impact of the landslides and poses a health hazard to local people (Knapen et al., 2006).

4.2.2 Impacts of Landslides on Forest Cover

Widespread stripping of forest and jungle cover by mass movements has been noted in many parts of the world but particularly in tropical areas as a result of large-scale, earthquake-induced landslide activity.

FIGURE 12 Debris flow along River Thingitungi, Magana area, Kijabe, during the April 2013 landslide. *Source: Kipseba et al. (2013).* (For colour version of this figure, the reader is referred to the online version of this chapter.)

4.2.3 Impacts of Landslides on Livestock and/or Wildlife

Although most kinds of wildlife are able to retreat fast enough to avoid injury from all but the fastest-moving landslides, all wild creatures are subject to landslide-caused habitat damage and destruction. Fish are probably most affected because they are dependent on stream access and water quality for their livelihood, both of which are often affected by landslides. In Kenya, a lot of cases have been reported where domestic livestock has borne the brunt of the landslides. Cows have been injured; goats and chicken pens have been destroyed. However, there is little else that is done to quantify the loss. Fish ponds have been reportedly destroyed by debris from landslides, thereby affecting the livelihoods of the directly affected families.

5 LANDSLIDE MITIGATION: HOW TO REDUCE THE EFFECTS OF LANDSLIDES

Vulnerability to landslide hazards is a function of location, type of human activity, use, and frequency of landslide events. The effects of landslides on people and structures can be lessened by total avoidance of landslide hazard areas or by restricting, prohibiting, or imposing conditions on hazard-zone activity. Local county governments can reduce landslide effects through land-use policies and regulations. Individuals can reduce their exposure to landslide hazards by educating themselves on the past hazard history of a site and by making inquiries to planning and engineering departments of local county governments. They can also obtain the professional services of an engineering geologist, a geotechnical engineer, or a civil engineer, who can properly evaluate the hazard potential of a site, built or unbuilt.

The hazard from landslides can be reduced by avoiding construction on steep slopes and existing landslides or by stabilizing the slopes. Stability increases when ground water is prevented from rising in the landslide mass by (1) covering the landslide with an impermeable membrane, (2) directing surface water away from the landslide, (3) draining ground water away from the landslide, and (4) minimizing surface irrigation. Slope stability is also increased when a retaining structure and/or the weight of a soil/rock debris is placed at the toe of the landslide or when mass is removed from the top of the slope.

In Kenya, mitigation measures that have been suggested in most cases involving landslide hazards include the following:

- Encouraging local communities to plant deep-rooted trees on steep slopes
- Discouraging cultivation on steep slopes
- Discouraging settlement on steep slopes
- Stopping vegetation clearing
- Putting up of soil conservation structures
- Constructing support structures on vulnerable areas
- Encouraging agroforestry in the areas
- Advising on putting up structures in safer areas

Areas that are generally prone to landslide hazards include those on existing old landslides or at the base of slopes, in or at the base of minor drainage channels, at the base or top of an old fill slope, at the base or top of a steep cut slope, and developed hillsides where leach field septic systems are used.

Areas that are typically considered safe from landslides are characterized by being on hard, non-jointed bedrock that has not moved in the past. They may also include those areas on relatively flat-lying areas away from sudden changes in slope angle and/or at the top or along the nose of ridges, set back from the tops of slopes.

Care should also be taken to look for and document any warning signs that may be occurring in an area. According to USGS (2012), such signs include and not limited to the following:

- Springs, seeps, or saturated ground in areas that have not typically been wet before
- New cracks or unusual bulges in the ground, street pavements, or sidewalks
- Soil moving away from foundations
- Ancillary structures such as decks and patios tilting and/or moving relative to the main house
- Tilting or cracking of concrete floors and foundations
- Broken water lines and other underground utilities
- Leaning telephone poles, trees, retaining walls, or fences
- Offset fence lines
- Sunken or down-dropped road beds
- Sudden decrease in creek water levels though rain is still falling or just recently stopped

- Sticking doors and windows and visible open spaces indicating jams (side posts of doors or windows) and frames out of plumb.

6 WAY FORWARD

Considering the damages so far being witnessed in the country, there is an urgent need to establish a National Disaster Management Authority, which will be tasked with among other activities, developing a comprehensive National Landslide Hazards Mitigation Strategy. Such a strategy would employ a wide range of scientific, planning, and policy tools to address various aspects of the landslide problem to effectively reduce losses from the ground failure. It should have the following major elements:

i. Carry out research to develop a predictive understanding of landslide processes and triggering mechanism.
ii. Hazard mapping and assessments. This would be done to delineate susceptible areas and different types of landslide hazards at a scale useful for planning and decision making.
iii. Carry out real-time monitoring of active landslides that pose substantial risk.
iv. Loss assessment. This would be achieved by compiling and evaluating information on the economic impacts of landslide hazards.
v. Establish an effective system for information collection, interpretation, and dissemination.
vi. Developing guidelines and training for scientists, engineers, and other decision makers.
vii. Engage in public awareness and education by developing information and education for the user community.
viii. Implementation of loss reduction measures by encouraging mitigation action.
ix. Emergency preparedness, response, and recovery geared towards building resilient communities.

In conclusion, therefore, it is clear that landslides are natural phenomena whose causative factors have substantial human contribution. Continuous monitoring, awareness campaigns, and mapping of landslide-prone areas will help to reduce losses resulting from landslides. It is encouraging to note that the government through the Mines and Geological Department has embarked on a programme to map out all landslide-prone areas, which will culminate in the production of a National Landslide Atlas—a useful tool to planners and other stakeholders. Collaboration among the stakeholders is a major factor in the management of landslides.

It has also been observed that at most of the sites of the tragic landslides, the residents live as squatters and/or peasants in highly densely populated area. In a recent landslide at Kijabe, there were 120 families living on about two acres land, making it overcrowded with homesteads. The land slopes at an angle of about 25°. This resulted in the families excavating foundations for constructing their houses very close to their neighbours, thus rendering the land unstable. In other areas, the fertile volcanic soil has encouraged the residents to practice subsistence cultivation, even on steep slopes. Most of the farms have poorly maintained soil conservation structures, mainly terraces. Many slope failures have occurred on such cultivated land, rendering the already peasant farmers even more destitute.

Population pressure and the need for more food have caused more disturbances in the land, hence the emergence of landslide occurrence. It is recommended that drastic efforts should be made to ensure the following interventions:

1. Deep-rooted trees especially eucalyptus should be planted especially on the slopy areas.
2. Construction of proper soil conservation measures.
3. Cultivation on steep slopes should be discouraged.
4. Detailed mapping needs to be done in the area to zone out susceptibility of the area to landslides since the slope instabilities are a new phenomenon in the area.
5. Sensitization of the local communities should be done through seminars by the Department of Mines and Geology and other stakeholders.

It is also proposed that the proposed National Disaster Management Authority establishes a landslide-reporting database where all the data collected would be stored and analyzed and then the end product be packaged for dissemination. Such integrated landslide-reporting databases have been used successfully in other countries of the world (Rego, 2001; Foster et al., 2008; Van Den Eeckhaut and Hervas, 2011) to provide early warnings and postdisaster activities, thereby becoming an indispensable service to the public.

ACKNOWLEDGEMENTS

The authors gratefully acknowledge the support given by the Commissioner of Mines and Geology for permission to use the photographs and reports done by the Department.

REFERENCES

Claessens, L., Knapen, A., Kitutu, M.G., Poesen, J., Deckers, J.A., 2007. Modelling landslide hazard, soil redistribution and sediment yield of landslides on the Ugandan footslopes of Mount Elgon. Geomorphology 90, 23–35.

Davies, T.C., 1996. Landslides research in Kenya. J. Afr. Earth Sci. 23, 41–49.

Foster, C., Gibson, A., Wildman, G., 2008. The New National Landslide Database and Landslide Hazard Assessment of Great Britain. British Geological Survey, Nottingham, UK.

http://pubs.usgs.gov/fs/2004/3072. Landslides types and processes (accessed 13.04.13).

Inganga, S.F., Ucakuwun, E.K., Some, D.K., 2001. Rate of swelling of expansive clays: a critical factor in the triggering of landslides and damage to structures. Doc. Nat. 136, 93–98.

Kenvironews, 2007. Landslides in Kenya—how do we preserve life and property? http://kenvironews.wordpress.com (accessed 30.04.13).

Kipseba, E.K., Ogora, M., Maina, G., Kotut, J., 2013. Preliminary reports on Kijabe Landslides, Lari District, Kiambu County. Mines and Geology Department, Ministry of Environment and Mineral Resources (Unpublished).

Kitutu, G.M., Muwanga, A., Poesen, J., Deckers, J.A., 2009. Influence of soil properties on landslide occurrences in Bududa district, Eastern Uganda. Afr. J. Agric. Res. 4 (7), 611–620.

Knapen, A., Kitutu, M.G., Poesen, J., Breugelmans, W., Deckers, J., Muwanga, A., 2006. Landslides in a densely populated county at the footslopes of Mount Elgon (Uganda): characteristics and causal factors. Geomorphology 73, 149–165.

Mines and Geology Department, 2012. Map of landslide prone areas in Kenya (Unpublished).

Mugagga, F., Kakembo, V., Buyinza, M., 2012. Land use changes on the slopes of Mount Elgon and the implications for the occurrence of landslides. Catena 90 (2012), 39–46.

Ngecu, W.M., Ichang'i, D.W., 1999. The environmental impact of landslides on the population living on the eastern footslopes of the Aberdare ranges in Kenya: a case study of Maringa Village landslide. Environ. Geol. 38 (3), 259–264.

Ngecu, W.M., Mathu, E.M., 1999. The El Nino triggered landslides and their socio-economic impacts on Kenya. Episodes 22 (4), 284–288.

Ngecu, W.M., Nyamai, C.M., Erima, G., 2004. The extent and significance of mass-movements in Eastern Africa: case studies of some major landslides in Uganda and Kenya. Environ. Geol. 46, 1123–1133. http://dx.doi.org/10.1007/s00254-004-1116-y.

Ogora, M., Kotut, J., 2013. Report on Yatta Landslide, Machakos County. Mines and Geology Department, Ministry of Environment and Mineral Resources (Unpublished), Nairobi.

Rego, J.A., 2001. National Disaster Management Information System & Networks: an Asian overview. Paper Presented on Global Disaster Information network (GDIN).

Rop, B.K., 2011. Landslide disaster vulnerability in Western Kenya and mitigation options: a synopsis of evidence and issues of Kuvasali landslide. J. Environ. Sci. Eng. 5 (1), 110.

Schuster, R.L., Highland, L.M., 2001. US Geological Survey Open-File Report 01-0276. US Geological Survey.

USGS, 2012. Landslide Hazards Program. http://landslides.usgs.gov/learning/prepare/ (accessed 16.08.12).

Van Den Eeckhaut, M., Hervas, J., 2011. State of the art of national landslide databases in Europe and their potential for assessing landslide susceptibility, hazard and risk. Geomorphology 139–140 (2012), 545–558.

Westerberg, L.O., 1999. Mass movements in East African highlands: processes, effects and scarp recovery. Dissertation 14. Department of Physical Geography, Stockholm University, Sweden.

Westerberg, L.O., Christianson, C., 1998. Landslides in East African highlands. Slope instability and its interrelation with landscape characteristics and land use. Adv. GeoEcol. 31, 317–325.

Chapter 21

Floods in Kenya

Alfred Opere
Department of Meteorology, University of Nairobi, Nairobi, Kenya

1 INTRODUCTION

Floods and droughts, associated with extreme climate events, have very devastating effects on almost all socioeconomic activities and are very common in many parts of Africa. Flooding in its most immediate form can inundate farms and villages and disrupt transportation networks, ultimately affecting food security and market distribution systems. Comprehensive, up-to-date information describing hydrologic conditions is therefore needed to anticipate and mitigate flood impacts on populations targeted for assistance. Available ground-based measurements of stream flow and rainfall are at present inadequate to provide comprehensive monitoring of regional flooding.

In Kenya, the hazards and impacts of floods were demonstrated by the 1997/1998 *El Niño* episodes. These floods led to severe loss of life (human and livestock) and property, destruction of infrastructure, disruption of the communication networks and large losses to the economy. They were also associated with land degradation (soil erosion), silting of hydropower dams and destruction of power lines. The recovery from the effects of these weather-related phenomena will take a long time, as national economies were seriously weakened. The recent Budalang'i floods in western Kenya also led to massive destruction of property and displacement of people.

Floods are a result of overflow in river banks and can cause enormous damage to loss of life and property including crops and infrastructure. These are common phenomena and are costly natural disasters. Floods are short-lived events that can happen suddenly, sometimes with little or no warning. They usually are caused by intense storms that produce more run-off than an area can infiltrate and store or a stream can carry within its normal channel. Rivers can also flood when dams fail or landslides temporarily block a channel.

Unlike droughts, defined by Palmer (1965), which has a slow onset, floods are rapid onset disasters. Flood hazards resulting from too much rainfall have resulted into disasters in Kenya. Examples are areas of Kano Plains in Kisumu County, Budalang'i in Busia County and the lower parts of the Tana River are

Developments in Earth Surface Processes, Vol. 16. http://dx.doi.org/10.1016/B978-0-444-59559-1.00021-9

susceptible to floods (The Republic of Kenya, LBDA & JICA, 1992). Arid and semi-arid areas of the country also experience flash floods. It is important to quantify the magnitude of floods that are often described by the size of a flood corresponding to a particular recurrence interval. By studying a long period of flow records for a stream, it is possible to estimate the size of a flood that would, for example, have a 5-year recurrence interval (called a 5-year flood). A 5-year flood would occur, statistically, once every 5 years.

Floods can occur at any time, but weather/climate patterns have a strong influence on when and where floods occur, Smith (2002). The land management practices in the upper catchments may also contribute to the enhancement of surface water run-off and flooding and an unplanned human occupation in the lowland can also enhance the damages due to flooding. In 1997/1998, the *El Niño*-associated floods affected many parts of Kenya, causing destruction to property, loss of lives, famine and waterborne disease epidemics. With inadequate preparation for the *El Niño* floods, national resources were overstretched in the response phase. The *El Niño*-induced floods of 1997–1998 caused some US $151.4 million in public and private property damage. This figure does not include the number of people who lost family members, savings, property and economic opportunities. Extreme climate events such as droughts and floods have adverse effects on all forms of life. The degree of the effects depends on the resistance and resilience of the affected communities.

Floods have adverse effects on health and health service in terms of disease outbreaks and the capability to reach people when roads are not passable. The congestion in the camps also poses a big health challenge in terms of communicable diseases. Flooding is associated with diseases such as malaria, typhoid and bilharzia (*schistosomiasis*). Interference with culture as seen in congestion of households with children of up to 15 years hurdled with their parents in the camping structures also contributes to social and health problems. Floods also enhance environmental degradation and destruction of homesteads. It also destroys the normal economic production, distribution of goods and food harvests.

In many low-lying areas around the mouths of the rivers and natural swamps, the inundation lasts for weeks, leading to total loss of crops. The worst affected are the poor who inhabit the flood plains and riverine lands to eke out a meager living from agriculture, livestock farming and fisheries. Because of poverty, lack of education and poor rural infrastructure, they are the most vulnerable to floods and post-flood consequences. The floods severely limit and hamper the developmental process, further increasing the vulnerability of the rural society and thereby perpetuating and increasing the incidence of poverty. Stagnant floodwater also causes vector-borne diseases, which result in high incidence of morbidity with consequent loss of alternative employment opportunities. An exception is made by farmers whose cropping practice relies on flood-recession agriculture like in the low lands of the

Tana delta and in other countries: for these communities the flooding is beneficial to their agricultural production and thus to their economy.

In most cases though people from inundated areas move to makeshift relief camps where they cluster together. Such makeshift homes soon become slums creating social problems and unhygienic conditions, which are conducive for the spread of contagious diseases and sexually transmitted diseases. Often, women and young girls are the worst sufferers.

2 HISTORY OF THE MAJOR FLOODS IN KENYA

Some of the past major floods in Kenya and associated impacts are highlighted in the following sections.

2.1 The *Uhuru Floods* of 1961

Uhuru Floods represents one of the early attempts of studies initiated for the purpose of studying the extent and magnitude of the flooding menace that inundated extensive low lying areas and caused widespread damage and destruction of existing infrastructures; dams, bridges, homesteads etc. as well as other facilities that existed then in Kenya. Other noteworthy flood events occurred in 1963/1964, 1968, 1977/1978, 1982, 1985 and 1990. These floods mainly hit the Lake Victoria basin and the coastal areas of the *Athi*, *Lamu* and *Tana* River basins; however, the detailed features are not well documented.

2.2 The El Niño Floods of 1997/1998

The hazards and impacts of the *El Niño* rains of 1997/8 demonstrated to Kenya the severe devastation of floods. The floods caused the loss of life of human and livestock, disrupted socio-economic activities due to the extensive damage to property, infrastructure and communication facilities, DMCN (2004). Further, the floods were also associated with land degradation, increased soil erosion with the consequent silting of hydropower dams and erosion of riverbanks, which affected water intake facilities and river gauging facilities, which are the basis for the operation of the National Hydrological Services in the country. The flow-measuring structures washed away by the floods included weirs, water-level gauges and data loggers. These facilities assist hydrologists to capture river level fluctuation data and this type of data forms the basis for the water sector resource management and development.

Among the intervention measures undertaken to ameliorate the impact of the *El Niño* were:

- The rehabilitation of river gauging stations washed away by floods. This was a project spearheaded by World Bank where several stations were rehabilitated.

- The installation of self-recording water-level data loggers at a number of sites to assist in the collection of river water level.

2.3 The April/May 2003 Floods

In this season, Kenya experienced massive flooding in the Western Province at Budalang'i and the lower reaches of the Nzoia River and in Nyanza Province at Ahero within the lower Nyando River basin. Further, the heavy rains that caused the flooding led to water supply disruption in urban and rural centres and washing away of river gauging facilities constructed along the rivers.

Among the most notable observations of the April/May 2003 flooding were:

- Complete inundation of the Budalang'i area and the Nyando River floodplain downstream of Ahero Town as a result of breaching of the dykes in the Nzoia and Nyando Rivers, respectively.
- Washing away of data loggers, water-level recorders, river gauging structures and bridges, affecting communication and flow monitoring activities in the field.

3 FLOOD HAZARDS IN KENYA

In Kenya, floods hazards are a recurrent phenomena. However, these flood hazards turn into disasters only in few places in the country. The most commonly affected places are the floodplains of the major rivers such as the lower Tana River, the lower Nzoia River at Budalang'i plains and the lower Nyando River at Kano Plains.

Figure 1 presents the significance of flood hazards in relation the other types of hazards in Kenya.

Epidemics (especially due to malaria but also due to HIV/AIDS, as explained before) have reached alarming levels in the recent past. There are, therefore, elaborate initiatives to curb the scourge of epidemic disasters in the country. On the other hand, while floods seem to have about the same prevalence in Kenya as droughts, droughts affect many more people than floods. This is illustrated in Figure 2.

Nevertheless, due to the existence of generally functional drought management strategies in the country, droughts disasters are not as devastating as flood disasters. This illustrated by Figure 3.

These figures confirm the fact that there is an urgent need for the development of flood disaster mitigation strategies in the country. Due to the key role that climate, and particularly climate change and variability, plays in the development of flood disasters, it is important to understand the changes in climate over the country and especially over western Kenya where flooding causes untold suffering and also acts as a poverty creation catalyst.

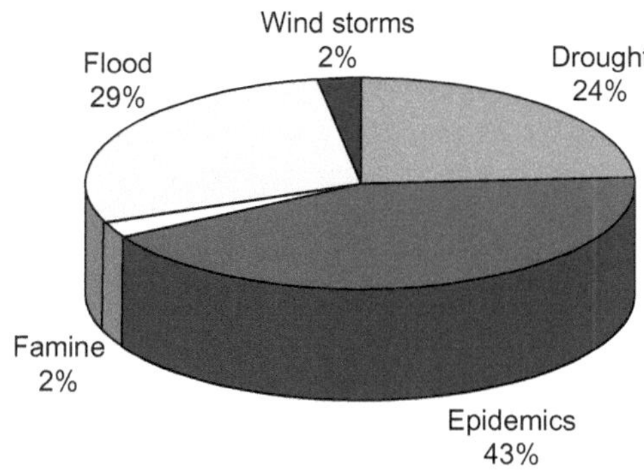

FIGURE 1 Comparison of the prevalence of hazards in Kenya. (For colour version of this figure, the reader is referred to the online version of this chapter.)

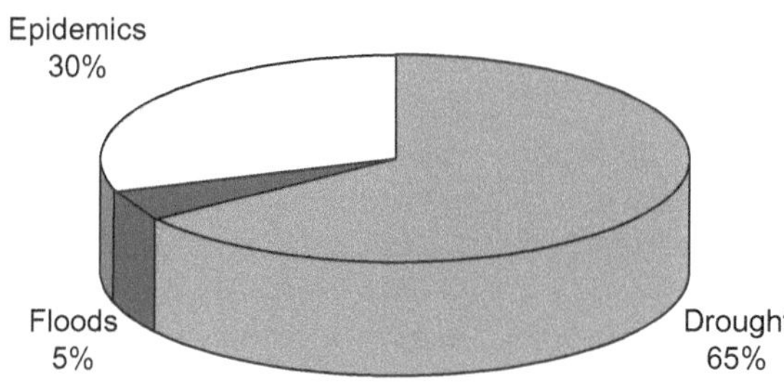

FIGURE 2 The percentage of people affected by the different types of hazards in Kenya. (For colour version of this figure, the reader is referred to the online version of this chapter.)

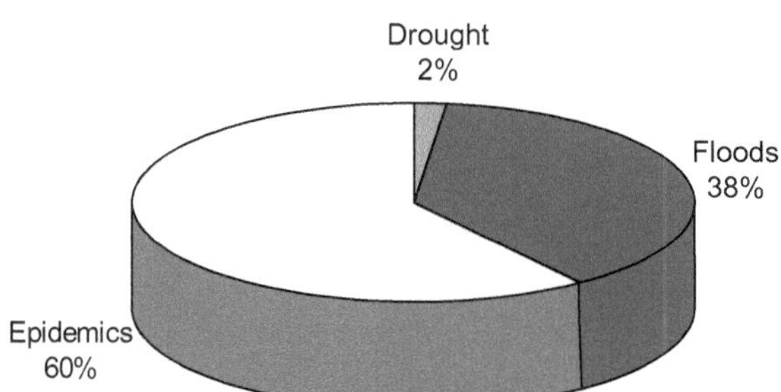

FIGURE 3 People killed or affected adversely by the different types of hazards in Kenya. (For colour version of this figure, the reader is referred to the online version of this chapter.)

4 IMPACTS OF FLOODS IN KENYA

Two case studies are here described in details, both linked to the April/May 2003 floods and both in Western Kenya: the case of floods in Budalang'i and the case of Kano Plain.

4.1 The Case of Floods in Budalang'i

Budalang'i division lies on the shores of Lake Victoria partly on the mouth of Yala River but mainly on the mouth of River Nzoia. River Nzoia is one of the largest rivers in western Kenya. The main stream of the river flows from the western side of the Elgeyo Escarpment (Sergoi, Sosiani and Kipkelion tributaries) and the Cherangani Hills (Chepkotet and Kaisungur tributaries) from an elevation of approximately 2286 m above mean sea level. Its tributaries, which flow from the high slopes of Mount Elgon, attain maximum elevation in the river's basin and are estimated at about 4300 m above mean sea level. The tributaries in Mount Elgon include Kuywa, Sioso, Ewaso, Rogai and Koitobos.

The average basin elevation is estimated at 1917 km above mean sea level, while the length of the mainstream is 275 km from source to mouth. The river follows the 275 km course with a mean slope of 0.010% from its source to discharge into Lake Victoria at about 1000 m above sea level. It enters the lake a short distance to the north of the Yala Swamp. The upper catchment is a high rainfall zone having a mean annual rainfall of between 1500 and 1700 mm per annum, while the flood plain is a low rainfall zone with mean annual rainfall below 800 mm per year.

The total catchment area drained by the basin's river network is about 12,950 km^2 when measured from Lumbwa Ferry area. Of this about 90.8% (11,667 km^2) is composed of a land area, which is relatively flat, with fairly deep soil, continuous vegetative cover, and shows a relatively stable landscape. The mean slope of the basin is 0.071%. The remaining land area is either covered by swamp (5.4%), areas of localized instability, sheet or gully eroded areas or steep sloping areas with rock outcrops.

Budalang'i has a total population of about 56,000 people (1999 Census), who inhabit the region mostly affected by floods. Of this, 54% of the total population is made up of women, while 60% is made up of children between 0 and 14 years. The population at the time of study was concentrated in 10 camps located in 7–8 sites. The entire Budalang'i has 6 administrative divisions with 6 locations and 18 sublocations. The areas affected by rivers Nzoia and Yala, which divide Budalang'i into two sections, suffer a lot during the over flooding of these rivers since the dykes no longer contain the floods as required.

The accumulation of many years of sediment in the river bed has made the channel of the river course to be above the general level of the flood plain as a result overbank flow across the 400–600-m-wide dykes causes massive flooding. The width of the channel decreases gradually from about 50 m at 140 km inland to around 40 m in the upper reaches of Kakamega, Bungoma, Trans Nzoia and Uasin Gishu Districts where the altitude is high and the slopes are steep. The sediment accumulates and reduces the capacity of river channel so that the river overflows the banks forming the delta.

The flow regime of the Nzoia is varied and is occasionally as low as 20 m^3/s and with extreme floods that may surpass 1100 m^3/s, which is the proposed protection level for the dykes for a 25-year, return flood. Siltation is heavy. Earlier estimates of siltation rates by Italconsult are in the order of 158 tonnes per day; however, recent assessments by Lake Basin Development Authority (LBDA) for the period 2000–2003 put it at 574 tonnes per day, which is thrice the Italconsult value. The discharge varies from a low flow of 2.8 m^3/s to a 100-year flood flow of 930 m^3/s.

The soils of the floodplain in the lower reaches of the Nzoia River are all alluvial. The river meanders in the floodplain depositing silt during seasonal floods. The major parts of the floodplain contain black cotton soils, while other areas have coarse textured sand silt mixture. In some places, there exist saline soils. The upper Nzoia Basin contains extensive seasonal swamp areas in the high and medium rainfall zones that are mainly utilized for grazing due to poor drainage. Once the river bursts its banks, the resultant flooding displaces close to 30,000 persons every rainfall season. People are usually evacuated to safer higher grounds during floods. Entire homesteads are swept away, property and crops worth hundreds of thousands of shillings are lost and many people perish as rivers break their banks, rendering large areas of land inaccessible. The floods enhance poverty, because crops and businesses are destroyed, (Oyugi et al., 2003).

However, basic needs are not readily available for those living in the camps. Moreover, the problem is further compounded by increased waterborne diseases like typhoid and malaria, increased shortage of food, limited environmental sanitation, low school attendance and increased poverty as most farmers have lost their livelihood. Livestock are weakened due to difficulties to access grazing grounds. Food shortages persist for a much longer time as the farming land remains underwater and is not possible to cultivate for another 6 months.

The lower reaches of River Nzoia are flat lands with low gradients. All the six locations in Budalang'i Division are perennially affected by flooding. These are, namely, Bunyala East, West, South, North and Central and Khajula location. Traditionally, the people have settled in the flood plain and this has meant that the people are directly affected by the perennial flooding. The April 2003 floods displaced 4000 families or about 24,000 people, covering an area of 60 km^2 with people moved to camps on higher areas and relying on food relief.

This situation has arisen after breaching of protection dykes that were constructed by the Ministry of Water from 1977 to 1984. The works involved construction of 32.8 km of earthen dykes (16.6 km in the southern side and 16.2 km) on the northern side that required a replacement of 690,000 m^3 of soil. They were designed for 25-year flood protection of 750 m^3/s. Now the dykes are occasionally overgrown with vegetation, breached at 20 points and have outlived their life of 20 years. The budgetary maintenance requirements are therefore high and have been diminishing annually.

A report given to the government in 1992 after the assessment of the area proposed an irrigation scheme as the solution to the unending flood menace.

The report said that a dam is needed to be constructed to trap the excess water that overflows the river. This would be followed by the construction of two canals, one in the northern riverbank measuring 40 km and the other on the southern side measuring 35 km, draining water onto the flat farms for irrigation. 'The project could cost about $20 million if implemented'. It is evident the present course of the river is highly unstable and could change from one moment to the next after a flood, if no protective works are provided.

The 2003 floods in Budalang'i saw nearly 24,000 out of a population of 56,000 people displaced (source: IFRC, 2003). Some 10,000 of these have been accommodated in the DO's camp, necessitating health emergency measures to control possible outbreaks of waterborne diseases. Scarcity of water sources and the contamination of pipes and borewells aggravated an already acute problem. Fishing and farming are the major economic activities in Budalang'i. Due to the floods and concentration in the camps, the economic activities are greatly disrupted.

4.2 The Case of Floods in Kano Plains

The main Nyando River, which traverses Kericho, Nyando and Kisumu Districts, has a total catchment area of 2606 km^2 and is 148 km long. Its main tributary, the Ainomutua River, drains 845 km^2. The flow regime of the Nyando is varied and has occasionally been as low as 2 m^3/s and with extreme floods above 850 m^3/s. Siltation is heavy. Earlier estimates of siltation rates by Italconsult are in the range of 685,000 tonnes per annum; however, recent assessments by LBDA for the period 2000–2003 are fourfold higher than the Italconsult findings.

A recent survey by a team of experts from the ministry has established that the peak discharges of the Nyando have reduced. The 1983 estimates obtained by Italconsult for the 25- and 50-year return period flows were 1100 and 1300 m^3/s but the 2003 estimates by the team are 776 and 879 m^3/s, respectively. This is in contrast to sediment loading, which has been found to increase threefold. The main rivers in the low plain are the Kibos and the Nyando. The other smaller rivers are Luanda, Nyaidho, Miriu and Awach.

The source of the river system is in Kericho and Nandi Districts, which lie above 2000 m above sea level with a mean annual rainfall of between 1800 and 2000 mm. The river basin has steep defined courses in their upper reaches, but on reaching the flat low-lying areas near the lake, they meander and periodically overflow their banks before terminating into swamps neighbouring the Lake Victoria. The vegetation of the upper catchment is mainly forest (*Tinderet* Forests), while the middle catchment may be classified as vegetative, that is, scattered trees and grass, which has greatly been modified by clearing, cultivation and burning due to human settlement. The total basin catchment area is 4484 km^2, which includes the Kano Plains 3356 km^2. The Kano Plains with an area of 73,000 ha comprise 13,000 ha of swamp and 25,000 ha proposed for paddy irrigation. The 45-km stretch inland from the lakeshore is rather flat.

The width of channel and the dykes is about 200–250 m. The agricultural irrigation schemes like the Ahero Scheme are found here. This is also the area where serious flooding occurs. The accumulation of many years of sediment in the riverbed has reduced the capacity of the river with time.

Drainage channels and ponding are common. Further inland near the Nyando and Kericho District border some 75 km away, the slope changes and the river width reaches 40 m, occasionally meandering. Few settlements are found in the valleys bottom. However, the people practise subsistence and livestock farming. Further inland are the highland areas of Kericho District where the channel is V-shaped and the altitude high and the slopes are steep. The low plain experiences a mean annual rainfall of 1260 mm, most of it falling between March and May and with a smaller peak between September and November. Extreme droughts occur in January and February. Severe convectional rains occur on lakeshores. The mean annual maximum temperature ranges from 25 to 30 °C, while the minimum is 9 to 18 °C. The soils are recent alluvial, medium to heavy clays, of poor drainage and structure and are therefore well suited for rice production. There have been two irrigation schemes managed by the National Irrigation Board, namely,

a. Ahero—860 ha under pumped water from river Nyando with 490 tenants and
b. West Kano scheme—810 ha with 500 tenants and irrigation water pumped from Lake Victoria and drainage water pumped back to Lake Victoria.

The Nyando River, which traverses the Kano Plains and covers a large area of the lower plains, is the most notorious for frequent flooding. The flooding results in social distress, health hazards, disruption of human settlement, loss of harvests, building and infrastructures and a general malaise of the population. This prevents any progress from being made in the development of this high potential Kano Plains. Before 1925, the so-called Nyalenda, Nyamware, Kawino, Kadhiambo, Bwanda and Middle Jimo areas were a stable lakeshore land with fishing and small-scale farming activities. In March of the same year, the lake regions experienced exceptionally heavy rains and many homes were abandoned and people settled in the upper lands. The situation did not last long and after a few years, life returned to normal along the lakeshore. The famous Uhuru rains of early 1961 saw the lake level remaining relatively high (approx. 1136 m above mean sea level), which combined with peak floods of 1963, 1968, 1977 and 1985 have had the cumulative effect of turning large shore lands at the deltas of the rivers into permanent swamps.

The Nyando River basin has extensive catchment areas that for the most part lie in zones of heavy rainfall. At the outlets of the major rivers are large swamps. The rivers begin to deposit their load of silt and suspended matter gathered from the catchment areas. Over the years, excessive siltation of the river in the basin has resulted in an overflow. Some limited embankments have recently been constructed but are contributing to the rise in the riverbed, increasing the risk of catastrophic floods. These annual floods result in inundation of large tracts of cropped land, displacement of thousands of people and damage to infrastructure.

The total area annually vulnerable to flooding is 15,000–20,000 ha. This area comprises the Kakola, Onjiko and Kochogo locations of Nyando District. Available data suggest that progressively greater flooding is being caused by smaller flow in the rivers concerned. For instance, a flow of the Nyando River of 850 m^3/s in 1962 apparently resulted in flood conditions no worse than was produced in 1970 by a flow of 600 m^3/s. Variation in the level of the Lake Victoria is a cause of flooding on the lower part of the Kano Plain. During the past 15 years, the level of the Lake Victoria has been up by about 1.40 m, impeding the discharge of the floods into the lake. Besides the regular flooding of the Kano Plain caused by the overflowing of rivers, the plain regularly suffers from impeded drainage.

According to records available from 1999 to date (2003), it has been observed that the lake fluctuates by about 4 cm and is on a rising limb and has permanently inundated a broad strip of land around the shore, thereby worsening the channel hydraulics and the hydrology of inflowing tributaries. The water table is shallow ranging from 3 to 10 m in some areas in the plains. Shallow wells and hand-dug wells are common sources of water supply.

5 VULNERABILITY INDICES FOR THE FLOOD-PRONE AREAS

The interrelationship between flood hazard, flood vulnerability and flood disaster is shown in Figure 4a, while the interrelationship between hazard, vulnerability and disaster is also illustrated in Figure 4b.

The vulnerability of a water resource system to climate change is a function of a number of physical features and social characteristics, Mutua (2001). The physical features associated with maximum vulnerability of water resources in a region include:

- the marginal hydrologic and climatic regime;
- high rates of sedimentation leading to reduction of reservoir storage;
- topography and land-use practices that promote soil erosion and flash flooding conditions; and
- deforestation, which allows increased surface run-off, increased soil erosion and more frequent significant flooding.

The social characteristics that increase vulnerability of water resources include:

- poverty and low income levels that prevent long-term planning and provisioning at the household level;
- lack of water control infrastructures;
- inadequate maintenance and deterioration of existing infrastructure;
- lack of human capital skills for system planning and management;

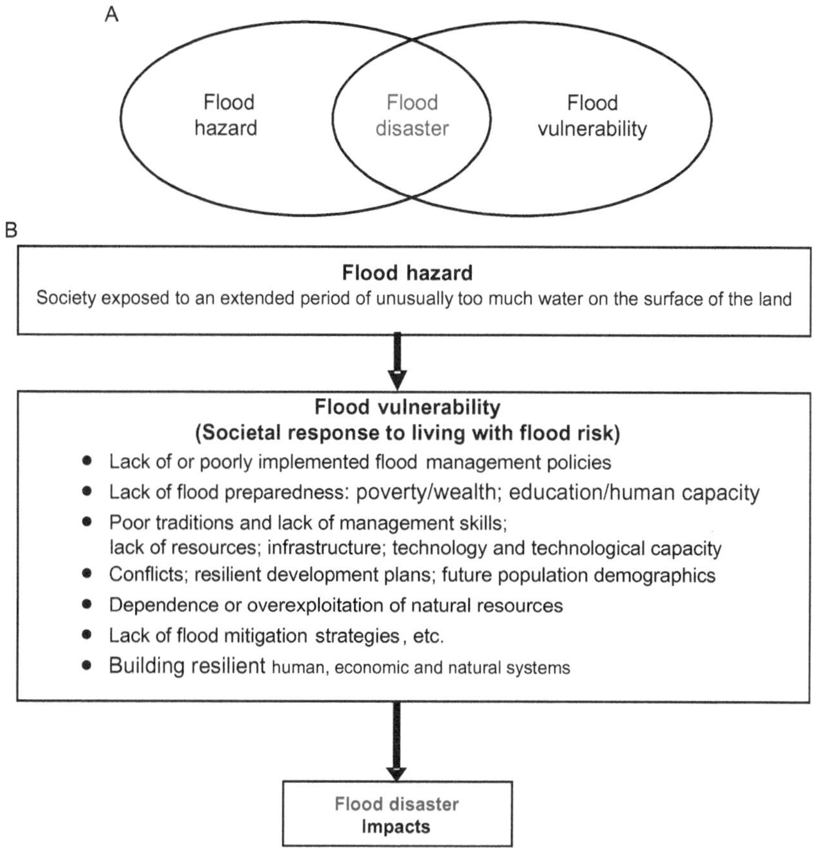

FIGURE 4 (a) A flood disaster occurs when a flood hazard happens within a community that is highly vulnerable. (b) Multiple indices representing different aspects of flood vulnerability can be derived from various physical, societal, economic and environmental factors. (For colour version of this figure, the reader is referred to the online version of this chapter.)

- lack of appropriate and empowered institutions;
- absence of appropriate land-use planning and management;
- high population densities and other factors that inhibit population mobility; and
- increasing demand for water because of rapid population growth and conservative attitudes towards risk, that is, unwillingness to live with some risks as a trade-off against more goods and services.

It should be noted that flooding poses serious threats to pollution of both surface and ground-water resources in an area.

The key indicators of vulnerability and risks in the water resources sector, as mentioned earlier, can be grouped as both natural and human-related factors. The natural factors include climate, topography and geology, while the human-related factors include settlement patterns, land-use patterns and pressures, migration patterns, population pressure, degradation of water catchment areas, unsustainable water supply, poverty and access to health and social services. These are briefly discussed in the sections below.

5.1 Climate

Of all the relevant factors in climate, precipitation is the main cause of disasters in the water resources sector. Abundant precipitation can lead to disasters such as flooding, water pollution, soil erosion, dam breaks, and water-related disease outbreaks and famine. On the other hand, scarcity of precipitation in areas, which ordinarily receive it, can lead to drought, water scarcity, loss of vegetation, loss of livestock and wildlife, famine and general suffering of people living in the affected areas. It is important that the characteristics and predictive potential of rainfall are factored in all the water resources management practices and policies in order to mitigate the adverse effects of disasters in the water resources sector.

5.2 Geology and Topography

The geologic characteristics of an area are known to influence the drainage patterns, the nature of soils and land-use patterns. For instance, heavy rainfall is more likely to cause flooding in low-lying regions occupied with clay formations because clay formations have low infiltration capacities and therefore surface run-off is generated rapidly. The same heavy rainfall in zones occupied by sandy soil will not cause significant flooding since rapid infiltration of rainfall reduces the likelihood of generation of high volume of surface run-off.

The topography of a region also determines land-use patterns and the patterns and intensity of erosion and landslides. It influences the gradient of rivers such that in areas with steep slopes, water flows rapidly into river channels. In such cases, the flood stage of a river is reached rapidly. In flat plains, the gentle slope reduces the velocity of flow leading to formation of meanders and flood plains. Water in such areas is not transported rapidly and tends to pile up making the low-lying zones more liable to flooding (e.g. Kano Plains).

5.3 Population Pressure and Settlement Patterns

Settlement patterns are partly influenced by population pressure. In urban areas, there is a tendency for the slums to develop in areas which have been designated as flood-prone zones. Settlement on steep slopes as well as cultivation on such lands also tends to increase the vulnerability of the community to landslides. This also increases possibility of increased rates of soil erosion particularly where overgrazing and deforestation have reduced vegetation cover. Modification of river channels through channel straightening can lead to rapid flow of water into streams, thus promoting rapid increase in water level in rivers. Flooding of the low-lying areas often follows this.

5.4 Poor Land-Use and Degradation of Catchment Areas

Poor land-use activities characterized by deforestation and clearance of bushes and other vegetation is the major cause of catchment degradation. Cultivation on steep slopes without applying soil conservation measures promotes soil erosion and rapid generation of surface runoff (Lake Basin River Catchment Conservation and Rehabilitation, 1987). Vegetation cover is essential since it retards the flow of surface run-off, thus encouraging more water to infiltrate into the soil and replenish soil moisture. The recharge of ground-water aquifers also take place through infiltration and deep percolation of rainwater.

Other causes of the degradation of catchment areas include poor construction of roads and footpaths, which are sources of sediments carried by the surface run-off to river channels. Lack of effective urban planning mechanisms promotes development of slums and other residential structures, which discharge sewage and domestic wastewater into river channels, thus degrading important water catchment areas.

5.5 Lack of Regulatory Systems

The enforcement of regulations governing settlement in zones designated as flood prone has been a major problem in Kenya in that the enforcement is weak, partly due to weak institutional capabilities to enforce regulations. Similarly, there is no effective coordination between different government departments and nongovernmental organizations, resulting into waste of resources and duplication of effort. The uneven development in the country particularly the huge difference in the living standard of urban and rural areas is encouraging the influx of rural poor into urban areas in search of better opportunities.

This has led to overcrowding in urban areas, severely overstraining the existing housing, health, water supply, sewage and educational and recreational facilities. Most of the present facilities in urban areas were designed cater for a small elite population during colonial times. Lack of housing and inability to access more decent housing has encouraged development of slums in areas, which are more liable to flooding. This makes the urban population more vulnerable to flooding.

5.6 Poverty

According to the 1997–2001 National Development Plan (GoK, 1997), out of the estimated population of 29 million people, 11 million people live below poverty line and 3 million are unemployed. This means that more than 40% of the population in Kenya lives below poverty line. The poverty in Kenya has increased in the recent past due to low productivity of the agricultural and industrial sectors. The dependency ratio is also extremely high. Studies show that areas prone to floods have high incidences of poverty.

Poverty is encouraging people to dwell in slums located in flood-prone zones of the urban areas. It has also contributed to the limited application of better land-use practices and soil conservation methods in rural areas. Cultivation in these areas is thus done by using unsustainable indigenous technology, which provides low yields and exposes soil to erosion hazards. Also, it has encouraged the conversion of swamps and wetlands into settlement and agricultural lands, thus subjecting the population to the risks of flooding.

5.7 Limited Access to Proper Health Care and Social Services

Poverty has also made it difficult for a large percentage of the urban and rural people who lack access to medical facilities. This has promoted high morbidity caused by waterborne and water-related diseases, which are otherwise treatable. This is made even worse by the inability of the government to provide highly subsidized access to public health facilities. Lack of an effective public health system is making people more liable to epidemics associated with the occurrence of extreme climatic events such as floods and droughts.

Lack of social amenities in both rural and urban areas is also thought to contribute to the low level of preparedness of the local population to handle disasters. This is particularly so given those facilities for dissemination of information to the general populace are often inadequate or completely lacking in certain areas. Thus, community mobilization is a difficult task, which is not easily achievable in periods of emergencies.

5.8 Poor Communication Infrastructure

Availability of good communication and transport network is essential in the management of disasters. These include telephone links, roads, railways, airports and airstrips. These are essential for evacuation of population during periods of emergencies. However, the current telephone network is poor and most rural areas are not linked to an efficient telephone system. In addition, road and railway network in most places is not in good condition and thus access to zones which are usually affected by droughts and floods is problematic.

The problem of communication has been made even worse by the fact that the present telephone and road network was designed cater for a smaller population. Road traffic has increased tremendously in the recent past, leading to rapid deterioration of roads which were designed for much smaller loads. Thus, lack of access to a reliable communication network has increased the risks and vulnerability of the population to extreme climatic events since the affected population cannot be mobilized rapidly. It should be noted that quantifying the vulnerability of the society is quite complex especially when one is looking for cumulative stresses induced by interannual variability and long-term climate changes. It will require multiple indices representing the different aspects of vulnerability including the coping mechanisms and adaptive capacity, hazards exposure, ecosystem sensitivity, land use and land-use changes.

The state of the current climate is, however, critical in planning the basic coping and adaptive capacities. Figure 5, in next section is a flood hazard map showing high potential areas for flood risks in Kenya.

6 FLOOD HAZARDS MAP FOR KENYA

A flood hazard map shows the areas over given regions that have the potential to flood. These are usually areas along riverbanks, lakeshores, wetland boundaries and arid and semi-arid areas which receive episodic rainfalls and which are generally flat and without adequate surface drainage. The spatial-extent of such areas depends on the severity of the flood event. The severity is often measured in terms of the return period of the flood under consideration. Return periods of 20, 50 and 100 years are often used in developing flood hazard maps. Physics-based rainfall run-off models are particularly useful in delineating flood prone areas in a given region.

In the western parts of Kenya, for example, serious flooding is usually experienced on the lower parts of the major rivers which flow into Lake

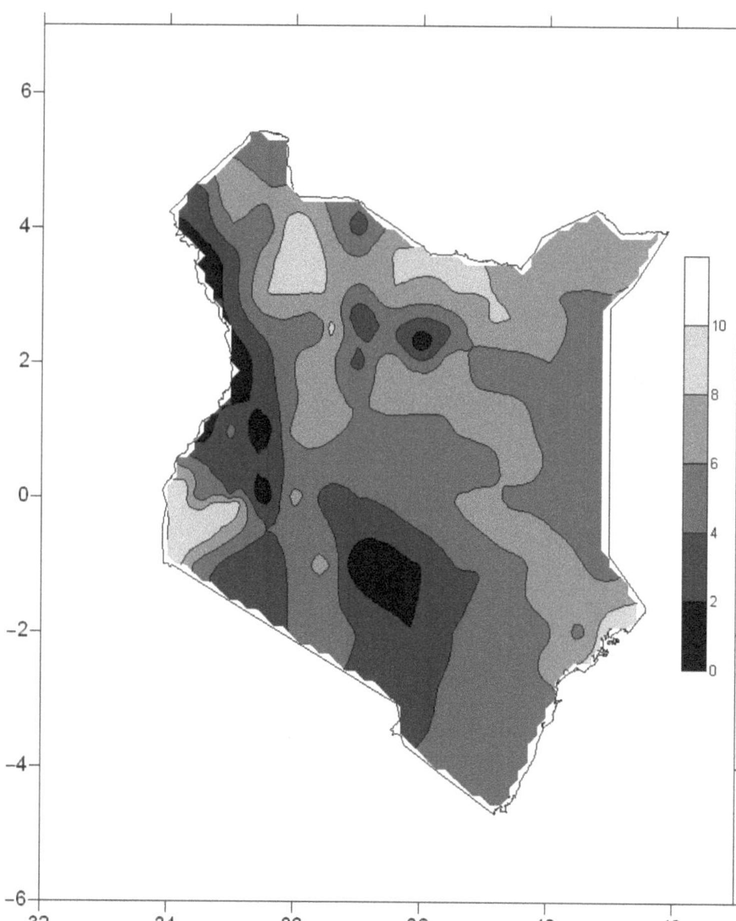

FIGURE 5 Flood vulnerability map of Kenya on a scale of 5 with low vulnerability at low values.

Victoria especially rivers Nzoia, Nyando, Kuja and to some extend near the mouth of river Yala. The Budalang'i and Kano plains are the most notorious and flood prone areas in western Kenya. In an attempt to reduce the flooding menace in the Budalang'i area, flood dykes were constructed on the banks of the Nzoia River just before it enters into Lake Victoria. The construction of the dykes created a sense of security amongst the riparian communities, who then have began settling more permanently in the areas where they had lived on sedentary basis before the construction of the dykes. This condition has therefore brought serious flooding impacts in this region considering the fact that the design of these dykes has proved to be faulty and fails very often.

REFERENCES

DMCN, 2004. Coping with Floods in Kenya: Vulnerability, Impacts and Adaptation Options for the Flood Prone Areas of Western Kenya, DMCN-UNEP project report, 2004.

Government of Kenya (GoK), 1997. National Development plan (1997–2001).

Lake Basin River Catchment Conservation and Rehabilitation, International Council for Research in Agro-forestry, Annual report, Nairobi, Kenya, 1987.

Management of flood control disaster in the Lake basin Development Authority (LBDA) with special reference to Kano Plains—a structural approach, LBDA, 2002.

Mutua, F.M., 2001. Report on the Impacts of climate change on Water Resources in the Kenya.

Oyugi, A., Ong'ang'a, O., Awange, J.L., 2003. Poverty reduction: a challenge for Lake Victoria basin. Osienala, Kisumu, Kenya.

Palmer, W.C., 1965. Meteorological drought. Research Paper No. 45, U.S. Department of Commerce Weather Bureau, Washington, DC.

Smith, L.A., 2002. What might we learn from climate forecasts? Proc. Natl. Acad. Sci. U.S.A. 99, 2487–2492.

The Republic of Kenya, LBDA, and JICA, 1992. Feasibility studies on Kano Plains Irrigation Project vol. 1, vol II-1, and vol II-2.

Chapter 22

Coastal Erosion

Christine A. Omuombo*, Daniel O. Olago*,† and Eric O. Odada*
*Department of Geology, University of Nairobi, P.O. Box 30197-00100, Nairobi, Kenya
†Institute for Climate Change and Adaptation (ICCA), University of Nairobi, P.O. Box 30197-00100, Nairobi, Kenya

1 INTRODUCTION

The natural 584 km of the Kenyan coastline comprises of sheltered bays, fringing reef, creeks, lagoons, mangrove swamps, exposed beaches, cliffs and rocky coasts and is vulnerable to coastal erosion (Mauffe, 1915; Ase, 1978, 1981; Cannon et al., 1981; Abuodha, 2003, 2004). Seasonal variation in the ocean circulation influences the sediment distribution, composition and wave action on the shoreline. Coastal erosion has been exacerbated by human occupation and intensive development of infrastructure that is concentrated along the coastal strip (Figure 1). In Diani–Bamburi area, shoreline recession of up to 10 m in 10 years is reported to have occurred over the period 1996–2006 (Ballot et al., 2006) due to modification to current flows caused by the construction of seawalls to protect the beachfront tourist hotels and resorts. Natural processes such as increase in sediment loading of, and deposition by, the Tana and Sabaki Rivers have resulted in recession along Malindi shoreline: about 500 m recession of shoreline has been observed from 1976 to 2003 due to sediment accretion leading to formation of dunes and flood plains (Odada, 1995; Abuodha 2003; McClanahan and Obura 1997; Kairu, 2003). In the Gazi area, longshore sediment transport has resulted in the erosion of Holocene and Pleistocene dune platforms forming new landforms such as sands spits (Arthurton, 1992; Kitheka, 1996; Kairu, 1997; Ohowa et al., 1997; Abuodha, 2003; Ballot et al., 2006; Kitheka et al., 2003).

2 CAUSES AND IMPACTS OF COASTAL EROSION

Although recent studies (Arthurton, 1992; Kairu, 1997, 2003; Kitheka et al., 2003; Ballot et al., 2006) show that coastal erosion has been prevalent in the short term (from the 1970s), it is still unclear how far back shoreline changes have occurred along the coast (Odada, 1995). However, the presence of relict dunes and beach platforms illustrates the dynamic nature of the coastal

Developments in Earth Surface Processes, Vol. 16. http://dx.doi.org/10.1016/B978-0-444-59559-1.00022-0

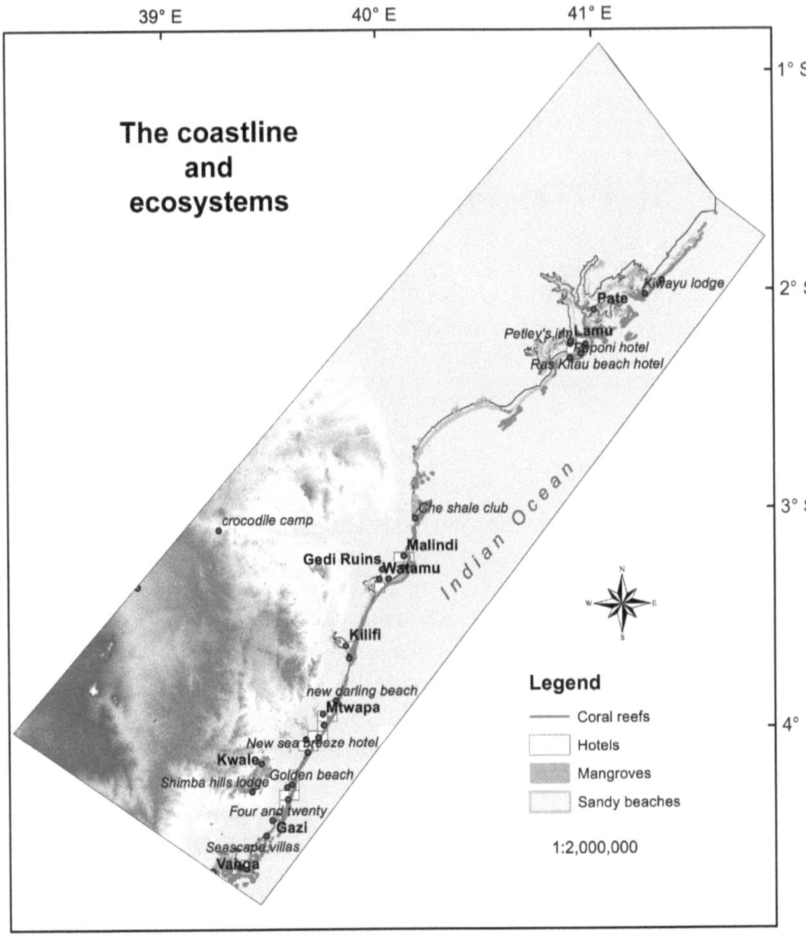

FIGURE 1 The configuration of the Kenyan coastline showing the major towns, hotels and ecosystems. *Hotels and sandy beach data from WRI (2007).* (For colour version of this figure, the reader is referred to the online version of this chapter.)

processes in the Quaternary period (Mauffe, 1915; Caswell, 1953, 1956; Thompson, 1954; Ase, 1978, 1981; Cannon et al., 1981; Braithwaite, 1984; Oosterom, 1988; Abuodha 2003; Abuodha, 2004). The configuration of the coastal zone is in such a way that the littoral zone is confined to the narrow continental shelf that is dominated by wave action. These waves contribute to the longshore transport and sediment down drift leading to sediment erosion and accretion with a net equilibrium transport of sediments (McClanahan, 1988).

The deltaic plains of the Sabaki–Tana Rivers, the old deltaic plains of the Lamu Archipelago and the fringing reef coast are in equilibrium with the dynamic longshore sediment transport alongside the provision of ecosystem goods and services (McClanahan and Obura, 1997; Kairu, 2003; Ballot et al., 2006). These wetlands are in most cases utilized for agriculture, water abstraction and livestock rearing with the associated changes in land use linked to clearing of vegetation cover and conversion of land to its current use. This has led to an influx in sediments from the catchment into the rivers, and this suspended and detrital sediment loading increases turbidity and siltation. During the monsoon rains, coastal and river flooding occurs with notable impact on the coastal ecosystems: the coral reefs (destruction and retardation of growth) and mangroves (sediment retention) (Leetma, 1973; van Katwijk et al., 1993; Kitheka, 1996, 1997; McClanahan and Obura, 1997; Obura, 2001). These ecosystems are natural stabilizers keeping the coast in equilibrium with the dynamic waves and tides that dominate the coastal waters. For instance, the influence of the Sabaki River is strongest in the Watamu–Malindi reef where although there is no consensus on the exact cause of the degradation of the coral ecosystem, it is postulated to be caused by the increase in sediment loading in the recent past due to frequent flooding (Odada, 1995; Leauthaud et al., 2013).

Harvesting of mangroves and coral reefs by the coastal communities, changing the landscape due to human settlement and economic activity along the shoreline increase its vulnerability to erosion and accretion resulting from tides and wave action (Kitheka, 1996, 1997). Measurements of total sediment flux from the Mwache mangrove forest wetland during the ebb and flood tides show an increased sediment input of about 50% more during the flood tide and a net sedimentation peak during high river discharge in spring tides during the dry season. A net accretion of 0.25 and 3.5 cm/year was estimated in the backwater and frontwater zone, respectively, of the mangrove forests (Kitheka et al., 2003). In the lagoonal systems, the bottom friction for sediment retention dissipates the wave energy along the coastline. The magnitude of protection depends on the tide level with a larger influence during the low tide than during the high tide (Pugh, 1978; Kitheka et al., 2003). In these systems, the fishermen are the key culprits to their destruction as trawling causes the dredging of the bottom sediments, which lowers the lagoon's buffering ability, thus allowing high-energy waves to reach the shore (Fulanda et al., 2011; Erftemeijer et al., 2012).

The coastal region is a key contributor to the National Gross Domestic Product with 45% from tourism and 15% from the shipping activities which are important to the coastal economy (WRI, 2007; Government of Kenya, 2009). Due to this, there has been an increase in real estate investment with the construction of beachfront apartments, hotels and resorts concentrated along the low-lying areas of the shoreline (Government of Kenya, 2010). In order to protect the prime beach properties from tide and wave action, seawalls have been erected. The construction of seawalls has led to erosion in some areas where sand accretion was the norm, resulting in relocation of some hotels and resorts further inland and away from the beach, while in others, there is sediment accretion and dune development that obstruct the aesthetic appeal of the beaches and sometimes leading to abandonment of the hotels (Odada, 1995; Mwanje, 1997; Kairu and Nyandwi, 2000; WRI, 2007; Government of Kenya, 2009, 2010).

The developments along the beaches are largely due to the unclear guidelines that govern and control the erection of engineering structures. Currently, there are no regulations that define beaches and buffer zones, and therefore, there are unplanned and sporadic tourist facilities being built very close to the beach and in some cases encroaching the beach, that is, that are built beyond the high water level (Odada, 1995; Mwanje 1997; WRI, 2007). During high tides, these establishments tend to flood and are thus reinforced by sand fillings, concrete walls and sea walls. These infrastructural developments increase shoreline instability and imperil the stability of seawalls or buildings due to high-energy waves from intensive erosion (Odada, 1995; Mwanje, 1997; Kairu and Nyandwi, 2000). These tourist hotels and resorts keep their beaches clean through the clearing of seaweed, and creation of beach gaming and relaxation areas has increased the human traffic in the beach, thus exposing the loose sands that are easily washed into the sea during the high tides. This has led to the characterization of these areas as coastal erosion 'hotspots' (Government of Kenya, 2009, 2010).

The change in the cultural perception of the local communities on the coastal forests, with special sacred enclaves termed as 'Kayas', which occupy a substantial stretch of the coastal lowlands, to potential resources for exploitation, has led to their encroachment and clearing for subsistence agriculture and human settlement. This destruction of forests has led to the exposure of unconsolidated soils, making the shoreline vulnerable due to the availability of material for erosion by tides, waves and run-off (Odada, 1995; Government of Kenya, 2009, 2010).

It is safe to conclude that the key actors in the coastal zone are the local communities, the fishermen, the hoteliers and tourist operators who cause notable changes in the natural coastal configuration, exposing the coast to erosion processes that lead to loss of beaches and corals. It should be noted that although not all areas of the coast are vulnerable to wave erosion, it mainly affects the unconsolidated and steep shorelines. The hydrodynamics of the coast is still poorly understood, and more research needs to be carried out focusing on the sediment dynamics along the coastal zone to distinguish the vulnerable areas.

3 EXTREME EVENTS

Extreme events such as El Niño, tsunamis, sea level rise and storm surges do impact the Kenyan coast. During the El Niño event of 1997/1998, high intense rainfall in East Africa following a long dry period led to increased sediment inflow into the ocean, increasing turbidity and detrital material deposition (van Katwijk et al., 1993; McClanahan et al., 2004). As a consequence of sediment loading and increasing ocean temperatures for a long period, coral mortality and bleaching occurred with varied intensity depending on the location (van Katwijk et al., 1993; Obura, 2001; McClanahan et al., 2004). It is estimated that 80% of the corals off the coast of Kenya were affected compared to 1% off the coast of South Africa (van Katwijk et al., 1993; Spencer et al., 2000; McClanahan et al., 2004). A greater temperature anomaly was experienced in East Africa compared to the Great Barrier Reef (GBR) of Australia; this is reflected in the taxa losses off the Kenyan coast (McClanahan et al., 2004). Tidal analyses of this event show that a substantial sea level rise in the western Indian Ocean contributed to the drowning of some of the corals and death of mangroves due to the freshening of the brackish water they occupy (Kitheka, 1997). The marine protected areas (MPAs) in the Malindi and Watamu areas were less affected by the global event and the biodiversity flourished in comparison to the already degraded corals on the coastline (van Katwijk et al., 1993; Obura, 2001; Kaunda-Arara and Rose, 2004; McClanahan et al., 2004; Muthiga, 2009).

A reconstruction of sea surface temperatures over the past two centuries has shown an estimated rise of 1.3 °C since 1880 on the western Indian Ocean (Cole et al., 2000); this is significant to the ecosystem changes witnessed before the El Niño event. The severity of the coral bleaching and mortality was seen at depths of 2 m, while the impact of coral bleaching off the GBR was much deeper at depths of 6 m. The significance of the temperature anomaly during the 1997/1998 El Niño event cannot be ignored if the predicted global warming projected at 6 °C temperature rise is to take place. The El Niño event provides an analogue of what is likely to happen to the coastal ecosystems off the Kenyan coast in a worst case scenario since reconstructions show decadal-scale increase in El Niño frequency and intensity since 1960 (van Katwijk et al., 1993; McClanahan et al., 2004; Nakamura et al., 2009).

Frequent storms are expected off the East African coast due to climate change-related global sea level rise. The low-lying areas of the Kenyan coast are in obvious threat of biodiversity, land and property losses and the disruption of economic activities beyond the country's borders since it has East Africa's largest seaport (Kabede et al., 2009; Brown et al., 2011). It is estimated that by 2100, 17% of Mombasa city could be submerged by a 30 cm sea level rise (Awuor et al., 2008; Kabede et al., 2009). The deltas of Tana and Sabaki Rivers are some of the vulnerable areas that have been identified as high risk with 5.7% expected loss of the Tana delta. The impacts of this sea level rise scenario intensify river flooding and sediment input into the ocean that will have unprecedented devastating impacts on the coastal ecosystem goods and services (Kabede et al., 2009; Brown et al., 2011).

On 26 December 2004, an earthquake of magnitude 9.0 Mw off the west coast of Sumatra generated a tsunami that led to the loss of approximately 70,000 lives and coastal erosion and subsidence, changing the geomorphological configuration of the coasts of the Indian Ocean (Borrero, 2005; Hori et al., 2007; Goto et al., 2011; Leone et al., 2011). The highest wave recorded at 14.1 m led to the deposition of large amounts of sediments from the ocean bottom to the shoreline with sequences of up to 4 m thick, and in some cases, the shoreline permanently moved about 1.5 km inland (Borrero, 2005; Goto et al., 2011; Leone et al., 2011). Kenya was struck by this tsunami 6 h later after the quake at low tide causing limited damage unlike in Somalia where over 200 lives were lost alongside property damage and losses (Government of Kenya, 2009, 2010).

4 MANAGEMENT OF THE SHORELINE

The oldest MPAs along the Kenyan coast are in Watamu and Malindi Marine Parks. An evaluation of the effectiveness of the MPAs was done after the 1997/1998 El Niño event where coral mortality and bleaching provided a measureable impact of the event on the coastal ecosystem. Results from the evaluation of the coral community off the Kenyan coast showed the preservation of up to 50% or more of the species within the MPAs, whereas there was up to 80% loss in the unprotected areas (McClanahan, 1997; Spencer et al., 2000; McClanahan et al., 2004). This and other successes seen in the MPAs have led to the emphasis on the need for the establishment of more protected areas along the coastline (McClanahan, 1997; Government of Kenya, 2009, 2010; Muthiga, 2009; Cinner et al., 2012).

Up until 2006, the management of the coastal resources in Kenya was centralized in the Ministry of Fisheries, but from then, the Ministry of Fisheries implemented the Beach Management Units for the fisher folk at the coast (McClanahan, 1997; Cinner et al., 2012). Through these units, the communities have been empowered to work alongside the national government to enforce fishing and conservation laws and have been effective in the conservation of the corals, lagoons and mangroves that are spawning grounds for fish (McClanahan, 1997). The other stakeholders especially from the tourism sector that is largely business-minded seem to rely on textbook explanations with regard to the coastal erosion and ecosystem changes due to the lack of firsthand long-term experience (Mayaka and Prasad, 2012).

Rapid modification of the coastline due to haphazard development of tourism infrastructure and subsequent construction of seawalls has modified the configuration of the coastline through altering sediment transport. Management of the coastal zone is still inefficient due to overlapping mandates, financial constraints and inadequate stakeholder participation that restricts achievement of the objectives of an integrated management (Mwanje, 1997; McClanahan, 1997). Currently, an Integrated Coastal Zone Management policy does not exist although the legal framework through the Environmental Management and Coordination Act No. 8 of 1999 provides a strong basis for the development of such a policy. It is due to this that the shoreline management strategy has been developed (Government of Kenya, 2010). This strategy divides Kenyan coast into 29 sediment cells based on the sediment transport and protruding morphological characteristics (Government of Kenya, 2010). The strategy addresses all the issues affecting the coastline including coastal erosion. The basis for this delineation and definition of sediment cells, although not clear-cut and straightforward, assigns priority management objectives to minimize risks of flooding, erosion, accretion and costal destabilization. This strategy aims at defining ambiguous terms such as beaches and buffer zones (60 m above mean high water mark of spring tide), restricting development in low-lying coastal areas, on eroding coastlines, unstable coasts and accreted land. The strategy recommends controlled, monitored and coordinated construction of future development while maintaining existing seawalls and revetments in highly developed areas.

REFERENCES

Abuodha, P.A.W., 2003. Geomorphology of the Kenyan coast: not as a result of sea-level change alone. Buletim Geologico, no. 43 Mozambique, 2–7. Paper presented during the 'Impact of Sea Level Rise: Past Present–Future' Inhaca Island, Mozambique 4–8 November 2002. Organized and sponsored by the Commission of the Holocene of the International Union for Quaternary Research (INQUA). http://www.oceandocs.org/bitstream/1834/407/1/1Abuodha.pdf (accessed 15 April 2013).

Abuodha, J.O.Z., 2004. Grain size distribution and composition of modern dune and beach sediments, Malindi Bay coast, Kenya. J. Afr. Earth Sci. 36, 41–45.

Arhturton, R.S., 1992. Beach erosion: case studies on the East Africa Coast. In: Proceedings of Bordomer 92, International Convention on Rational Use of Coastal zone. Intergovernmental Oceanographic Commission, Bordeaux, pp. 91–95. http://erepository.uonbi.ac.ke/bitstream/handle/123456789/31642/scan0001.pdf?sequence=1.

Ase, L.E., 1978. Preliminary report on studies of shore displacement at the southern coast of Kenya. Geogr. Ann. 60A (3–4), 209–221.

Ase, L.E., 1981. Studies of shores and shore displacement on the southern coast of Kenya—especially in Kilifi District. Geogr. Ann. 63A (3–4), 303–310.

Awuor, C.B., Orindi, V.A., Adwera, A.O., 2008. Climate change and coastal cities: the case of Mombasa, Kenya. Environ. Urban. 20 (1), 231–242. http://dx.doi.org/10.1177/0956247808089158.

Ballot, J., Hoyng, C., Kateman, I., Smits, M., de Winter, R., 2006. Report on coastal erosion project. Master's Thesis, Delft University of Technology, The Netherlands.

Borrero, J.C., 2005. Field data and satellite imagery of tsunami effects in Banda Aceh. Science 308, 1596.

Braithwaite, C.J.R., 1984. Depositional history of the late Pleistocene limestones of the Kenya coast. J. Geol. Soc. London 141, 685–699. http://dx.doi.org/10.1144/gsjgs.141.4.0685.

Brown, S., Kebede, A.S., Nicholls, R.J., 2011. Sea Level Rise and Impacts in Africa, 2000 to 2100. School of Engineering and Environment, University of Southampton, Southampton, UK.

Cannon, R.T., Simiyu Siambi, W.M.N., Karanja, F.M., 1981. The Proto-Indian Ocean and a probable Paleozoic/Mesozoic Triradial rift system in East Africa. Earth Planet. Sci. Lett. 52, 419–426. http://www.sciencedirect.com/science/article/pii/0012821X81901941.

Caswell, P.V., 1953. The geology of the Mombasa—Kwale area. Report No. 24. Geological Survey of Kenya.

Caswell, P.V., 1956. The geology of the Kilifi—Mazeras area. Report No. 34. Geological Survey of Kenya.

Cinner, J.E., Daw, T.M., McClanahan, T.R., Muthiga, N., Abunge, C., Hamed, S., Mwaka, B., Rabearisoa, A., Wamukota, A., Fisher, E., Jiddawi, N., 2012. Transitions toward co-management: the process of marine resource management devolution in three east African countries. Glob. Environ. Chang. 22, 651–658.

Cole, J.E., Dunbaar, R.B., McClanahan, T.R., Muthiga, N.A., 2000. Tropical Pacific forcing of decadal SST variability in the Western Indian Ocean over the past two centuries. Science 287, 617–619.

Erftemeijer, P.L.A., Reigl, B., Hoeksema, B.W., Todd, P.A., 2012. Environmental impacts of dredging and other sediment disturbances on corals: a review. Mar. Pollut. Bull. 64, 1737–1765.

Fulanda, B., Ohtomi, J., Mueni, E., Kimani, E., 2011. Fishery trends, resource-use and management system in Ungwana Bay fishery Kenya. Ocean Coast. Manag. 54, 401–414.

Goto, K., Kahashi, J., Ole, T., Imamura, F., 2011. Remarkable bathymetric change in the nearshore zone by the 2004 Indian Ocean tsunami: Kirinda Harbor, Sri Lanka. Geomorphology 127, 107–116.

Government of Kenya, 2009. State of the Coast Report: Towards Integrated Management of Coastal and Marine Resources in Kenya. National Environment Management Authority (NEMA), Nairobi, 88 pp.

Government of Kenya, 2010. Shoreline Management Strategy for Kenya. National Environment Management Authority (NEMA), Nairobi, 87 pp.

Hori, K., Kuzumoto, R., Hirouchi, D., Umitsu, M., Janjirawuttikul, N., Patanakanog, B., 2007. Horizontal and vertical variation of 2004 Indian tsunami deposits: example of two transects along the western coast of Thailand. Mar. Geol. 239, 163–172.

Kebede, A.S., Hanson, S., Nicholls, R.J., Mokrech, M., 2009. Impacts of Climate Change and Sea-Level Rise: A Case Study of Mombasa, Kenya. School of Engineering and Environment, University of Southampton, Southampton, UK.

Kairu, K.K., 1997. Vulnerability of the Kenyan shoreline to coastal instability. In: UNESCO-Kenya National Seminar of Sustainable Coastal Development Through Integrated Planning and Management Focused on Mitigating the Impacts of Coastline Instability. Whitesands Hotel, Mombasa, 23–25 June. Summary Report, 13–25. http://unesdoc.unesco.org/images/001383/00138383/138391eo.pdf.

Kairu, K., Nyandwi, N. (Eds.), 2000. Guidelines for the study of shoreline change in the Western Indian Ocean region. IOC Manuals and Guides No. 40. UNESCO 2000.

Kairu, K., 2003. Coastal erosion in Kenya: Malindi-Ungwana Bay Case Study. A report prepared for UNEP/GPA, UNEP and WIOMSA.

Kaunda-Arara, B., Rose, G.A., 2004. Effects of marine reef National Parks on fishery CPUE in coastal Kenya. Biol. Conserv. 118, 1–13. http://www.sciencedirect.com/science/article/pii/S0006320703002994.

Kitheka, J.U., 1996. Water circulation and coastal trapping of brackish water in a tropical mangrove-dominated bay in Kenya. Limnol. Oceanogr. 41 (1), 169–176.

Kitheka, J.U., 1997. Coastal tidally—driven circulation and the role of water exchange in the linkage between tropical coastal ecosystems. Estuarine Coastal Shelf Sci. 45, 177–187.

Kitheka, J.U., Ongwenyi, G.S., Mavuti, K.M., 2003. Fluxes and exchange of suspended sediment in tidal inlets draining a degraded mangrove forest in Kenya. Estuarine Coastal Shelf Sci. 56, 655–667.

Leauthaud, C., Duvail, S., Hamerlynck, O., Paul, J.-L., Cochet, H., Nyunja, J., Albergel, J., Grünberger, O., 2013. Floods and livelihoods: the impact of changing water resources on wetland agro-ecological production systems in the Tana River delta, Kenya. Glob. Environ. Chang. 23, 252–263.

Leetma, A., 1973. The response of the Somali current at 20S to the Southwest monsoon of 1971. Deep Sea Res. 20, 397–400.

Leone, F., Lavigne, F., Paris, R., Denain, J.-C., Vinet, F., 2011. A spatial analysis of the December 26th, 2004 tsunami-induced damages: lessons learned for a better risk assessment integrating buildings. Appl. Geogr. 31, 363–375.

Mauffe, H.B., 1915. The coastal series of sediments in the East Africa Protectorate. Geol. Mag. 2 (6), 274–277 (Decade VI).

Mayaka, M.A., Prasad, H., 2012. Tourism in Kenya: an analysis of strategic issues and challenges. Tourism Manage. Perspect. 1, 48–56.

McClanahan, T.R., 1988. Seasonality of East Africa's coastal waters. Mar. Ecol. Prog. Ser. 44, 191–199.

McClanahan, T.R., 1997. Primary succession of coral-reef algae: differing patterns on fished versus unfished reefs. J. Exp. Mar. Biol. Ecol. 218, 77–102.

McClanahan, T.R., Obura, D., 1997. Sedimentation effects on shallow coral communities in Kenya. J. Exp. Mar. Biol. Ecol. 209, 103–122.

McClanahan, T.R., Baird, A.H., Marshall, P.A., Toscano, M.A., 2004. Comparing bleaching and mortality responses of hard corals between southern Kenya and the Great Barrier Reef, Australia. Mar. Pollut. Bull. 48, 327–335.

Muthiga, N.A., 2009. Evaluating the effectiveness of management of the Malindi–Watamu marine protected area complex in Kenya. Ocean Coast. Manag. 52, 417–423.

Mwanje, J.I., 1997. Socio economic impacts of coastal instability (erosion) in Kenya: a case study. In: UNESCOKenya National Seminar of Sustainable Coastal Development Through Integrated Planning and Management Focused on Mitigating the Impacts of Coastline Instability. Whitesands Hotel, Mombasa, 23–25 June. Summary Report, pp. 26–40. http://unesdoc.unesco.org/images/0013/001383/138391eo.pdf.

Nakamura, N., Kayanne, H., Iijima, H., McClanahan, T.R., Behera, S.K., Yamagata, T., 2009. Mode shift in the Indian Ocean climate under global warming stress. Geophys. Res. Lett. 36 (1–5): http://dx.doi.org/10.1029/2009GL040590, L23708.

Obura, D.O., 2001. Kenya. Mar. Pollut. Bull. 42 (12), 1264–1278.

Odada, E.O., 1995. Coastal Erosion and Flooding in East Africa and the Western Indian Ocean. In: Proceedings of the Workshop and Policy Conference on Integrated Coastal Zone Management in Eastern Africa Including the Island States, 21–13 April 1993, ISBN: 971-9014-00-8.

Ohowa, B.O., Mwashote, B.M., Shimbira, W.S., 1997. Dissolved inorganic nutrient fluxes from two seasonal rivers into Gazi Bay, Kenya. Estuarine Coastal Shelf Sci. 45, 189–195.

Oosterom, A.P., 1988. The geomorphology of southeast Kenya. Ph.D. ThesisWageningen Agricultural University, The Netherland, 227 pp.

Pugh, D., 1978. Sea levels at Aldabra Atoll, Mombasa and Mahé, western equatorial Indian Ocean, related to tides, meteorology and ocean circulation. Deep Sea Res. 26A, 237–258.

Spencer, T., Teleki, K.A., Bradshaw, C., Spalding, M.D., 2000. Coral bleaching in Southern Seychelles during the 1997–1998 Indian Ocean Warm event. Mar. Pollut. Bull. 40 (7), 569–586.

Thompson, A.O., 1954. The geology of the Malindi area. Report No. 36. Geological Survey of Kenya.

Van Katwijk, M.M., Meier, N.F., van Loon, R., van Hove, E.M., Giesen, W.B.J.T., van der Velde, G., den Hartog, C., 1993. Sabaki river sediment load and coral stress: correlation between sediments and condition of the Malindi–Watamu reefs in Kenya (Indian Ocean). Mar. Biol. 117, 675–683.

WRI, 2007. Nature benefits in Kenya: an atlas of ecosystem and human well-being. World Resource Institute, Department of Resource Surveys and Remote Sensing, Ministry of Environment and Natural Resource, Kenya Central Bureau of Statistics, Ministry of Planning and National Development, Kenya and International Livestock Research Institute.

Chapter 23

Monitoring Drought with the Combined Drought Index in Kenya

Zoltan Balint*[,1], Francis Mutua[†], Peris Muchiri* and Christian T. Omuto[‡]
*FAO-SWALIM, Nairobi, Kenya
[†]Department of Meterology, University of Nairobi, Nairobi, Kenya
[‡]Department of Environmental and Biosystems Engineering, University of Nairobi, P.O. Box 30197-00100, Nairobi, Kenya

1 INTRODUCTION

Drought is a climate characteristic with features such as unfavourable weather conditions that lead to scarcity of freshwater sources, high temperature and strong winds. It is generally recognised by the climate research community as an integral part of the environment as well as a recurrent feature of the climate (Gommes and Petrassi, 1994; Jones and Hulme, 1996; Lloyd-Hughes and Saunders, 2002; Herweijer and Seager, 2008). Although it is a common phenomenon of climate variability, its occurrence is often a devastating and complex natural hazard. In Africa, many countries have repeatedly been affected by droughts over the last decades, resulting in considerable environmental, social and economic damage as well as in worsening food security (Hulme, 1992; Prospero and Lamb, 2003). Climate change studies further indicate a trend towards increasing climate variability in the African continent, most likely resulting in more frequent drought events (Hansen and Lebedeff, 1987; Druyan and Hall, 1996; Dai, 2010). In the Horn of Africa, including Kenya, drought occurrence and its impacts have triggered actions at various political and management level (http://www.economist.com/node/14506436, accessed on 17 April 2010).

Numerous researches and the general opinion in the region agree that there is a need for adequate and up-to-date information on the occurrence and severity of drought episodes, probability of their duration, and their possible impacts (Moron, 1997; Mutai et al., 1998; Shanko and Camberlin, 1998). This is important for the design of appropriate mitigation and coping strategies for

[1]Retired

Developments in Earth Surface Processes, Vol. 16. http://dx.doi.org/10.1016/B978-0-444-59559-1.00023-2

drought. This need calls for a scientific approach for the development of a relevant drought monitoring framework that is appropriate for the region.

Droughts and their impacts have been studied for many decades in the Horn of Africa. Most of these studies have been largely based on one or two variables that are linked to the characterization of the climate such as rainfall, temperature, vegetation and soil moisture (Shanko and Camberlin, 1998; Ntale and Gan, 2003; Hastenrat et al., 2007). None of these approaches have been able to fully trace the drought footprints and offer reliable prediction in the Horn of Africa. This is particularly true because drought is a composite natural phenomenon whose monitoring methodology should be able to include at least six variables simultaneously, namely, rainfall deficit and its persistence, soil moisture deficit and its persistence, and temperature excess and its persistence. The objective of this study was to develop a drought monitoring methodology for Kenya and the Horn of Africa that is able to measure the natural components of drought by comparing the prevailing situation to the multiyear average situation at a given time in a year at a given place.

2 DROUGHT DEFINITIONS, TYPES AND ESTIMATION

Drought is a natural phenomenon that is triggered and sustained by inadequate availability of freshwater supplies for human and ecosystem's needs over an extended period of time. It has many facets in any single region: starting with the deficiency of precipitation and subsequently developing into drought conditions if the deficiency persists for a long time, depending on the requirements of the affected system. Besides rainfall, droughts are associated with a decline of soil moisture, stream flow and a drop of groundwater table. Unfortunately, the term drought is often used for food shortages, which are not necessarily caused or linked to drought in a true sense. Food shortage can be caused by many other factors, such as population pressure on the land, types of seeds used, overgrazing, plant disease, pest problems, armed conflicts or volatile market conditions, among others. These characteristics make the definition of drought quite complex. A universal definition of drought is therefore an unrealistic expectation. Nonetheless, there are some attempts in the literature, which can be broadly categorised as either conceptual or operational (Wilhite and Glantz, 1985). Conceptual definitions are of the 'dictionary' type, which generally define the boundaries of the concept of drought. They can be seen as generic descriptions of the drought phenomenon, for example, 'a long period with no rain, especially during a crop-growing season'.

Operational definitions attempt to identify the onset, severity, continuation and termination of drought episodes. They are often used in an 'operational' mode in many disciplines to analyse sectoral drought attributes such as frequency, severity and duration. Each discipline incorporates different physical, biological and/or socioeconomic factors in its definition of drought. As an example, an operational definition of agricultural drought might be one that compares daily or monthly precipitation to the corresponding evapotranspiration rates for determining soil moisture depletion rate. The resultant comparison is then used to relate drought effects to the plant behaviour at various stages of drought development (WMO, 2010).

The following definition was developed by this study for drought monitoring purposes based on various drought definitions in the literature:

Drought is an extended period, during which, fresh water availability and accessibility for the ecosystem at a given time and place is below normal, due to unfavourable spatial and temporal distribution of rainfall, temperature, soil moisture and wind characteristics.

Different drought 'types' are effectively different stages of the same process. Thus, meteorologic droughts attempt to explain the primary causes, while agricultural and hydrologic droughts attempt to explain the secondary impacts of the meteorologic droughts. This is illustrated in Figure 1, based on and further developed from Wilhite and Glantz (1985).

The process of monitoring the evolution of the drought is complex. Ideally it should include measurement and evaluation of all variables incorporated in

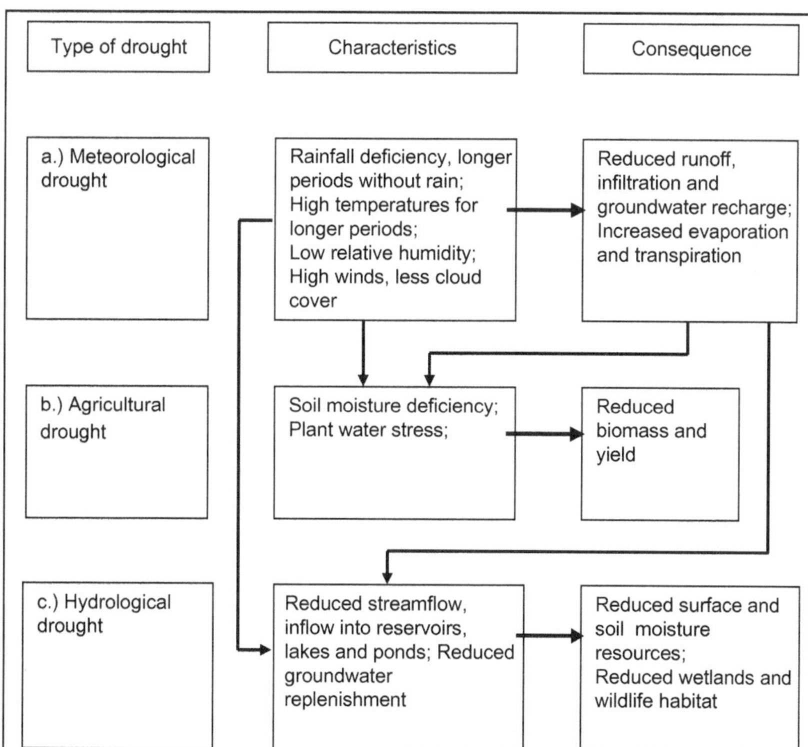

FIGURE 1 Schema for the relationship between drought types, characteristics and consequences. (Modified chart of Wilhite, 1993.)

the definition of drought in the preceding text. At present, there are many indices that have been developed to measure drought magnitudes, in different parts of the world (Szinell et al., 1998; Wu et al., 2001; Morid et al., 2006; Shakya and Yamaguchi, 2010). Although none of these indices is inherently superior to the rest in all circumstances, some indices are better suited than others for specific country/regional uses. These drought indices can be broadly put into two categories: (a) drought indices based on water balance calculation and (b) statistical drought indices based on time series analysis.

The water balance methodology requires application of several climatic and physical variables at a given time and space. Some of these variables might be calculated using some time series analysis, but all in all, their final goal is to determine the water deficit of the crop at a given time and space based on a distributed parameter model. Examples of these types of indices include the Palmer Drought Severity Index, the Palmer Hydrological Drought Index, the Palmer Z-Index, the Crop Moisture Index, the Surface Water Supply Index and the Reclamation Drought Index.

Most statistical indices are based on one or maximum two parameters, mostly rainfall and sometimes temperature deficiency/excess. To date, the most commonly used drought indices in this category include the Percent Normal Drought Index, the Standardised Precipitation Index, the Precipitation Decile Index and the Weighted Anomaly Standardised Precipitation.

The indices in the preceding text were analysed for suitability in practical monitoring of drought in Kenya.

3 METHODOLOGY FOR CALCULATING THE COMBINED DROUGHT INDEX

3.1 General Considerations and Data Used

Drought is conceived in this study as a combination of the following: a precipitation component, which considers rainfall deficits and dryness persistence; a vegetation component, which is used as a proxy for soil moisture deficit and which considers NDVI deficits and deficit persistence; and a temperature component, which considers temperature excesses and persistence of high temperatures. The drought index calculated using the precipitation component is referred to in the study as Precipitation Drought Index (PDI), while the index based on temperature is named as Temperature Drought Index (TDI) and that based on the vegetation component is named as Vegetation Drought Index (VDI). The drought index that combines the three drought components is named as Combined Drought Index (CDI).

Based on the considerations in the preceding text, the CDI was developed by Balint et al. (2011). The new index is a statistical index comparing the present hydrometeorological conditions with the long-term average characteristics in the same interest period within the year.

The index for the various components in simple words can be expressed as

$$\text{Drought Index} = \frac{\text{actual average for IP}}{\text{LTM for IP}} * \sqrt{\frac{\text{actual length of continuous deficit/excess in the IP}}{\text{LTM length of continuous deficit/excess in the IP}}}$$

where IP is the interest period, LTM is the long-term mean, and deficit applies to rainfall or NDVI, while excess applies to temperature.

The following six different time series were used for the calculation of the drought indices in the preceding text:

a. The dekadal (10-day) or monthly rainfall record.
b. The time series of the rainfall run lengths ($RL^{(P)}$) for the specific interest period (IP).
c. The dekadal (10-day) or monthly temperature record.
d. The time series of the temperature run lengths ($RL^{(T)}$) for the specific IP.
e. The dekadal (10-day) or monthly NDVI record.
f. The time series of the NDVI run lengths ($RL^{(NDVI)}$) for the specific IP.

The run length in these time series characterises the persistence of the unfavourable weather conditions, during which the uninterrupted drought pressure exists. In the case of PDI and VDI, the run length is the period within the IP, in which the rainfall or NDVI is continuously below the long-term mean value. In the case of TDI, run length is the time period within the IP, in which the temperature is continuously above the long-term mean value characteristic for the same time unit (e.g. month).

The time series in the preceding text can be grouped in two categories. In group A, small values of the data indicate dry conditions and larger values indicate wet conditions. Rainfall and NDVI data fall into this group. In group B, large values in the time series indicate conditions that can contribute to drought, while the small values indicate better than drought conditions. Temperature data fall into this category.

Since the application of the drought indices in the preceding text invariably involves the comparison of actual drought conditions with the long-term average conditions, some sort of input data standardisation may be inevitable for homogeneous mathematical interpretation. In this study, this was done by moving the x-axis to the level of ($T_{max}+1$) for temperature, ($RL_{max}+1$) for run length and ($NDVI_{min}-0.01$) for NDVI, where T_{max} is the maximum temperature, RL_{max} is the longest run length in the whole dataset used and $NDVI_{min}$ is the minimum NDVI for the station under consideration. Adding 1.0 to the highest temperature and run-length values and subtracting 0.01 from the minimum NDVI serves the purpose of avoiding division by 0. The x-axis shift results into modified data series as given in Equation (1). This modification only serves to ease the calculation process. We found that the modification does not change the nature of the data series and that of the results in any way:

$$\begin{aligned} T^* &= (T_{\max} + 1) - T \\ \mathrm{RL}^* &= (\mathrm{RL}_{\max} + 1) - \mathrm{RL} \\ \mathrm{NDVI}^* &= \mathrm{NDVI} - (\mathrm{NDVI}_{\min} - 0.01) \end{aligned} \tag{1}$$

We found it advisable working with modified rainfall data specifically in countries that have distinct, long dry seasons without any rain. In these cases, a whole period is characterised by exclusively zero values, including the LTM, which can result into unrealistically large values when dividing by very small values close to zero. Shifting the x-axis of the coordinate system by 1 mm tended to solve such problems. In countries where some rainfall can be expected any time throughout the year, and the original data series work well, we still recommend using modified precipitation data as shown in Equation (2):

$$P^* = P + 1 \tag{2}$$

where P is the precipitation amount.

3.2 Calculation of PDI, TDI and VDI

The equations for calculating the PDI, the VDI and the TDI for year i and time unit (dekad/month) m are given in Equations (3–5):

$$\mathrm{PDI}_{i,m} = \frac{\frac{1}{\mathrm{IP}}\sum_{j=0}^{\mathrm{IP}-1} P^*_{i,(m-j)}}{\frac{1}{(n*\mathrm{IP})}\sum_{k=1}^{n}\left[\sum_{j=0}^{\mathrm{IP}-1} P^*_{(m-j),k}\right]} * \sqrt{\left(\frac{\mathrm{RL}^{(P^*)}_{m,i}}{\frac{1}{n}\sum_{k=1}^{n}\mathrm{RL}^{(P^*)}_{m,k}}\right)} \tag{3}$$

$$\mathrm{VDI}_{i,m} = \frac{\frac{1}{\mathrm{IP}}\sum_{j=0}^{\mathrm{IP}-1} \mathrm{NDVI}^*_{i,(m-j)}}{\frac{1}{(n*\mathrm{IP})}\sum_{k=1}^{n}\left[\sum_{j=0}^{\mathrm{IP}-1} \mathrm{NDVI}^*_{(m-j),k}\right]} * \sqrt{\left(\frac{\mathrm{RL}^{(\mathrm{NDVI}^*)}_{m,i}}{\frac{1}{n}\sum_{k=1}^{n}\mathrm{RL}^{(\mathrm{NDVI}^*)}_{m,k}}\right)} \tag{4}$$

$$\mathrm{TDI}_{i,m} = \frac{\frac{1}{\mathrm{IP}}\sum_{j=0}^{\mathrm{IP}-1} \left[T^*_{i,(m-j)}\right]}{\frac{1}{(n*\mathrm{IP})}\sum_{k=1}^{n}\left[\sum_{j=0}^{\mathrm{IP}-1} T^*_{(m-j),k}\right]} * \sqrt{\left(\frac{\mathrm{RL}^{(T^*)}_{m,i}}{\frac{1}{n}\sum_{k=1}^{n}\mathrm{RL}^{(T^*)}_{m,k}}\right)} \tag{5}$$

where IP is the interest period (e.g. 3, 4, 5, ... dekads or months), $\mathrm{RL}^{(P)}$ (run length) is the maximum number of successive dekads or months below long-term average rainfall in the IP, $\mathrm{RL}^{(T)}$ is the maximum number of successive dekads or months above long-term average temperature, $\mathrm{RL}^{(\mathrm{NDVI})}$ is the maximum number of successive dekads or months below long-term average NDVI in the IP, n is the number of years where relevant data are available,

j is the summation running parameter covering the IP and k is the summation parameter covering the years where relevant data are available.

Equations (3)–(5) are separately dimensionless and measure the drought severity in a given IP. The first term is the ratio of average modified rainfall, modified NDVI or modified temperature during the IP to the long-term mean of the same parameters during the same IP. The numerator is a high-frequency component that measures the current conditions, while the denominator is a low-frequency term that measures the long-term average. The second term is a measure of persistence of dryness whose square root is intended to scale down the variability of the second term (i.e. to decrease its impact on the final value of the drought index). In this study, it was assumed that the effect of the first and second terms is multiplicative. The low values in Equations (3)–(5) indicate strong drought conditions and vice versa. In using the equations, the actual drought index represents the severity of drought for the IP ending in time unit m. For example, if the IP $=3$ months, then the drought index (say PDI) of 0.35 for January 2006 implies actual drought for November and December 2006 and January 2006.

3.3 Calculation of the CDI

The CDI is computed as the weighted average of the VDI and the 2-dekad lagged PDI and TDI, as shown for decadal calculation on Equation (6):

$$\mathrm{CDI}_{i,m} = w_{\mathrm{PDI}} * \mathrm{PDI}_{i,m-2} + w_{\mathrm{TDI}} * \mathrm{TDI}_{i,m-2} + w_{\mathrm{VDI}} * \mathrm{VDI}_{i,m} \tag{6}$$

The lags on the PDI and TDI demonstrate that the effects of rainfall and temperature on vegetation are slightly delayed. In general, the present-day rainfall or temperature in a given area is likely to affect the vegetation of the area in about two-dekad time. It is important to note that this lagging is not necessarily valid for time units of the order of 1 month. For larger time units, Equation (6) becomes

$$\mathrm{CDI}_{i,m} = w_{\mathrm{PDI}} * \mathrm{PDI}_{i,m} + w_{\mathrm{TDI}} * \mathrm{TDI}_{i,m} + w_{\mathrm{VDI}} * \mathrm{VDI}_{i,m} \tag{7}$$

where w are the weights that are 50% for w_{PDI} and $w_{\mathrm{TDI}} = w_{\mathrm{VDI}} = 25$–25%. Where either temperature or NDVI data are missing, w_{PDI} is assigned a weight of 67% and the others assigned 33%.

It is important to note that the CDI does not measure the physical parameters of either the vegetation or the soil. It does not attempt to simulate either the physical phenomena or the water balance. It is an index that measures how much the present conditions deviate from the reference level, which is the multiyear long-term average as characterised by the time interval.

3.4 Interpretation of the Results

The severity (scale) of the drought is measured by the value of the CDI.

If the computed drought index:

- equals 1.00, then conditions in the actual time unit can be considered 'normal';
- is greater than 1.00, then they can be considered better than the 'normal'; and
- is less than 1.00, then they can be considered worse than the 'normal'.

An initial classification of drought categories based on the value of the CDI was set up below.

CDI Value	Drought Severity
>1.0	No drought
1.0–0.8	Mild drought
0.8–0.6	Moderate drought
0.6–0.4	Severe drought
<0.4	Extreme drought

The classification earlier can be modified, based on socioeconomic analysis or geographic considerations and experience. The impact of the drought depends on a number of factors: on the crop type, soil type, climatic zone, availability of groundwater, vulnerability of the ecosystem, etc. Therefore, the methodology gives the researcher the flexibility to analyse various lengths of drought duration, starting from a few dekads (10-day periods) to several months, seasons or even years. The basic period of analysis is called the 'interest period'. If the plants are sensitive even to a few weeks of drought conditions, dekadal data can be used with an IP of at least 3 dekads. As the CDI is calculated similarly to a moving average, in this case, the CDI values will characterise always the 3 dekads immediately before the selected date. In a more general situation, monthly data are recommended with an IP of 3–6 months; however, as seen in the succeeding text, sometimes the 12-month IP reflects best the longer-term characteristics of the drought. The shorter the IP, the more oscillation is shown in the graphs and the longer the IP the smoother are the graphs, reflecting longer-term trends but loosing short-term drought characteristics.

Besides all the features in the preceding text, the calculations later also clearly demonstrate that the drier the area, the more volatile the CDI values become, reflecting more abrupt changes in the behaviour of the climate there.

4 DROUGHTS IN KENYA, MEASURED BY THE CDI

4.1 Calculation CDI for Measuring Drought Severity

Monthly rainfall, temperature and NDVI data have been collected and used to calculate the CDI at the following stations.

	Name	Climate Zone
1	Dagoretti	Semi-arid
2	Embu	Semi-arid
3	Lodwar	Arid
4	Wajir	Arid
6	Kakamega	Subhumid

The results of the CDI calculations with monthly data and a 3-month IP are shown in Figure 2, and they clearly reflect the theoretical considerations in Section 3. There are two stations in semi-arid, one in subhumid and two in arid climatic zones. Figure 3 is a comparison between dekadal and monthly calculation at the Embu station. The oscillation of the CDI values is the most intensive with the 5-dekad analysis and much smoother with the 3-month analysis. Short-term droughts are detected well by the 5-dekad analysis;

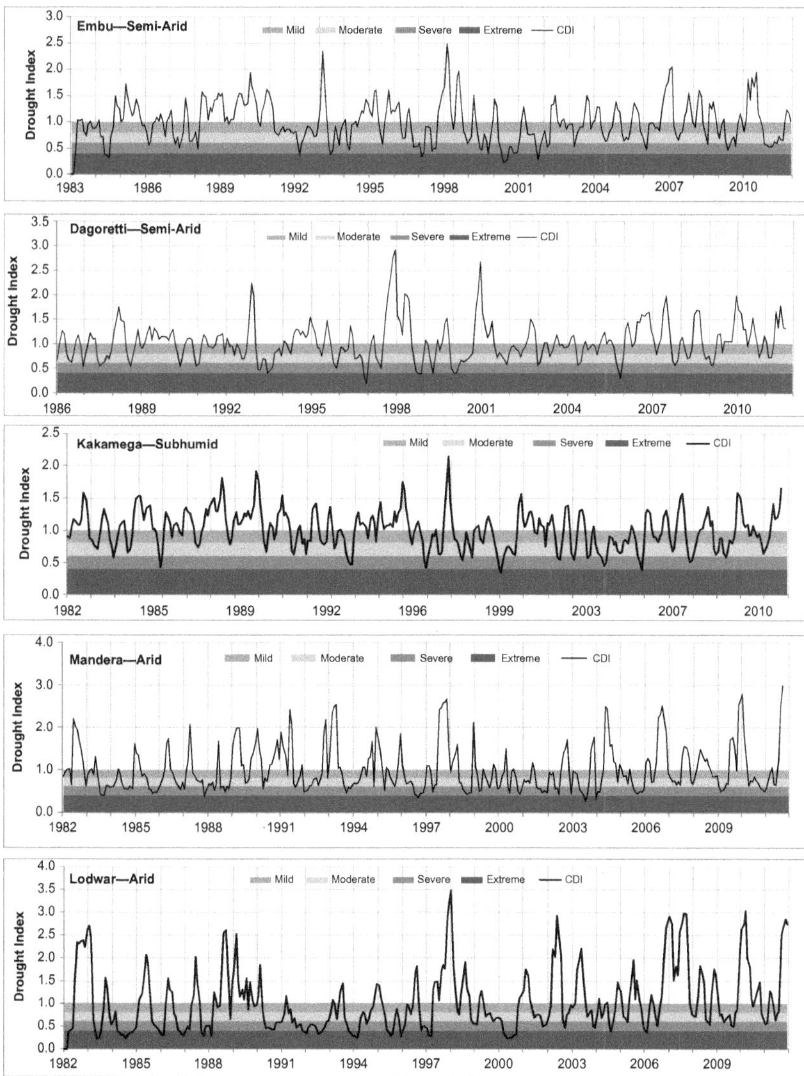

FIGURE 2 CDI calculations with monthly input data and a 3-month interest period. (For colour version of this figure, the reader is referred to the online version of this chapter.)

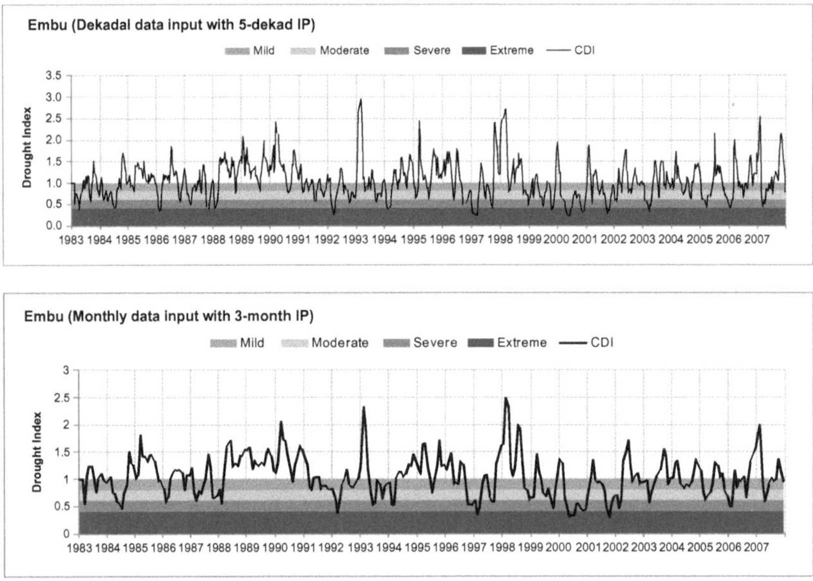

FIGURE 3 Comparison of decadal and monthly CDI calculation with 3-month/5-dekad interest-periods. (For colour version of this figure, the reader is referred to the online version of this chapter.)

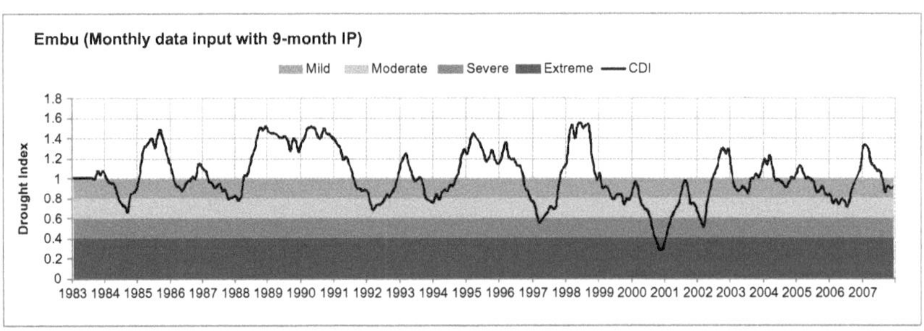

FIGURE 4 Example for a 9-month analysis for detecting longer droughts. (For colour version of this figure, the reader is referred to the online version of this chapter.)

however, the more severe droughts causing wider-ranging impacts and national catastrophes are better shown with the 9-month analysis in Figure 4.

4.2 Comparison of CDI Calculations with Drought Reports

'Verification' of the CDI results is not easy, because there are no exact, objective, detailed reports about the impacts of droughts in various years. Drought obviously starts in different months in different parts of the country, and national disaster is declared when its impact exceeds a certain extent. Some regions might not experience drought at all, while in other regions, the severity of drought might be exaggerated purely on political grounds. However, analysing CDI results together with drought reports, we found that they reinforce each other and paint a more detailed and more realistic picture about temporal and spatial development of the drought.

As shown in Table 1, there is evidence of drought impacts in Kenya emphasising the country's vulnerability to climate variability and the limitations of adaptive capacity. The table shows the drought events that the government declared as disasters from 1980 to 2011. During the 1980s, there were two major nationwide drought events in 1983/1984 and 1987; this has been well captured by the CDI as seen on Figure 2 for all stations. 1992–1994 and 1999/2000 also witnessed severe droughts in many parts of the country as shown in Tables 1 and 2 as well as the CDI graphs. The last decade has seen an increased number of drought occurrences. Devastating droughts during that period include that of 2001, 2002, 2004, 2006 and the most recent and much publicised one in 2010/2011. This again has been well captured by the CDI calculations.

TABLE 1 Recent History of Drought Incidences in Kenya (1980–2011)

Year	Region	Remark
1980	Widespread	40,000 People affected
1983/1984	Central, Rift Valley, eastern and northeastern	Severe food shortages in eastern province and less in central
1987	Eastern and central provinces	4.7 Million people dependent on relief power and water rationing
1991/1992	Northeastern, Rift Valley, eastern and coast provinces	1.5 Million people affected
1993/1994	Northern, central and eastern provinces	
1995/1996	Widespread	1.41 Million people affected
1997	Northern parts of Kenya	2 Million people affected
1999/2000	Countrywide except west and coast provinces	4.4 Million people affected (worst drought in 37 years)
2004	Widespread	2.3 Million people affected
2005	Northern parts of Kenya	2.5 Million people affected
2010/2011	Widespread	3.5 Million people affected

Sources: Republic of Kenya (2004), Republic of Kenya (2011).

TABLE 2 CDI Trend from the End of 1999–2000

Year	Month	Mandera	Lodwar	Embu	Dagoretti	Kakamega
1999	Oct	0.7	0.61	0.39	0.87	1.22
1999	Nov	0.58	0.67	0.73	1.06	1.08
1999	Dec	0.78	0.68	1.05	1.41	0.96
2000	Jan	1.14	0.67	1.44	1.52	0.8
2000	Feb	1.02	0.60	1.30	1.05	0.69
2000	Mar	0.56	0.36	0.63	0.50	0.43
2000	Apr	0.58	0.27	0.48	0.41	0.35
2000	May	0.86	0.23	0.23	0.41	0.49
2000	Jun	0.96	0.24	0.25	0.55	0.63
2000	Jul	1.51	0.25	0.29	0.66	0.74
2000	Aug	0.6	0.3	0.53	0.63	0.75
2000	Sep	0.47	0.31	0.54	0.66	0.71
2000	Oct	0.97	0.90	0.42	0.69	0.64
2000	Nov	1.02	1.13	0.41	0.75	0.62
2000	Dec	0.79	1.23	0.43	0.81	0.91

Drought severity legend					
Category	No Drought	Mild	Moderate	Severe	Extreme
colour code					
CDI	>1.0	1.0–0.8	0.8–0.6	0.6–0.4	<0.4

5 LONG-TERM TRENDS OF DROUGHT EVENTS IN KENYA

Is severity of drought increasing in Kenya? Is it going to cause more harm than in the previous years? Is the frequency of drought events growing? These are all questions that are important to answer in the course of preparing any drought management plan. FAO-SWALIM took the first steps in trying to answer these questions. Simple statistical analysis on various characteristics of the CDI graphs was performed for selected stations in Kenya with the following questions:

- Is there a general, increasing trend in severity of drought?
- Is the number of extreme/severe droughts increasing within a decade?

Results of the analysis are shown in Figure 5 and Table 3, and the summary conclusions are as follows:

1. To answer the first question, the complete time series were divided into three decades: years 1980–1989, 1990–1999 and 2000–2009 (Figure 5). The averages of the CDI values of less than 1.0 (mild to extreme

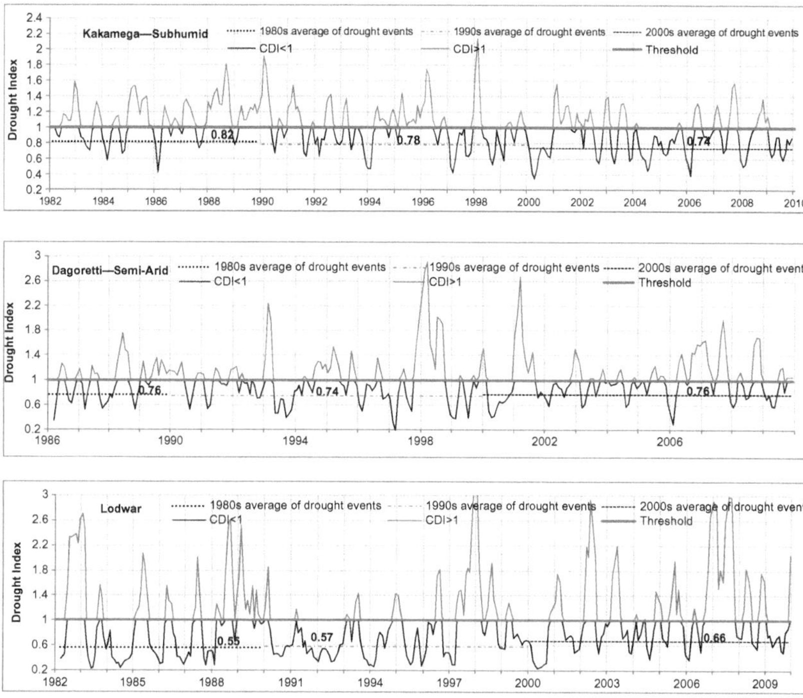

FIGURE 5 Decadal linear trends of below average CDI values at three Kenyan stations (CDI < 1.0).

TABLE 3 Number of Moderate to Extreme Drought Occurrences Per Decade (CDI < 0.8)

Station	1980	1990	2000
Dagoretti	23	67	71
Lodwar	54	86	75
Mandera	66	67	91
Kakamega	31	57	85
Embu	35	76	88

droughts) calculated for the periods in the preceding text show a shift towards more severe droughts from the 1980s to 2000s in most stations, while a few of them showed an improvement in drought severity in the 2000s.

2. Table 3 speaks for itself. The number of the severe/extreme droughts per decade is increasing in all stations.

3. While the drought severity is not necessarily getting worse in all stations, there is evidence of an increasing trend of the frequency of drought occurrences in the country.

Note: The ones in the preceding text are just some examples of many more possible statistical analyses that can be performed by the researchers according to the particular questions.

6 SHORT-TERM FORECAST OF DROUGHT BY THE CDI

First attempts have been made in this study to perform 'what-if' analysis, a simplified ensemble forecast for the evolution of the drought until the end of the season. The calculation is performed on a monthly basis. CDI values are calculated at the end of each month, and three scenarios are assumed about the future development of the climatic conditions in the remaining months of the season.

The assumed input scenarios are

1. future rainfall conditions 25% better than normal with normal temperature and NDVI,
2. future conditions around normal,
3. future rainfall conditions 25% worse than normal with normal temperature and NDVI.

The results are calculated, and hypothetical series of future CDI values can help decision-makers evaluate the anticipated drought situation at the end of the season.

To test the early warning methodology, a simulation example for Embu for 2007 is presented in Figure 6.

Short-term early warning is given for the three scenarios and compared with the actual situation in 2007. The example shows quite clearly that the 25% below average rainfall assumption captured very closely the actual situation in the first 6 months of the year.

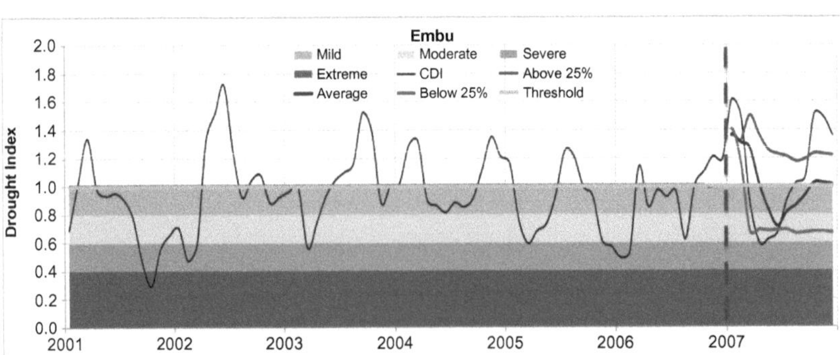

FIGURE 6 Short-term forecast (simulation) of the drought in 2007 at the Embu station. (For colour version of this figure, the reader is referred to the online version of this chapter.)

7 CONCLUSIONS

This chapter uses a statistical approach that combines different parameters to an index herein represented as the CDI. This index:

1. Can clearly trace the footprints of droughts in Kenya.
2. Has the potential to give short-term early warning up to the end of the season.
3. Has the potential for use in climate trends and climate change analysis.
4. The results are supported with drought reports in the country.
5. The authors recommend that the CDI should be tested worldwide.

REFERENCES

Balint, Z., Mutua, F.M., Muchiri, P., 2011. Drought Monitoring with the Combined Drought Index. Methodology and Software, FAO-SWALIM Nairobi, Kenya.

Dai, A., 2010. Drought under global warming: a review. Interdiscip. Rev. 2, 45–65.

Druyan, L.M., Hall, T.M., 1996. The sensitivity of African wave disturbances to remote forcing. J. Appl. Meteorol. 35, 1100–1110.

Gommes, R., Petrassi, F., 1994. Rainfall Variability and Drought in Sub-Saharan Africa Since 1960Agrometeorology Series Working Paper No. 9. Food and Agriculture Organization, Rome, Italy.

Hansen, J., Lebedeff, S., 1987. Global trends of measured surface air-temperature. J. Geophys. Res. Atmos. 92, 345–372.

Hastenrat, S., Polzin, D., Mutai, C.C., 2007. Diagnosing the 2005 drought in Equatorial East Africa. Int. J. Climatol. 20, 4628–4637.

Herweijer, C., Seager, R., 2008. The global footprint of persistent extra-tropical drought in the instrumental era. Int. J. Climatol. 28, 1761–1774.

Hulme, M., 1992. Rainfall changes in Africa: 1931–60 to 1961–90. International Journal of Climatology 12, 685–699.

Jones, P.D., Hulme, M., 1996. Calculating regional climatic time series for temperature and precipitation: methods and illustrations. Journal of Climatology 16, 361–377, (R).

Lloyd-Hughes, B., Saunders, M.A., 2002. A drought climatology for Europe. Int. J. Climatol. 22, 1571–1592.

Morid, S., Smakhtin, V., Moghaddasi, M., 2006. Comparison of seven meteorological indices for drought monitoring in Iran. Int. J. Climatol. 26, 971–985.

Moron, V., 1997. Trend, decadal and interannual variability in annual rainfall of subequatorial and tropical North Africa (190–1994). Int. J. Climatol. 17, 785–805.

Mutai, C.C., Ward, M.N., Colman, A.W., 1998. Towards the prediction of the East Africa short rains based on sea-surface temperature–atmosphere coupling. Int. J. Climatol. 18, 975–997.

Ntale, H.K., Gan, T.W., 2003. Drought indices and their application to East Africa. Int. J. Climatol. 23, 1335–1357.

Prospero, J.M., Lamb, P.J., 2003. African droughts and dust transport to the Caribbean: climate change implications. Science 302, 1024–1027.

Shakya, N., Yamaguchi, Y., 2010. Vegetation, water and thermal stress index for study of drought in Nepal and central Northeastern India. Int. J. Remote. Sens. 31, 903–912.

Shanko, D., Camberlin, P., 1998. The effects of the southwest Indian Ocean tropical cyclones on Ethiopian drought. Int. J. Climatol. 18, 1373–1388.

Szinell, C.S., Bussay, A., Szentimrey, T., 1998. Drought tendencies in Hungary. Int. J. Climatol. 18, 1479–1491.

Wilhite, D.A., 1993. Understanding the phenomenon of drought. Hydrol. Rev. 12 (5), 136–148.

Wilhite, D.A., Glantz, M.H., 1985. Understanding the drought phenomenon: the role of definitions. Water Int. 10 (3), 111–120.

WMO, 2010. Experts Recommend Agricultural Drought Indices for improved understanding of food production conditions. Press Release No. 887, Geneva/Murcia, 8 June 2010.

Wu, H., Hayes, M.J., Weiss, A., Hu, Q., 2001. An evaluation of the standardized precipitation index, the China-z index and the statistical z-score. Int. J. Climatol. 21, 745–758.

Conclusions

The contributions of this book, each in a different way, point toward four main big topics for a wise and sustainable protection and use of the natural resources of this country: (a) challenges due to lack of data or lack of access to data; (b) challenges due to a lack of integrated and controlled exploitation of the environment; (c) opportunities provided by the potential for data gap filling, the potential for renewable power generation, and by a sound water management in the country; and (d) finally, some caution and realism that needs to be applied when converting the knowledge of natural resources into policy and policy implementation.

1 CHALLENGES DUE TO LACK OF DATA

Despite years of rising attention by the scientific, humanitarian, and civil society, still Kenya is lacking an adequate base of data on its natural resources and also an easy access to the data. In different thematic and geographic contexts (Deloitte, 2011; Heather, 2012), it has been proven that open access to data generated by public tax payers has increased the scientific knowledge of a country, and as a cascade also the private initiative, the possibility for civil society organisation to monitor and pressure for a better use and management of the environment and also a dialogue between various public agencies and governmental bodies. In the specific case of Kenya, we still face several challenges in these regards. Examples have been put by McClain and Subalusky (2013) highlighting the gaps in hydro-meteorological (rainfall, temperature, insolation, river discharge) discharge data for the Mara River basin, by Kuria (2013) regarding the limited access to or the poor metadata accompanying the almost 13,000 borehole data that he has analysed for a groundwater assessment of Kenya, and by Omuto (2013) who highlights how almost half of the 540 soil profiles that are available for Kenya are hosted by an international organisation outside Kenya (ISRIC, www.isric.org) and just less than the other half is contained in reports and academic outputs that are not publicly available. Challenges are posed also by lack of resources for maintaining medium- to long-term monitoring networks that would serve as a crucial part of baseline generation for any sort of development projects (Maina-Gichaba et al., 2013; Opere, 2013). As a final example, coastal waters and tidal mechanisms are still poorly researched upon (Oludhe, 2013; Omuombo et al., 2013), while they already represent a substantial part of the country's economy and livelihoods.

2 CHALLENGES POSED BY LACK OF INTEGRATED APPROACH IN EXPLOITATION OF NATURAL RESOURCES

Kenya censuses through the recent decades (KNBS, 2008) show an unstoppable growing population and an increasing urbanisation pattern, common to the global pattern. This trend has an inevitable impact on the natural resource of Kenya coupled with the still too low crop yield, and consequently an expansion of farmland at the expense of natural vegetation (as highlighted by Brink and Eva, 2009; Brink et al., 2013; Pellika et al., 2013) and charcoal production as the main source of energy (Pellika et al., 2013) especially in the medium- to small-size towns as opposed to other sources of energy like geothermal or other renewable energy sources such as wind (Oludhe, 2013). Challenges are also posed by the present practices of wildlife and freshwater ecology management as highlighted by Ogada and Nyingi (2013) and Nyingi et al. (2013), respectively. The lack of involvement or marginalisation of local communities in the management of wildlife has proved to be an unsustainable way to manage wildlife in Kenya. While this may have deep roots in the history of wildlife management, it needs to make a U-turn in order to get an integrated approach to the management of this important source of national revenue and also to counterbalance the present-day conflicts linked to it. This is even more an urgent issue in view of the increasing trends of poaching for large mammals (elephants and rhinos) registered in recent years.

Groundwater resource regulations have a well-written legislation as spelled out in the Water Act of 2002 and draft Water Act of 2012 (GoK, 2002, 2012). Despite this, the practice of reporting incomplete borehole information during well construction and of water abstraction above the set limits hampers the sustainable use of this crucial resource (Kuria, 2013) especially in view of climatic changes and their implications for a higher reliability on groundwater. It is possible to argue that also here it is missing the link between groundwater table and ecosystem and the benefits in the mid- and long term that these ecosystems provide to a large portion of population, often also to the ones that abstract water above the limits.

3 OPPORTUNITIES

Three main areas of opportunities are here highlighted.

Data gaps filling: Remote sensing techniques are not anymore recent being in place for the past 40 years or so (http://landsat.usgs.gov/about_mission_history.php). Before remote sensing, we also have other aerial resources like aerial photography. Although all the old and most of the new remote sensing methods do not penetrate below ground or below water surfaces, they offer a wide range of opportunities for filling the gaps in those data series that from the early 1990s started to become hole filled instead of data filled. Particularly in the fields of climatology, hydrology, and vegetation studies, these can be of extreme help and there is an opportunity for combining field observation and calibration with remote sensing data gap filling (Kiema, 2013). This would be useful if accompanied by a restructuring of the monitoring networks across the country and especially in the areas so far mostly neglected like the arid and semi-arid lands: it is here that a large part of the pastoral population will suffer from most of the impacts of climate change, and it is here that we need to strengthen the observation of natural processes, above and below ground, in an integrated manner. Recent efforts are made in this direction, and the presence of SERVIR Africa in Kenya (http://www.servir.net/africa/) could lead as a leverage for this kind of activities. Also, generations of scholars and technicians have been formed in African and international colleges, and there is a bulk of brains in Kenya that could be put into work to provide the much needed appropriate solutions for data gap filling.

Power generation: Kenya has an enormous potential for renewable energy production. Recent studies highlight that geothermal power alone could provide the foreseen 15,000 MW of energy that the country aims to in the Vision 2030 (Oludhe, 2013). This would also free Kenya from spending for oil import and using polluting sources of power generation. Needless to say that it would also have the potential of generating a leadership role (in both public policy and private sector) for other African countries with similar potential (most of the ones along the young East African Rift System). The development of renewable energy solutions would leverage the production of power for the private sector, infrastructure, and public uses, and will set the base for a more efficient use of this resource.

Better management of water resources as a cascade from improved data and sharing, improved knowledge and also as a reduction in hydropower in favour of geothermal power generation, would increase the environmental flows throughout the country. These are extremely important for sustaining the various ecosystems, and the concepts of keeping the inter- and intra-annual variability of water flows have been well exposed by various contributions to this book (Kuria, 2013; McClain and Subalusky, 2013; Nyingi et al., 2013). New opportunities arise in Kenya in this regard as new dams are planned and a growing micro-hydropower construction is under way for private sector power generation. A deep dialogue between upstream and downstream users of ecosystem services is needed.

4 KNOWLEDGE APPLICATION AND SHARING

Looking at the knowledge of the country that was available almost 40 years ago and at the present situation emerging also from this book, it is clear that there is not only an improved and increased knowledge on Kenya's natural resources but also a modern attitude by most scholars pointing to the need for an integrated natural resource management. Most of the studies here presented in one way or another the need for sustainable natural resource exploitation, given the rapidly growing urban expansion, climate change effects, and the inevitable fact that an increasing population will use more resources.

Recently, a report by Transparency International (http://www.transparency.org/gcb2013/report, last visited on 17 August 2013) highlighted several indicators on the patterns of corruption worldwide. When asking the respondent if they see corruption increased compared to the previous year, almost all countries worldwide answered yes. This is an alarming factor because better governance (including governance of natural resources) can only pass through a better policy implementation, and corruption pulls the line to the opposite direction. The good news is that more than 50% of the global population agrees that the ordinary citizen can make a difference in this trend.

Also in the management of natural resources, there is a stringent need for fair policy implementation, especially in context where the legislation is—on paper—already good in many cases, like in Kenya. There is an opportunity here for scholars and civil society involved in the natural resources world: they have the potential to make a change and to inform a wider audience and make them aware of the linkages between causes and effects in natural resource management at different levels. Several examples are already available in this direction like the achievements made by the Green Belt movement and by other unconventional organisations that work for the environment and with the people in need of a better environment.

We hope that the readers of this book will gain a more in-depth understanding of the challenges and will appreciate the opportunities deriving from a sound natural resource management in Kenya as opposed to a short-term, selfish attitude that will leave no more resource to the subsequent generations.

REFERENCES

Brink, A.B., Eva, H.D., 2009. Monitoring 25 years of land cover change dynamics in Africa: a sample based remote sensing approach. Appl. Geogr. 29 (4), 501–512.

Brink, A.B., Bodart, C., Buchanan, G., Clerici, M., Donnay, F., Eshiamwata, G., Gregoire, J.-M., Kirui, K.B., Lupi, A., Palumbo, I., Pekel, J.-F., 2013. Vegetation. In: Paron, P., Olago, D.O., Omuto, C.T. (Eds.), Kenya: A Natural Outlook. Elsevier.

Deloitte, 2011. Unlocking growth: how open data creates new opportunities for the UK, , Available at: http://www.deloitte.com/view/en_GB/uk/market-insights/deloitte-analytics/2d376cdde7e14310VgnVCM1000001a56f00aRCRD.htm.

GoK, 2002. The Water Act, 2002. Available at: http://www.water.go.ke/index.php?option=com_docman&Itemid=89.

GoK, 2012. The Water Act, 2012, Draft. Available at: http://worldbank.org%2FINTAFRICA%2FResources%2F257994-1335471959878%2Fdraft-water-bill-2012.pdf&ei=VOERUubrEOas0QWdn4DICQ&usg=AFQjCNGZ9-khEFO9eS5WAxwWT6FzSePgDg&sig2=izvmxs540okEZ60AnUwfsw.

Heather, J., 2012. The impact of open access on research and scholarship. Reflections on the Berlin 9 Open Access Conference. Coll. Res. Libr. News 73 (2), 83–87.

Kiema, J.B.K., 2013. Remote sensing applications for water data in Kenya case study. In: Paron, P., Olago, D.O., Omuto, C.T. (Eds.), Kenya: A Natural Outlook. Elsevier.

KNBS, 2008. Kenya Demographic and Health Survey. Available at: http://www.knbs.or.ke/surveys.php.

Kuria, Z., 2013. Groundwater distribution and aquifer characterisation in Kenya. In: Paron, P., Olago, D.O., Omuto, C.T. (Eds.), Kenya: A Natural Outlook. Elsevier.

Maina-Gichaba, C., Kipseba, E.K., Masibo, M., 2013. Overview of landslide occurrences in Kenya. Causes, mitigation and challenges. In: Paron, P., Olago, D.O., Omuto, C.T. (Eds.), Kenya: A Natural Outlook. Elsevier.

McClain, M.E., Subalusky, A.L., 2013. Surface flows for people and wildlife in the transboundary Mara River Basin. In: Paron, P., Olago, D.O., Omuto, C.T. (Eds.), Kenya: A Natural Outlook. Elsevier.

Nyingi, D.W., Gichuki, N., Ogada, M., 2013. Freshwater ecology of Kenyan highlands and lowlands. In: Paron, P., Olago, D.O., Omuto, C.T. (Eds.), Kenya: A Natural Outlook. Elsevier.

Ogada, M., Nyingi, D.O., 2013. The management of wildlife and fisheries resources in Kenya: origins, present challenges and future perspectives. In: Paron, P., Olago, D.O., Omuto, C.T. (Eds.), Kenya: A Natural Outlook. Elsevier.

Oludhe, C., 2013. Renewable energy resources in Kenya. In: Paron, P., Olago, D.O., Omuto, C.T. (Eds.), Kenya: A Natural Outlook. Elsevier.

Omuombo, C.A., Olago, D.O., Odada, E.O., 2013. Coastal waters. In: Paron, P., Olago, D.O., Omuto, C.T. (Eds.), Kenya: A Natural Outlook. Elsevier.

Omuto, C.T., 2013. Major soil and data types in Kenya. In: Paron, P., Olago, D.O., Omuto, C.T. (Eds.), Kenya: A Natural Outlook. Elsevier.

Opere, A., 2013. Floods in Kenya. In: Paron, P., Olago, D.O., Omuto, C.T. (Eds.), Kenya: A Natural Outlook. Elsevier.

Pellikka, P.K.E., Clark, B.J.F., Gonsamo Gosa, A., Himberg, N., Hurskainen, P., Maeda, E., Mwang'ombe, J., Omoro, L.M.A., Siljander, M., 2013. Agricultural expansion and its consequences in the Taita Hills, Kenya. In: Paron, P., Olago, D.O., Omuto, C.T. (Eds.), Kenya: A Natural Outlook. Elsevier.

Index

Note: Page numbers followed by "*f*" indicate figures, and "*t*" indicate tables.

D

L

M

N

S

T

FSC
www.fsc.org
MIX
Papier | Fördert
gute Waldnutzung
FSC® C083411

Kolibri 360 GmbH
Ferdinand-Jühlke-Straße 7
99095 Erfurt, Deutschland
produktsicherheit@kolibri360.de